CONTENTS

About Catch-Up Maths iii
How To Use the QR Codes in Catch-Up Maths .. iv
Australian Curriculum Correlations............ v

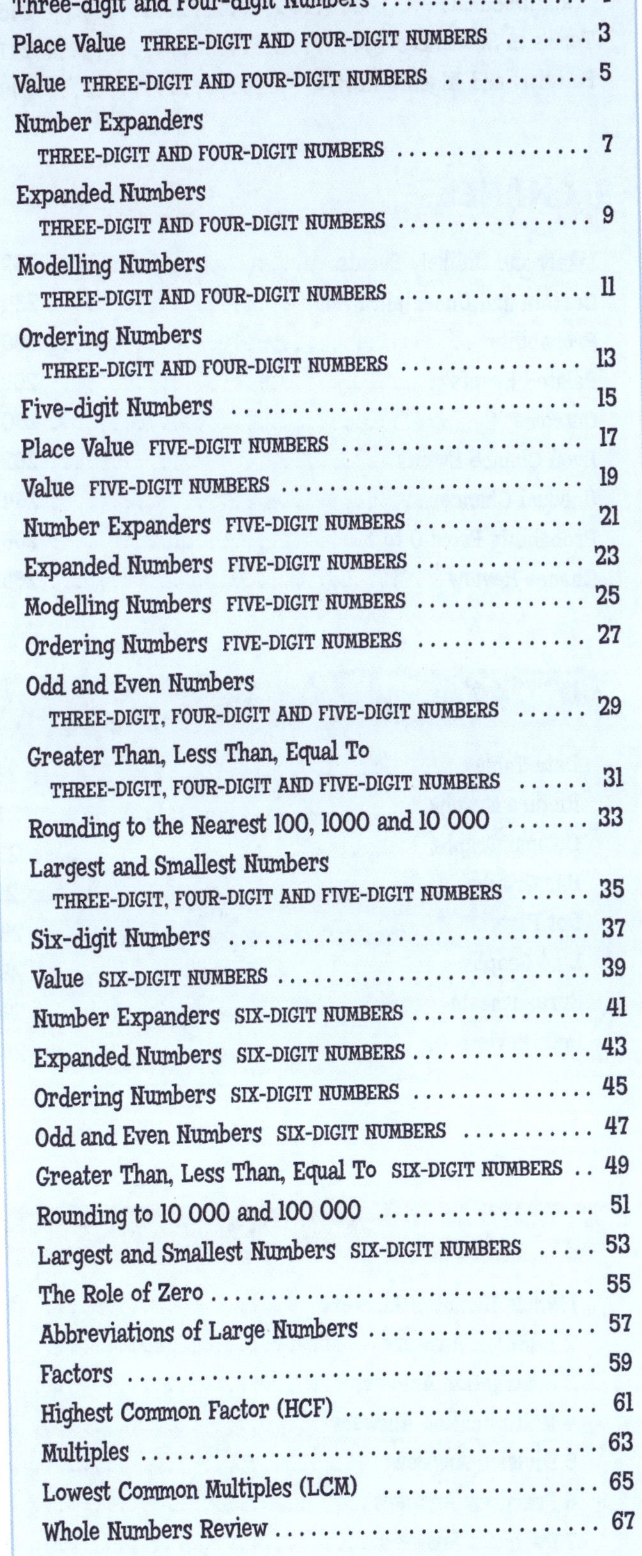

1 WHOLE NUMBERS

Three-digit and Four-digit Numbers 1
Place Value THREE-DIGIT AND FOUR-DIGIT NUMBERS 3
Value THREE-DIGIT AND FOUR-DIGIT NUMBERS 5
Number Expanders THREE-DIGIT AND FOUR-DIGIT NUMBERS 7
Expanded Numbers THREE-DIGIT AND FOUR-DIGIT NUMBERS 9
Modelling Numbers THREE-DIGIT AND FOUR-DIGIT NUMBERS 11
Ordering Numbers THREE-DIGIT AND FOUR-DIGIT NUMBERS 13
Five-digit Numbers 15
Place Value FIVE-DIGIT NUMBERS 17
Value FIVE-DIGIT NUMBERS 19
Number Expanders FIVE-DIGIT NUMBERS 21
Expanded Numbers FIVE-DIGIT NUMBERS 23
Modelling Numbers FIVE-DIGIT NUMBERS 25
Ordering Numbers FIVE-DIGIT NUMBERS 27
Odd and Even Numbers THREE-DIGIT, FOUR-DIGIT AND FIVE-DIGIT NUMBERS 29
Greater Than, Less Than, Equal To THREE-DIGIT, FOUR-DIGIT AND FIVE-DIGIT NUMBERS 31
Rounding to the Nearest 100, 1000 and 10 000 33
Largest and Smallest Numbers THREE-DIGIT, FOUR-DIGIT AND FIVE-DIGIT NUMBERS 35
Six-digit Numbers 37
Value SIX-DIGIT NUMBERS 39
Number Expanders SIX-DIGIT NUMBERS 41
Expanded Numbers SIX-DIGIT NUMBERS 43
Ordering Numbers SIX-DIGIT NUMBERS 45
Odd and Even Numbers SIX-DIGIT NUMBERS 47
Greater Than, Less Than, Equal To SIX-DIGIT NUMBERS .. 49
Rounding to 10 000 and 100 000 51
Largest and Smallest Numbers SIX-DIGIT NUMBERS 53
The Role of Zero 55
Abbreviations of Large Numbers 57
Factors 59
Highest Common Factor (HCF) 61
Multiples 63
Lowest Common Multiples (LCM) 65
Whole Numbers Review 67

2 ADDITION

Adding Three or More Single-digit Numbers 76
Sum 78
Relating Addition and Subtraction 80
Addition Without Trading TWO-DIGIT AND THREE-DIGIT NUMBERS 82
Adding With Trading TWO-DIGIT AND THREE-DIGIT NUMBERS 84
Adding With and Without Trading FOUR-DIGIT NUMBERS .. 86
Rounding to Estimate Addition Answers 88
Using the Jump Strategy to Solve Addition TWO-DIGIT AND THREE-DIGIT NUMBERS 90
Using the Jump Strategy THREE-DIGIT AND FOUR-DIGIT NUMBERS 92
Using the Split Strategy to Solve Addition TWO-DIGIT AND THREE-DIGIT NUMBERS 94
Using the Split Strategy THREE-DIGIT AND FOUR-DIGIT NUMBERS 96
Rounding Money 98
Working Out Change 100
Money Problems – Spending and Saving 102
Addition Review 108

3 SUBTRACTION

Using the Jump Strategy to Solve Subtraction TWO-DIGIT AND THREE-DIGIT NUMBERS 114
Using the Jump Strategy THREE-DIGIT AND FOUR-DIGIT NUMBERS 116
Using the Split Strategy to Solve Subtraction TWO-DIGIT AND THREE-DIGIT NUMBERS 118
Using the Split Strategy THREE-DIGIT AND FOUR-DIGIT NUMBERS 120
Subtraction Without Trading TWO-DIGIT AND THREE-DIGIT NUMBERS 122
Subtraction Without Trading THREE-DIGIT AND FOUR-DIGIT NUMBERS 124
Subtraction Without Trading FOUR-DIGIT AND FIVE-DIGIT NUMBERS 126
Subtraction With Trading TWO-DIGIT AND THREE-DIGIT NUMBERS 128
Subtraction With Trading THREE-DIGIT AND FOUR-DIGIT NUMBERS 130
Subtraction With Trading FIVE-DIGIT NUMBERS 132
Trading From Larger Place Values 134
What's the Difference? 136
Rounding to Estimate Subtraction Answers 138
Subtraction Review 140

CONTENTS

4 MULTIPLICATION

Skip Counting 146
Product, Factors and Multiples 148
Multiplying 2-digit by 1-digit Numbers 150
Formal Algorithms 152
Multiplying 3-digit and 4-digit Numbers by 1-digit Numbers 155
Multiplying 2-digit and 3-digit Numbers by 2-digit Numbers 159
Multiplication Review 162

5 DIVISION

Groups and Equal Rows 168
Relating × to ÷ 170
Quotient, Divisor and Dividend 172
Formal Division 174
Different Ways to Write Division 176
Division With Remainders 178
Division of 2-digit Numbers 180
Division of 3-digit Numbers 182
Recording Remainders as Fractions and Decimals 184
Division Review 186

6 FRACTIONS

Parts of a Fraction 194
Halves and Halves of Collections 196
Quarters and Eighths – Fractions and Collections 198
Thirds and Fifths – Fractions and Collections 200
Equivalent Fractions 202
Comparing Fractions 204
Proper and Improper Fractions 206
Mixed Numbers 208
Improper Fractions and Mixed Numbers 210
Add and Subtract Fractions with the Same Denominator 212
Fractions Review 214

7 DECIMALS

Writing Decimals 218
Relating Tenths to Hundredths 220
Decimal Fractions and Fractions in Words 222
Thousandths 224
Place Value 226
Partitioning Decimals 228
Decimals Review 231

8 PATTERNS AND ALGEBRA

Number Patterns 238
Pattern Grids 241
Equivalent Number Sentences 243
Number Sentences and Patterns with Fractions and Decimals 245
Terms in Sequences 247
Patterns and Algebra Review 249

9 CHANCE

Likely and Unlikely Events 252
Certain and Uncertain Events 254
Probability 256
Related Events 258
Outcomes 260
Even Chance Events 262
Unequal Chances 264
Probability From 0 to 1 266
Chance Review 268

10 DATA

Data Tables 272
Picture Graphs 275
Column Graphs 278
Bar Graphs 281
Dot Plots 283
Line Graphs 288
Spreadsheets 293
Data Review 296

ANSWERS

1 Whole Numbers Answers 302
2 Addition Answers 308
3 Subtraction Answers 311
4 Multiplication Answers 313
5 Division Answers 314
6 Fractions Answers 316
7 Decimals Answers 318
8 Patterns and Algebra Answers 321
9 Chance Answers 322
10 Data Answers 323

ABOUT CATCH-UP MATHS

The Catch-Up Maths series enables students to start from scratch when they are struggling with their year level maths. Each book takes maths back to the foundation and ensures that all basic concepts are consolidated before moving forward. Lots of revision and opportunities to practise and build confidence are provided before moving on to new topics.

Each new strand and sub-strand of the primary maths curriculum is introduced clearly with simple explanations, examples and trial questions (with answers), before children move to the Practice section. To ensure concepts are understood fully, videos of the author working through and explaining new concepts are included in every chapter.

This book has 10 chapters divided between the Year 5 Number & Algebra and Statistics & Probability strands of the Australian Maths Curriculum. The chapters are:

1 Whole Numbers
2 Addition
3 Subtraction
4 Multiplication
5 Division
6 Fractions
7 Decimals
8 Patterns & Algebra
9 Chance
10 Data

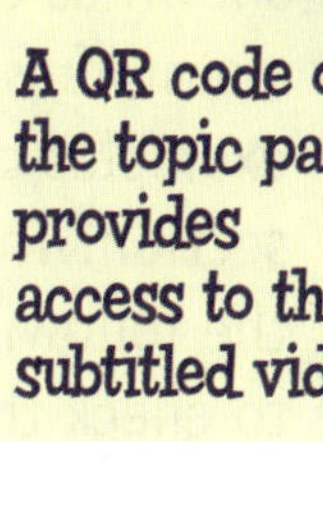

A QR code on the topic page provides access to the subtitled video.

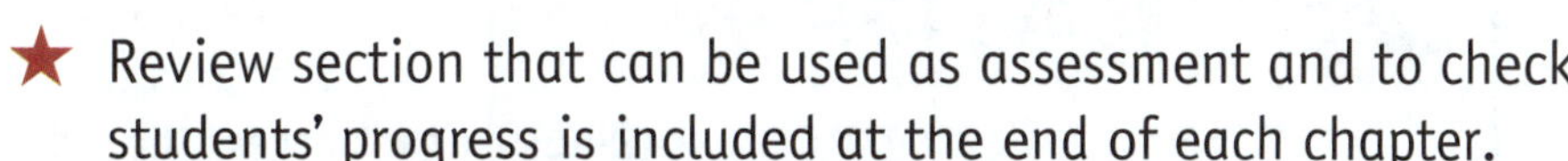

★ Review section that can be used as assessment and to check students' progress is included at the end of each chapter.

★ Answers are at the back of the book.

How to use this book

Children can work through the pages from front to back, or choose individual topics to reinforce areas where they are struggling.

The topics are introduced with:

- clear instructions, using simple language
- completed examples and incomplete examples for students to tackle before moving on to the **Your Turn** section
- a video linked by QR code that shows the incomplete example and has the author giving extra instruction about the page.

Each Your Turn section contains a SELF CHECK for students to reflect and give self-assessment on their understanding.

HOW TO USE THE QR CODES IN CATCH-UP MATHS

A unique aspect of the **Catch-Up Maths** series is the **instructional video** created by the author for every new topic.

The videos give a step-by-step explanation of the examples and work through them to provide a helpful lesson for the student. The videos are simply accessed via the QR code on the same page and can be watched on a phone, tablet, computer or interactive whiteboard.

Access the video using a QR code reader app

Each video shows the page from the book. The author talks through the concepts and examples, and demonstrates what students need to do. Her words are shown as easy-to-read captions so that the audio can be turned down in noisy classrooms, and for students with hearing difficulties. The solutions to the examples are presented before students are expected to tackle the 'Your Turn' section. This careful instruction ensures that students can confidently move on to the following Practice questions. Those assisting students should encourage them to check their 'Your Turn' answers before moving on.

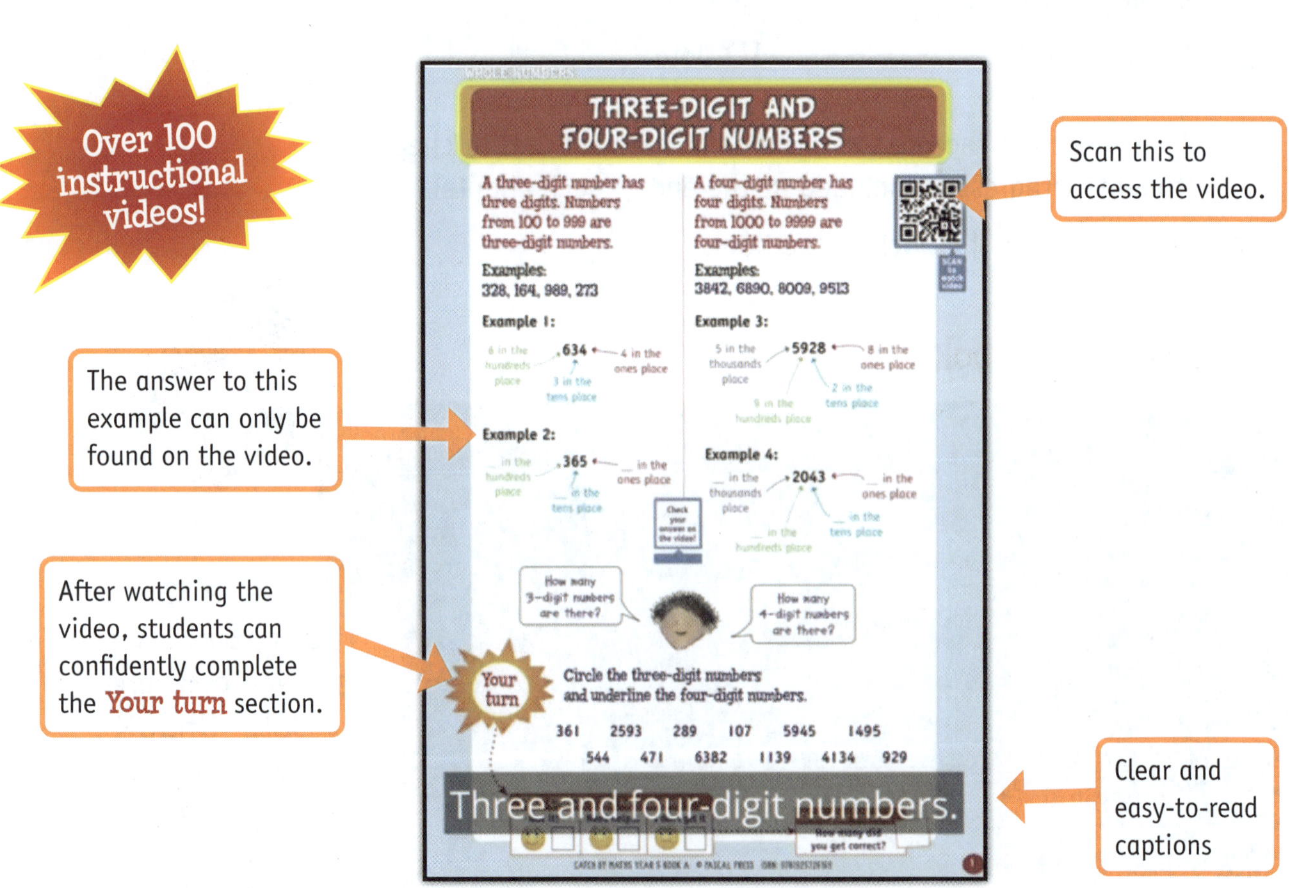

CATCH UP MATHS YEAR 5 BOOK A © PASCAL PRESS ISBN: 9781925726169

AUSTRALIAN CURRICULUM CORRELATIONS

ACARA CODE	Content Description	Pages
1. Whole Numbers		**PAGES 1–75**
ACMNA027 Year 2	Recognise, model, represent and order numbers to at least 1000	1, 2, 11, 12, 13, 14
ACMNA028 Year 2	Group, partition and rearrange collections up to 1000 in hundreds, tens and ones to facilitate more efficient counting	3, 4, 5, 6, 7, 8, 9, 10
ACMNA029 Year 2	Explore the connection between addition and subtraction	70, 71
ACMNA030 Year 2	Solve simple addition and subtraction problems using a range of efficient mental and written strategies	66, 67, 68, 69, 72, 73
ACMNA051 Year 3	Investigate the conditions required for a number to be odd or even and identify odd and even numbers	29, 30
ACMNA052 Year 3	Recognise, model, represent and order numbers to at least 10 000	1, 2, 11, 12, 13, 14, 31, 32, 35, 36
ACMNA053 Year 3	Apply place value to partition, rearrange and regroup numbers to at least 10 000 to assist calculations and solve problems	3, 4, 5, 6, 7, 8, 9, 10, 72, 73, 74, 75
ACMNA054 Year 3	Recognise and explain the connection between addition and subtraction	70, 71
ACMNA055 Year 3	Recall addition facts for single-digit numbers and related subtraction facts to develop increasingly efficient mental strategies for computation	66, 67, 68, 69
ACMNA071 Year 4	Investigate and use the properties of odd and even numbers	29, 30, 47, 48
ACMNA072 Year 4	Recognise, represent and order numbers to at least tens of thousands	13, 15, 16, 25, 26, 27, 28, 31, 32, 33, 34, 35, 36, 37, 38, 45, 49, 50, 53, 54, 55, 56, 57, 58
ACMNA073 Year 4	Apply place value to partition, rearrange and regroup numbers to at least tens of thousands to assist calculations and solve problems	3, 17, 18, 19, 20, 21, 22, 23, 24, 27, 39, 40, 41, 42, 43, 44, 51, 52
ACMNA098 Year 5	Identify and describe factors and multiples of whole numbers and use them to solve problems	33, 34, 59, 60, 61, 62, 63, 64, 65, 66,
ACMNA099 Year 5	Use estimation and rounding to check the reasonableness of answers to calculations	33, 51, 52
2. Addition		**PAGES 76–113**
ACMNA030 Year 2	Solve simple addition and subtraction problems using a range of efficient mental and written strategies	104, 105, 108, 109
ACMNA053 Year 3	Apply place value to partition, rearrange and regroup numbers to at least 10 000 to assist calculations and solve problems	76, 77
ACMNA054 Year 3	Recognise and explain the connection between addition and subtraction	114, 115
ACMNA055 Year 3	Recall addition facts for single-digit numbers and related subtraction facts to develop increasingly efficient mental strategies for computation	76, 77, 78, 80, 81, 82, 83, 84, 85, 86, 87, 104, 105, 106, 107, 110, 111, 112, 113
ACMNA059 Year 3	Represent money values in multiple ways and count the change required for simple transactions to the nearest five cents	88, 89

ACARA CODE	Content Description	Pages
ACMNA073 Year 4	Apply place value to partition, rearrange and regroup numbers to at least tens of thousands to assist calculations and solve problems	90, 92
ACMNA080 Year 4	Solve problems involving purchases and the calculation of change to the nearest five cents with and without digital technologies	88, 89, 90, 91, 92, 93
ACMNA099 Year 5	Use estimation and rounding to check the reasonableness of answers to calculations	78, 79, 88, 89, 90, 91
ACMNA106 Year 5	Create simple financial plans	92, 93, 94, 95, 96, 97, 98, 100, 102
ACMNA291 Year 5	Use efficient mental and written strategies and apply appropriate digital technologies to solve problems	100, 102
3. Subtraction		**PAGES 114–145**
ACMNA054 Year 3	Recognise and explain the connection between addition and subtraction	114, 115
ACMNA055 Year 3	Recall addition facts for single-digit numbers and related subtraction facts to develop increasingly efficient mental strategies for computation	114, 115, 116, 117, 118, 119, 120, 121, 122, 123, 124, 125, 126, 127
ACMNA074 Year 4	Investigate number sequences involving multiples of 3, 4, 6, 7, 8, and 9	144, 145
ACMNA075 Year 4	Recall multiplication facts up to 10 × 10 and related division facts	144, 145
ACMNA099 Year 5	Use estimation and rounding to check the reasonableness of answers to calculations	128, 129
4. Multiplication		**PAGES 146–167**
ACMNA032 Year 2	Recognise and represent division as grouping into equal sets and solve simple problems using these representations	166, 167
ACMNA056 Year 3	Recall multiplication facts of two, three, five and ten and related division facts	144, 145, 146, 147, 168, 169
ACMNA057 Year 3	Represent and solve problems involving multiplication using efficient mental and written strategies and appropriate digital technologies	148, 149
ACMNA075 Year 4	Recall multiplication facts up to 10 × 10 and related division facts	146, 147, 168, 169
ACMNA076 Year 4	Develop efficient mental and written strategies, and use appropriate digital technologies for multiplication and for division where there is no remainder	148, 149, 150, 151
ACMNA098 Year 5	Identify and describe factors and multiples of whole numbers and use them to solve problems	146, 147
ACMNA100 Year 5	Solve problems involving multiplication of large numbers by one- or two-digit numbers using efficient mental, written strategies and appropriate digital technologies	150, 151, 152, 153, 154, 155, 156, 157, 158, 159, 160, 161

Australian Curriculum Correlations continued

ACARA CODE	Content Description	Pages
5. Division		**PAGES 168–193**
ACMNA076 Year 4	Develop efficient mental and written strategies, and use appropriate digital technologies for multiplication and for division where there is no remainder	170, 171, 174, 175, 178, 179, 180, 181
ACMNA079 Year 4	Recognise that the place value system can be extended to tenths and hundredths. Make connections between fractions and decimal notation	184, 185
ACMNA101 Year 5	Solve problems involving division by a one digit number, including those that result in a remainder	170, 171, 172, 173, 174, 175, 176, 177, 178, 179, 182, 183
6. Fractions		**PAGES 194–217**
ACMNA033 Year 2	Recognise and interpret common uses of halves, quarters and eighths of shapes and collections	194, 195, 196, 197, 198, 199
ACMNA058 Year 3	Model and represent unit fractions including $\frac{1}{2}$, $\frac{1}{4}$, $\frac{1}{3}$, $\frac{1}{5}$ and their multiples to a complete whole	200, 201
ACMNA077 Year 4	Investigate equivalent fractions used in contexts	202, 203
ACMNA102 Year 5	Compare and order common unit fractions and locate and represent them on a number line	204, 205
ACMNA103 Year 5	Investigate strategies to solve problems involving addition and subtraction of fractions with the same denominator	206, 207, 208, 209, 210, 211, 212, 213
7. Decimals		**PAGES 218–237**
ACMNA079 Year 4	Recognise that the place value system can be extended to tenths and hundredths. Make connections between fractions and decimal notation	218, 219, 220, 221, 222, 223
ACMNA104 Year 5	Recognise that the place value system can be extended beyond hundredths	224, 225, 228, 229
ACMNA105 Year 5	Compare, order and represent decimals	226, 227
8. Patterns & Algebra		**PAGES 238–251**
ACMNA035 Year 2	Describe patterns with numbers and identify missing elements	238, 239, 240
ACMNA036 Year 2	Solve problems by using number sentences for addition or subtraction	243, 244
ACMNA060 Year 3	Describe, continue, and create number patterns resulting from performing addition or subtraction	238, 239, 240, 241, 242
ACMNA074 Year 4	Investigate number sequences involving multiples of 3, 4, 6, 7, 8, and 9	247, 248
ACMNA078 Year 4	Count by quarters, halves and thirds, including with mixed numerals. Locate and represent these fractions on a number line	245, 246
ACMNA081 Year 4	Explore and describe number patterns resulting from performing multiplication	247, 248
ACMNA083 Year 4	Find unknown quantities in number sentences involving addition and subtraction and identify equivalent number sentences involving addition and subtraction	243
ACMNA107 Year 5	Describe, continue and create patterns with fractions, decimals and whole numbers resulting from addition and subtraction	245, 246
ACMNA121 Year 5	Find unknown quantities in number sentences involving multiplication and division and identify equivalent number sentences involving multiplication and division	243, 244
9. Chance		**PAGES 252–271**
ACMSP047 Year 2	Identify practical activities and everyday events that involve chance. Describe outcomes as 'likely' or 'unlikely' and identify some events as 'certain' or 'impossible'	252, 253, 254, 255, 256, 257
ACMSP067 Year 3	Conduct chance experiments, identify and describe possible outcomes and recognise variation in results	260, 261
ACMSP092 Year 4	Describe possible everyday events and order their chances of occurring	256, 257
ACMSP093 Year 4	Identify everyday events where one cannot happen if the other happens	258, 259
ACMSP094 Year 4	Identify events where the chance of one will not be affected by the occurrence of the other	258, 259
ACMSP116 Year 5	List outcomes of chance experiments involving equally likely outcomes and represent probabilities of those outcomes using fractions	260, 261, 262, 263, 264, 265
ACMSP117 Year 5	Recognise that probabilities range from 0 to 1	266, 267
10. Data		**PAGES 272–301**
ACMSP048 Year 2	Identify a question of interest based on one categorical variable. Gather data relevant to the question	272, 273, 274
ACMSP049 Year 2	Collect, check and classify data	272, 273, 274
ACMSP050 Year 2	Create displays of data using lists, table and picture graphs and interpret them	275, 276, 277
ACMSP068 Year 3	Identify questions or issues for categorical variables. Identify data sources and plan methods of data collection and recording	278, 279, 280
ACMSP069 Year 3	Collect data, organise into categories and create displays using lists, tables, picture graphs and simple column graphs, with and without the use of digital technologies	278, 279, 280
ACMSP096 Year 4	Construct suitable data displays, with and without the use of digital technologies, from given or collected data. Include tables, column graphs and picture graphs where one picture can represent many data values	275, 276, 277, 278, 279, 280
ACMSP118 Year 5	Pose questions and collect categorical or numerical data by observation or survey	272, 273, 274
ACMSP119 Year 5	Construct displays, including column graphs, dot plots and tables, appropriate for data type, with and without the use of digital technologies	281, 282, 283, 284, 285, 286, 287, 288, 289, 290, 291, 292, 293, 294, 295
ACMSP120 Year 5	Describe and interpret different data sets in context	293, 294, 295, 296

THREE-DIGIT AND FOUR-DIGIT NUMBERS

A three-digit number has three digits. Numbers from 100 to 999 are three-digit numbers.

Examples:
328, 164, 989, 273

Example 1:

634

6 in the hundreds place

3 in the tens place

4 in the ones place

Example 2:

A four-digit number has four digits. Numbers from 1000 to 9999 are four-digit numbers.

Examples:
3842, 6890, 8009, 9513

SCAN to watch video

Example 3:

Example 4:

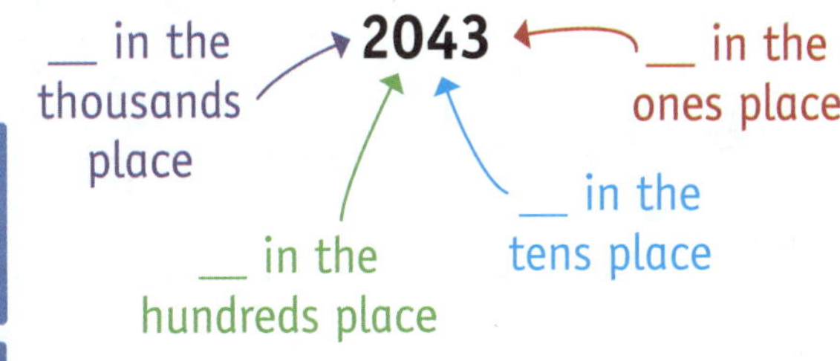

Check your answer on the video!

How many 3-digit numbers are there?

How many 4-digit numbers are there?

Your turn

Circle the three-digit numbers and underline the four-digit numbers.

361 2593 289 107 5945 1495

544 471 6382 1139 4134 929

SELF CHECK Tick how you feel

Got it!	Need help...	I don't get it
☐	☐	☐

Check your answers

How many did you get correct? ☐

PRACTICE

1 Fill in the table.

	Number	Thousands	Hundreds	Tens	Ones
●	357	0	3	5	7
a	924				
b	561				
c	426				
d	840				
e	1384				
f	8972				
g	6259				

2 Write these numbers in words.

● 6259 six thousand, two hundred and fifty-nine

a 783 ______________________

b 248 ______________________

c 510 ______________________

d 1963 ______________________

e 2490 ______________________

f 5721 ______________________

3 Match the number to the correct label.

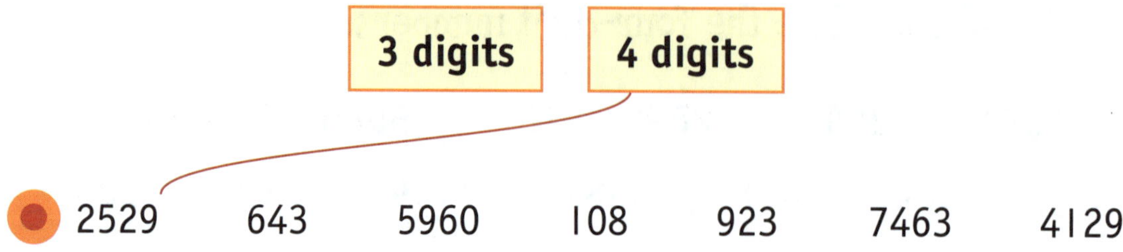

CATCH UP MATHS YEAR 5 BOOK A © PASCAL PRESS ISBN: 9781925726169

PLACE VALUE
THREE-DIGIT AND FOUR-DIGIT NUMBERS

Where a digit is in a number is called the place value.

A three-digit number has a hundreds place, a tens place and a ones place.

Example 1:

HTO
726

The place value of the 7 is hundreds.
The place value of the 2 is tens.
The place value of the 6 is ones.

Example 2:

HTO
308

The place value of the 3 is ____________.
The place value of the __ is tens.
The place value of the 8 is ________.

A four-digit number has a thousands place, a hundreds place, a tens place and a ones place.

Example 3:

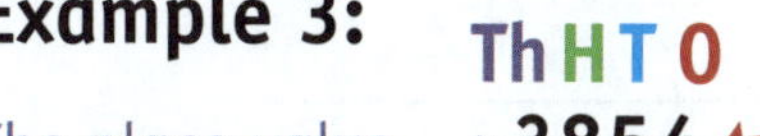

ThHTO
3854

The place value of the 3 is thousands.
The place value of the 8 is hundreds.
The place value of the 5 is tens.
The place value of the 4 is ones.

Example 4:

ThHTO
8614

The place value of the 8 is ____________.
The place value of the 6 is ____________.
The place value of the 1 is ______.
The place value of the __ is ones.

Check your answer on the video!

1 What is the place value of the 5?

● 523 hundreds
a 3275 ____________
b 5961 ____________
c 152 ____________
d 2053 ____________
e 1295 ____________

2 Circle the thousands in purple, hundreds in green, tens in blue and ones in red.

● 2897 a 5921 b 253 c 320 d 4815

SELF CHECK Tick how you feel

Got it!	Need help...	I don't get it
☐	☐	☐

Check your answers
How many did you get correct? ☐

PRACTICE

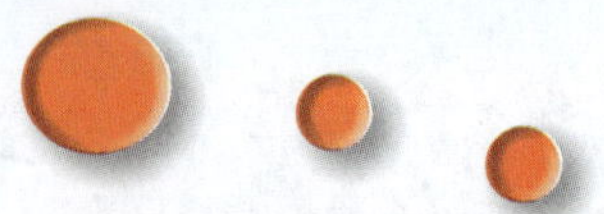

1 Write the place value of the underlined digit.

●	$37\underline{2}4$	tens	g	$\underline{1}08$	______
a	$5\underline{2}36$	______	h	$64\underline{2}$	______
b	$\underline{7}314$	______	i	$\underline{7}398$	______
c	$25\underline{7}$	______	j	$24\underline{5}0$	______
d	$12\underline{9}3$	______	k	$37\underline{0}6$	______
e	$1\underline{0}36$	______	l	$\underline{5}2$	______
f	$51\underline{3}0$	______	m	$325\underline{4}$	______

2 Write the numbers.

● 6 thousands, 2 tens, 3 hundreds, 0 ones 6320

a 5 tens, 8 ones, 3 thousands, 4 hundreds ______

b 9 hundreds, 0 tens, 0 ones ______

c 6 ones, 5 tens, 7 hundreds ______

d 1 thousands, 4 ones, 8 hundreds, 3 tens ______

e 2 hundreds, 3 tens, 5 ones ______

3 Colour the digits in the thousands purple, hundreds green, tens blue and ones red.

● 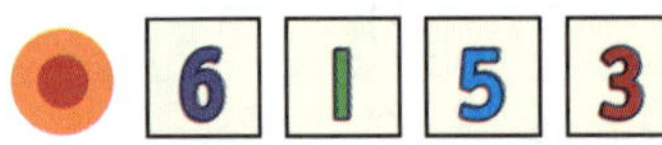

●	6 1 5 3	e	8 3 7	j	2 1 9
a	8 4 3 5	f	4 5 8 7	k	2 0 1 3
b	4 7 5 6	g	9 2 6	l	3 2 4
c	6 6 5	h	8 1 7 0	m	5 9 6
d	7 5 6	i	6 1 0 9	n	4 9 2 3

CATCH UP MATHS YEAR 5 BOOK A © PASCAL PRESS ISBN: 9781925726169

VALUE
THREE-DIGIT AND FOUR-DIGIT NUMBERS

The value of a number is how much a number is worth.
To find the value of a digit, look at where it is in a number.

**For example, what is the value of the 3 in 7310?
The 3 is in the hundreds column, so the value of 3 is 300.**

Remember, the place value is where the digit is in a number.

The value is how much the digit is worth.

Example 1:

In the number 593,
the value of 5 is 500
the value of 9 is 90
and the value of 3 is 3.

Example 2:

In the number 1729,
the value of 1 is 1000
the value of 7 is 700
the value of 2 is 20
and the value of 9 is 9.

Example 3:

In the number 471,
the value of 4 is ____,
the value of __ is 70
and the value of 1 is __.

Example 4:

In the number 8235,
the value of 8 is ____
the value of 2 is ____
the value of __ is 30
and the value of 5 is 5.

Circle the numbers where the value of 3 is 30.

- (537) 634 3921 8231
- 1325 1813 432 13

SELF CHECK Tick how you feel

Got it!	Need help...	I don't get it
☐	☐	☐

Check your answers
How many did you get correct? ☐

PRACTICE

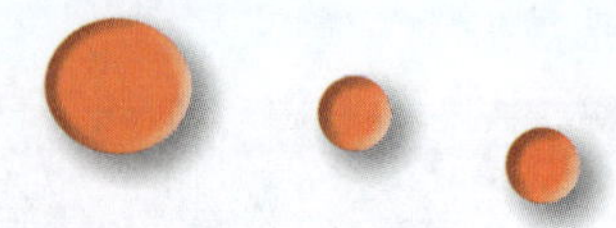

1 What is the value of 4 in these numbers?

● 3247	40	e 6842	______	j 4832	______
a 1354	______	f 1473	______	k 743	______
b 436	______	g 4825	______	l 4903	______
c 134	______	h 7415	______	m 849	______
d 543	______	i 6497	______	n 5243	______

2 Write a number that has a 5 with the given value.

● 50	4953	b 500	______	d 5000	______
a 5	______	c 50	______		

3 Circle the digit with the least value and underline the digit with the most value.

● 5382	d 2589	h 582	l 873
a 530	e 9803	i 471	m 684
b 1037	f 1009	j 4936	n 3781
c 265	g 8632	k 5419	o 9275

4 Circle the numbers with the matching value.

● The value of 5 is 50: 457 255 386 857 59 2581

a The value of 4 is 400: 342 439 5543 477 1472 415

b The value of 6 is 6000: 6358 4600 1362 6426 47 6156

c The value of 7 is 7: 1297 1732 407 372 5470 1357

d The value of 3 is 30: 735 32 4534 351 937 358

CATCH UP MATHS YEAR 5 BOOK A © PASCAL PRESS ISBN: 9781925726169

NUMBER EXPANDERS

THREE-DIGIT AND FOUR-DIGIT NUMBERS

A number expander for the three-digit number 386 has hundreds, tens and ones.

3	hundreds	8	tens	6	ones

A number expander for the four-digit number 2849 has thousands, hundreds, tens and ones.

2	thousands	8	hundreds	4	tens	9	ones

Number expanders can be folded:

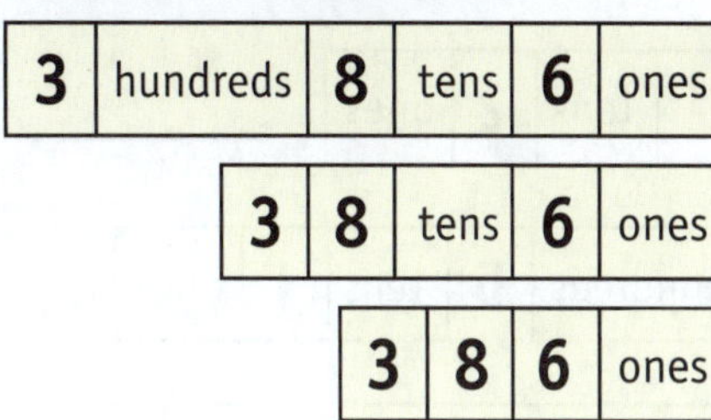

3	hundreds	8	tens	6	ones
		3	8	tens	6	ones
			3	8	6	ones

SCAN to watch video

Example 1: Show 2849 on the number expander.

2	thousands	8	hundreds	4	tens	9	ones
		2	8	hundreds	4	tens	9	ones
				2	8	4	tens	9	ones
					2	8	4	9	ones

Example 2: Show 897 on the number expander.

	hundreds		tens		ones
				tens		ones
						ones

Check your answer on the video!

Your turn

Show the numbers on the number expanders.

● 846

8	hundreds	4	tens	6	ones
		8	4	tens	6	ones
			8	4	6	ones

a 478

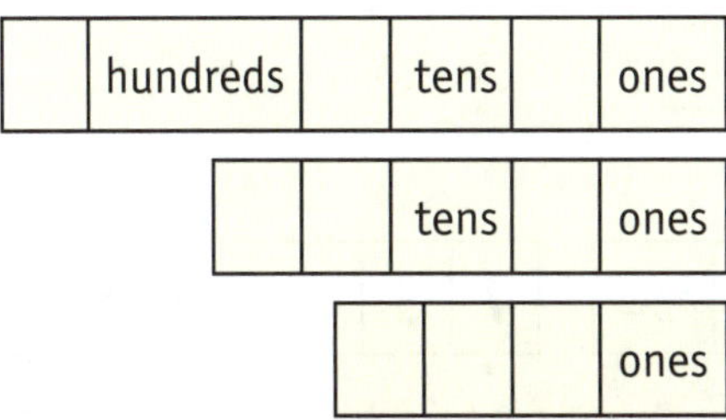

	hundreds		tens		ones
				tens		ones
						ones

b 9321

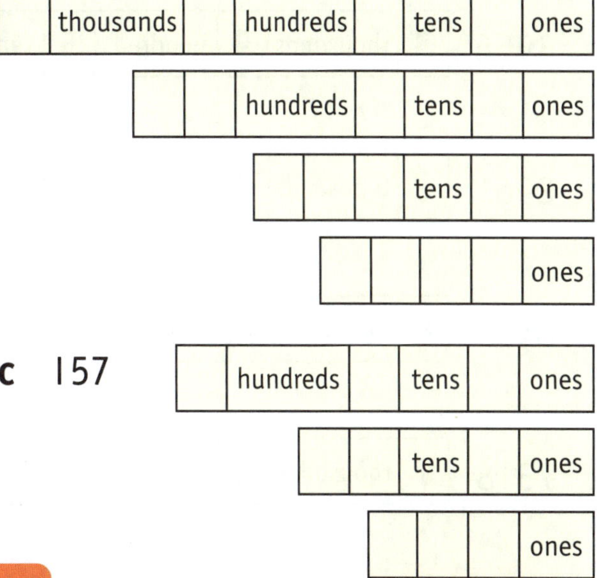

	thousands		hundreds		tens		ones
				hundreds		tens		ones
						tens		ones
							ones

c 157

	hundreds		tens		ones
				tens		ones
						ones

SELF CHECK Tick how you feel

Got it! ☐  Need help... ☐  I don't get it ☐

Check your answers

How many did you get correct? ☐

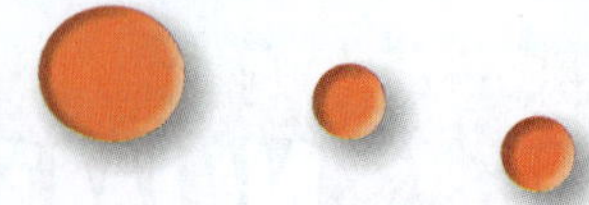

1 Write the number shown on the number expanders.

- | 6 | thousands | 0 | hundreds | 9 | tens | 3 | ones | 6093

a | 7 | thousands | 1 | hundreds | 4 | tens | 3 | ones | ______

b | 6 | 7 | tens | 2 | ones | ______

c | 2 | hundreds | 5 | tens | 1 | ones | ______

d | 5 | 7 | hundreds | 4 | tens | 9 | ones | ______

e | 3 | 4 | 8 | 6 | ones | ______

f | 3 | 2 | hundreds | 5 | tens | 9 | ones | ______

2 Cross out the incorrect number expander.

- 2583 | 2 | thousands | 5 | hundreds | 8 | tens | 3 | ones | ~~| 5 | 2 | hundreds | 8 | tens | 3 | ones |~~

a 652 | 6 | hundreds | 5 | tens | 2 | ones | | 5 | 2 | tens | 6 | ones |

b 3956 | 3 | thousands | 9 | hundreds | 5 | tens | 6 | ones | | 3 | 9 | 6 | 5 | ones |

c 2961 | 2 | thousands | 6 | hundreds | 9 | tens | 1 | ones | | 2 | 9 | 6 | tens | 1 | ones |

d 843 | 8 | 4 | tens | 3 | ones | | 8 | 3 | 4 | ones |

e 7318 | 7 | thousands | 3 | hundreds | 1 | tens | 8 | ones | | 7 | 3 | hundreds | 8 | tens | 1 | ones |

f 259 | 2 | hundreds | 9 | tens | 5 | ones | | 2 | 5 | tens | 9 | ones |

CATCH UP MATHS YEAR 5 BOOK A © PASCAL PRESS ISBN: 9781925726169

EXPANDED NUMBERS
THREE-DIGIT AND FOUR-DIGIT NUMBERS

We expand a number by writing the value of each digit.

Example 1: Write 763 in expanded form.

763 = 700 + 60 + 3

(7 hundreds) (6 tens) (3 ones)

When a number is written this way, it is in expanded form.

Example 2: Write 2852 in expanded form.

2852 = 2000 + 800 + 50 + 2

(2 thousands) (8 hundreds) (5 tens) (2 ones)

When a number has a zero in it, leave out that place value.

Example 3:

3092 = 3000 + 90 + 2

(3 thousands) (9 tens) (2 ones)

Leave out the hundreds.

Example 4:

Complete the expanded form.

8243 = 8000 + ____ + 40 + __

(8 ______________) (2 hundreds) (__ tens) (3 ones)

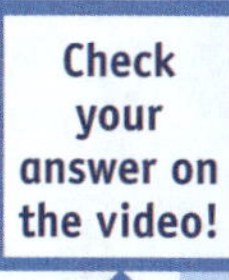

Write these numbers in expanded form.

- 347 = 300 + 40 + 7
  = 3 hundreds + 4 tens + 7 ones

a 432 = ____ + ____ + __
  = __ hundreds + __ tens + __ ones

b 3065 = ______ + ___ + __
  = __ thousands + __ tens + __ ones

Check your answers
How many did you get correct?

PRACTICE

1 Write these numbers in expanded form.

- 2538 = 2000 + 500 + 30 + 8

a 857 = ____ + ____ + ____

b 329 = ____ + ____ + ____

c 1093 = ______ + ____ + ____

d 1834 = ______ + ____ + ____ + ____

e 506 = ____ + ____

f 4370 = ______ + ____ + ____

2 Write these numbers in expanded form.

- 1464 = 1000 + 400 + 60 + 4

a 302 = ______________________

b 1642 = ______________________

c 2037 = ______________________

d 4060 = ______________________

e 850 = ______________________

3 Match the numbers with their expanded form.

• 282	6000 + 800 + 9
a 1463	200 + 80 + 2
b 707	400 + 20 + 4
c 1490	1000 + 400 + 60 + 3
d 540	1000 + 400 + 90
e 6809	500 + 40
f 424	700 + 7

CATCH UP MATHS YEAR 5 BOOK A © PASCAL PRESS ISBN: 9781925726169

MODELLING NUMBERS
THREE-DIGIT AND FOUR-DIGIT NUMBERS

You can use Base 10 blocks and an abacus to model numbers.

Example 1:
Model 436 with base 10 blocks and on the abacus.

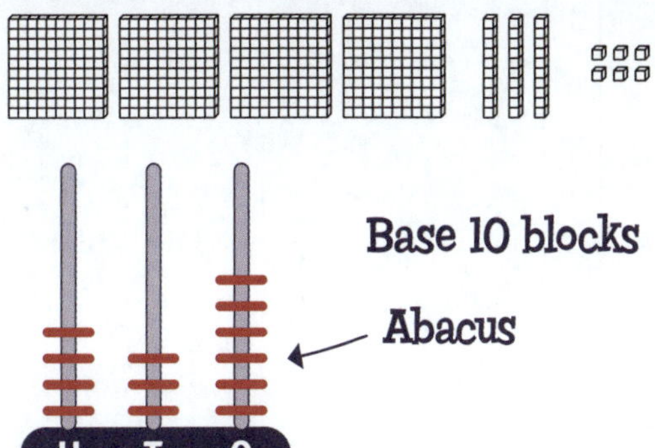

Example 3: Complete the number models for 256.

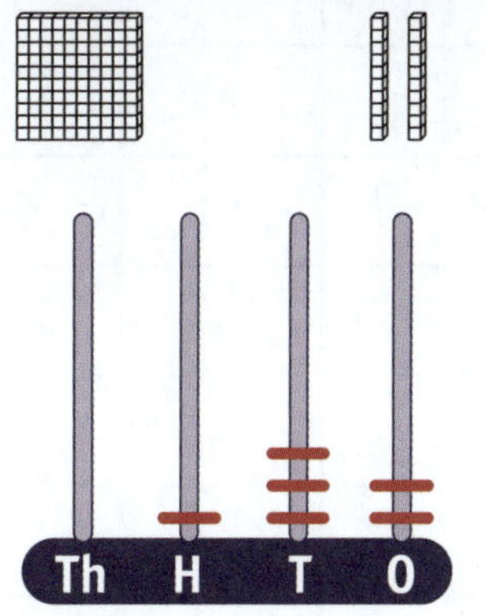
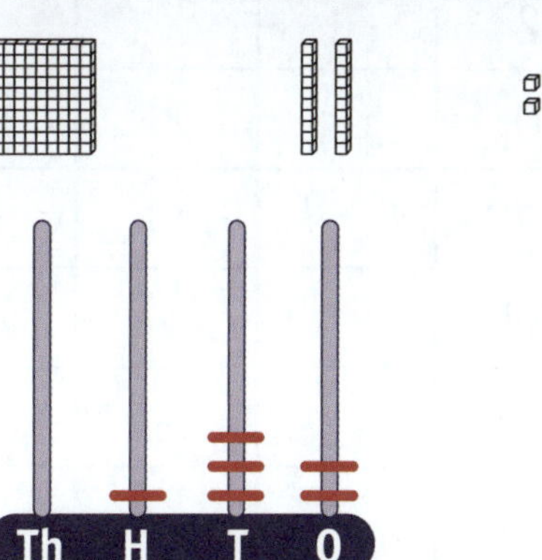

Example 2: Model 3051.

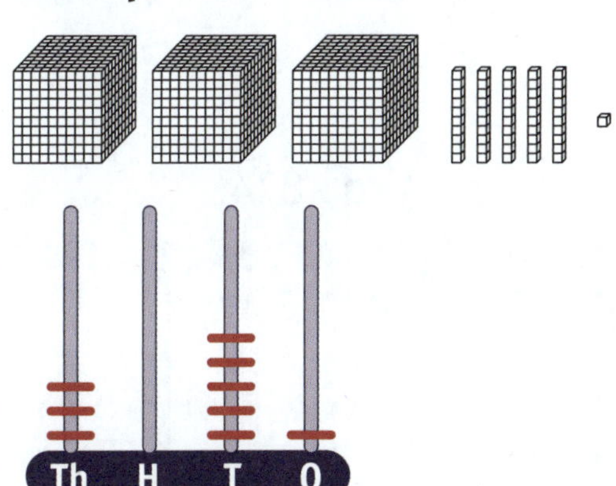

Example 4: Complete the number models for 1213.

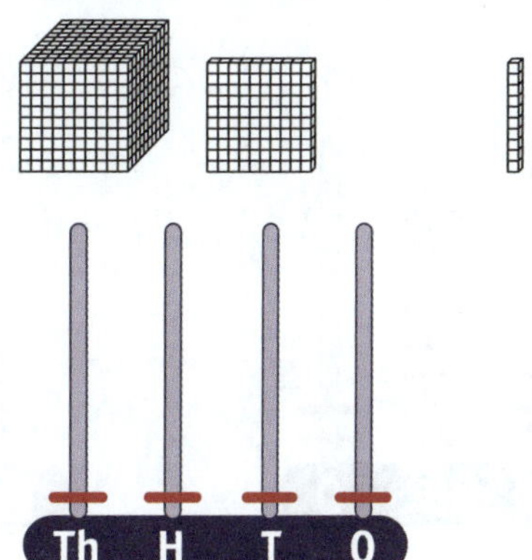

Your turn

Write the numbers.

● 4253

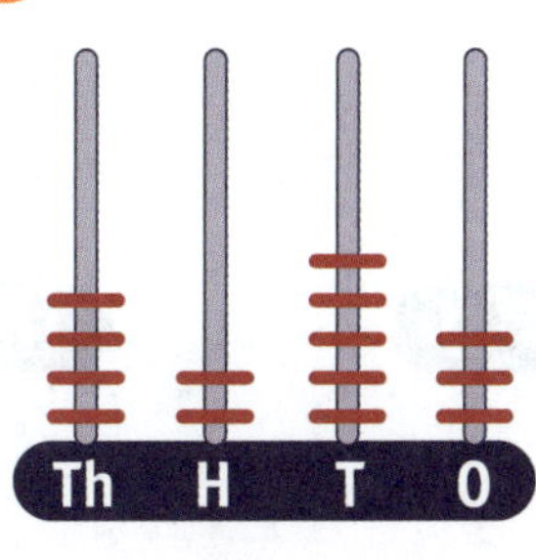

a ______

b ______

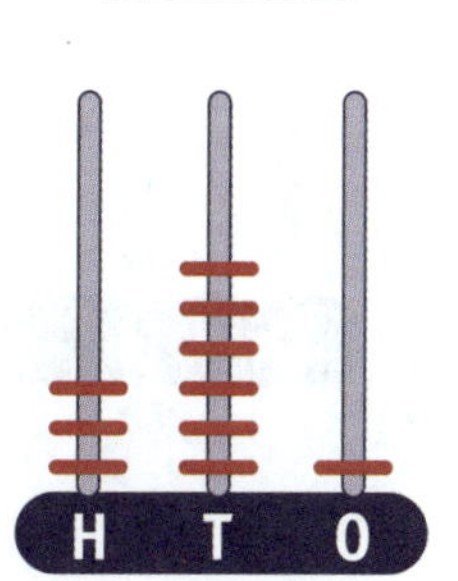

SELF CHECK Tick how you feel

Got it!	Need help...	I don't get it
☐	☐	☐

Check your answers

How many did you get correct?

PRACTICE

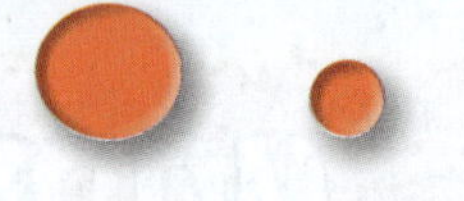

1 How many of each block do you need to make the numbers?

	Number	thousands block	hundreds block	tens block	ones block
●	356	0	3	5	6
a	732				
b	410				
c	503				
d	1083				
e	4860				
f	3609				
g	5917				

2 Show the number on the abacus.

● 4382

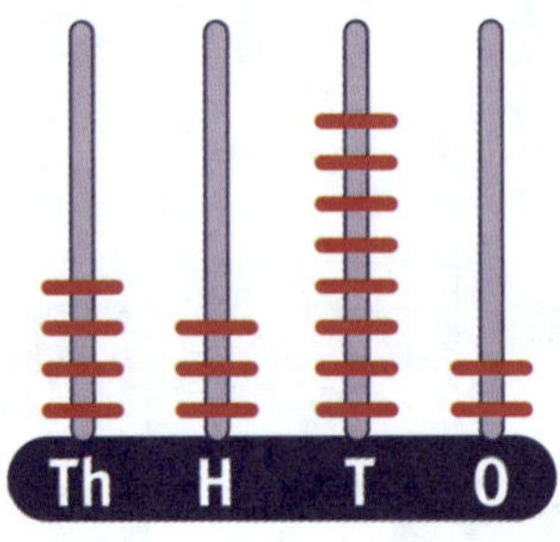

b 340

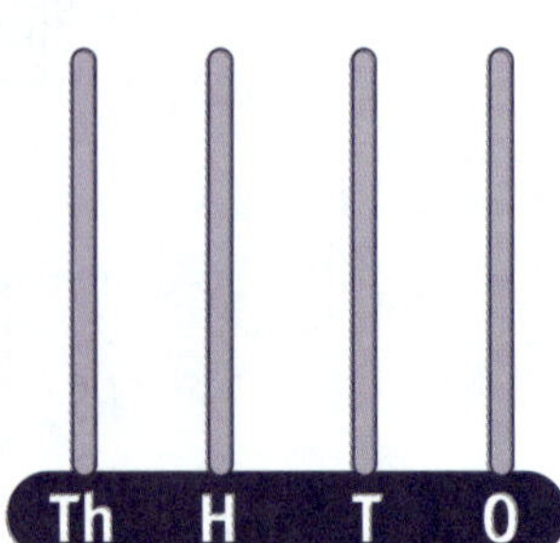

d 874

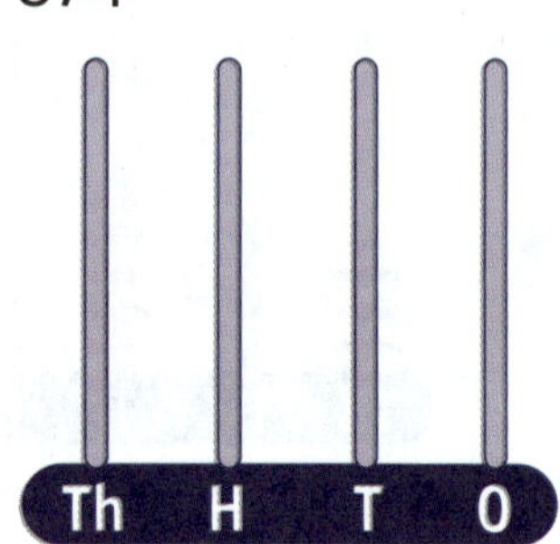

a 7685

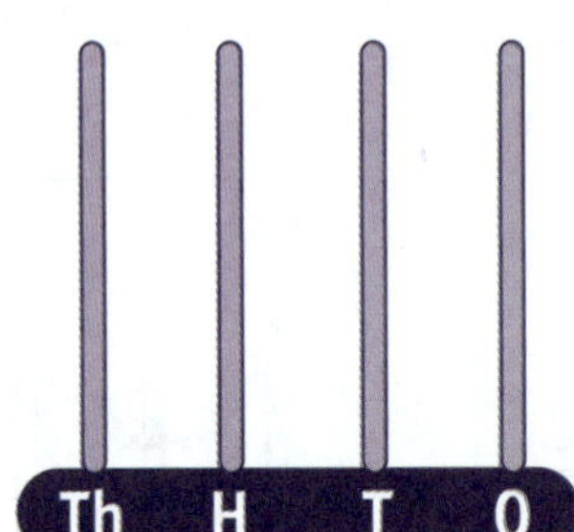

c 2708

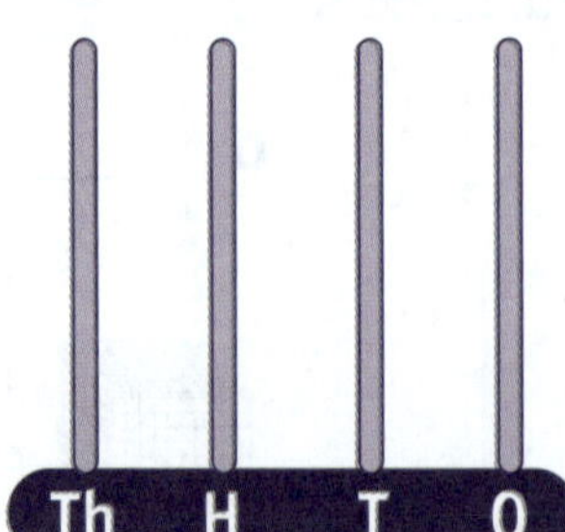

e 5983

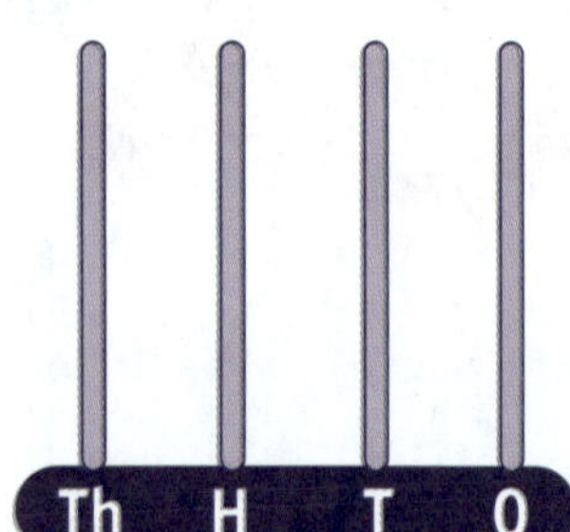

CATCH UP MATHS YEAR 5 BOOK A © PASCAL PRESS ISBN: 9781925726169

ORDERING NUMBERS
THREE-DIGIT AND FOUR-DIGIT NUMBERS

When numbers are ordered from smallest to largest, it is called ascending order.
When numbers are ordered largest to smallest, it is called descending order.

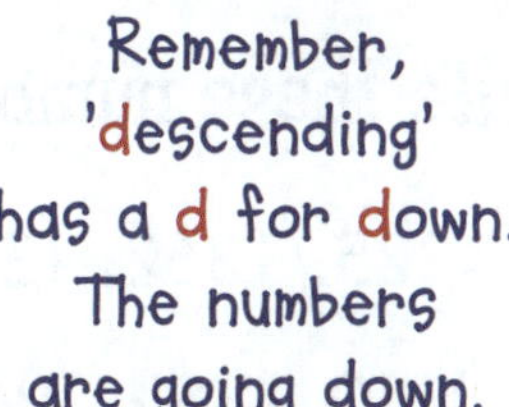

Example 1:
These 3-digit numbers are in ascending order:
125, 192, 195, 205

Example 2:
These 3-digit numbers are in descending order:
205, 195, 192, 125

Example 3:
These 4-digit numbers are in ascending order:
1625, 1792, 1826, 1983

Example 4:
These 4-digit numbers are in descending order:
1983, 1826, 1792, 1625

Example 5:
Write a number that fits into this list of ascending numbers.
249, 327, _____, 409

Example 6:
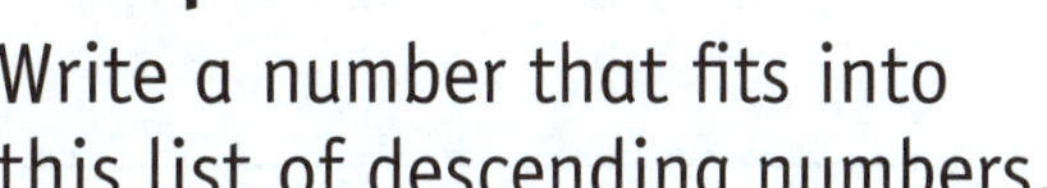
Write a number that fits into this list of descending numbers.
8219, 6467, _______, 5124

Write A for ascending or D for descending to describe how the numbers are ordered.

- ● 253, 297, 1358 **A**
- **a** 372, 419, 536 ☐
- **b** 1324, 1493, 2974 ☐
- **c** 957, 942, 810 ☐
- **d** 5382, 5625, 5789 ☐
- **e** 6352, 6125, 6005 ☐

SELF CHECK Tick how you feel		
Got it! ☐	Need help... ☐	I don't get it ☐

Check your answers
How many did you get correct? ☐

PRACTICE

1 Write these numbers in ascending order.

- 523, 632, 143, 510
  143, 510, 523, 632

a 101, 536, 217, 437

b 2597, 2417, 2560, 2650

c 243, 342, 423, 432

d 1791, 1197, 1719, 1179

e 6560, 6650, 5660, 5060

2 Write these numbers in descending order.

- 1325, 1362, 1341, 1105, 1610
  1610, 1362, 1341, 1325, 1105

a 110, 635, 712, 743, 620

b 2795, 2714, 2605, 2560, 2597

c 342, 423, 603, 204, 362

d 3752, 3257, 3725, 3275, 3572

e 4901, 4119, 4911, 4191, 4109

3 Match these groups of numbers with the correct label to describe the order of the numbers.

- 6243, 6124, 1264, 1064 → descending

a 249, 942, 1249, 1942

b 7260, 7000, 4382, 2876

ascending

descending

c 126, 621, 1026, 1626

d 439, 934, 1037, 1255

e 5541, 5411, 1455, 1154

FIVE-DIGIT NUMBERS

A five-digit number is a number from 10 000 to 99 999.
A number with 5 digits is in the tens of thousands.
A 5-digit number has five digits.

25 235 62 349 92 106 57 965 19 000

Example 1: The number 25 345 in words is twenty-five thousand, three hundred and forty-five.

Example 2: The number 92 304 in words is ninety-two thousand, three hundred and four.

Example 3: The number 84 372 in words is eighty-four ________________, three ______________ and seventy-_____.

Example 4: The number 73 205 in words is seventy-three ________________, ______ hundred and _____.

1 Underline the numbers that have five digits.

47 236 24 37 651 6 1523 10 350

721 29 734 73 497 13 659 42 715

2 Write these numbers in words.

83 264

eighty-three thousand, two hundred and sixty-four

a 40 395

____

b 62 536

____

Check your answers
How many did you get correct?

PRACTICE

Match the numbers to the words.

●	17 490	forty-two thousand and ten
a	50 286	twenty-four thousand, one hundred
b	93 275	seventeen thousand, four hundred and ninety
c	87 340	thirty-six thousand and nine
d	67 038	eighty-seven thousand, three hundred and forty
e	24 100	sixty-seven thousand and thirty-eight
f	71 854	ninety-three thousand, two hundred and seventy-five
g	36 009	seventy-one thousand, eight hundred and fifty-four
h	42 010	fifty thousand, two hundred and eighty-six

Write these numbers in words.

● 99 382 ninety-nine thousand, three hundred and eighty-two

a 74 056 __________

b 80 090 __________

c 32 840 __________

d 62 008 __________

Circle all the five-digit numbers.

● (38 252) 68 560 4300 22 870 32 590

3247 28 490 30 003 7214 45 310

CATCH UP MATHS YEAR 5 BOOK A © PASCAL PRESS ISBN: 9781925726169

PLACE VALUE
FIVE-DIGIT NUMBERS

The place value is the value of a digit based on where it is in a number.

Example 1: This place value chart shows 89 374.

Ten Thousands	Thousands	Hundreds	Tens	Ones
8	9	3	7	4

The number 89 374 has:

8 ten thousands
9 thousands
3 hundreds
7 tens
4 ones

Example 2:
The number 74 351 has:

__ ten thousands
__ thousands
__ hundreds
__ tens
__ ones

Example 3:
The number 15 113 has:

1 ______________
5 ______________
1 ______________
1 __________
3 __________

Your turn

Trace the numbers in the ten thousands place in orange, the thousands place in purple, the hundreds place in green, the tens place in blue and the ones place in red.

● 25 361   43 240   329   7267   927

1352   672   93   3289   14   63 053

SELF CHECK Tick how you feel

Got it!	Need help...	I don't get it
☐	  ☐	☐

Check your answers
How many did you get correct?

  ISBN: 9781925726169

PRACTICE

1 Circle ten thousands in orange, thousands in purple, hundreds in green, tens in blue and ones in red.

- (3)(6) (2)(8)(5)
- a 42 560
- b 95 390
- c 59 936
- d 99 363
- e 38 342
- f 67 429
- g 84 132
- h 53 931
- i 19 150
- j 10 000
- k 35 003
- l 62 415
- m 71 317
- n 59 020

2 Fill in the chart.

	Number	Ten Thousands	Thousands	Hundreds	Tens	Ones
●	52 059	5	2	0	5	9
a	63 000					
b	42 618					
c	86 254					
d	38 913					
e	71 352					

3 What is the place value of the underlined digits?

- 36 $\underline{2}$58 hundreds
- a $\underline{2}$4 736 ____________
- b 42 82$\underline{0}$ ____________
- c 3$\underline{0}$ 745 ____________
- d 65 8$\underline{1}$2 ____________
- e 19 $\underline{0}$56 ____________
- f $\underline{2}$2 490 ____________
- g 7$\underline{3}$ 592 ____________
- h 38 $\underline{2}$50 ____________
- i 53 2$\underline{5}$2 ____________
- j 84 25$\underline{6}$ ____________
- k 9$\underline{1}$ 029 ____________
- l 83 $\underline{2}$43 ____________

CATCH UP MATHS YEAR 5 BOOK A © PASCAL PRESS ISBN: 9781925726169

VALUE
FIVE-DIGIT NUMBERS

The value of a number is how much it is worth.

Example 1:

TT Th H T O

25 967

The value of the 2 is 20 000.

The value of the 5 is 5000.

The value of the 9 is 900.

The value of the 6 is 60.

The value of the 7 is 7.

Example 2:

TT Th H T O

78 496

The value of the 7 is ________.

The value of the 8 is ______.

The value of the 4 is _____.

The value of the 9 is ___.

The value of the 6 is __.

1 Write a number where 6 has the given value.

●	60	363	c	6000	________
a	600	________	d	60 000	________
b	6	________	e	60	________

2 Circle the numbers where 5 has a value of 50.

● (652) 5632 54 64 357 15

2593 3156 3205 50 326 25 937

SELF CHECK Tick how you feel

Got it!	Need help...	I don't get it
 ☐	 ☐	☐

Check your answers
How many did you get correct? ☐

PRACTICE

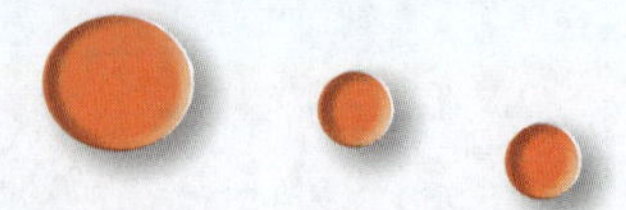

1 What is the value of the underlined digit?

- ● 32 651 — 30 000
- a 36 324 ______
- b 80 357 ______
- c 71 352 ______
- d 99 062 ______
- e 14 271 ______
- f 52 437 ______
- g 81 950 ______
- h 13 313 ______
- i 95 449 ______
- j 57 595 ______
- k 25 311 ______

2 Use the digits 4, 9, 8, 5, 2 to make a five-digit number with:

- ● 9 ten thousands — 94 852
- a 2 hundreds ______
- b 5 ones ______
- c 4 thousands ______
- d 9 ones ______
- e 8 thousands ______
- f 2 tens ______
- g 4 tens ______
- h 8 hundreds ______

3 Write the five-digit number in different ways.

a 53 428

- ● 53 thousands + 428
- 534 hundreds + ______
- 53 428 + ______
- 5342 tens + ______
- 5 ten thousands + ______

b 56 320

- ● 563 hundreds + 20
- 56 320 + ______
- 56 thousands + ______
- 5 ten thousands + ______
- 5632 tens + ______

CATCH UP MATHS YEAR 5 BOOK A © PASCAL PRESS ISBN: 9781925726169

NUMBER EXPANDERS
FIVE-DIGIT NUMBERS

A number expander can show different ways of writing a number.

6	ten thousands	9	thousands	5	hundreds	7	tens	2	ones

SCAN to watch video

Example 1: We can make 69 572 using only thousands, hundreds, tens and ones:

6	9	thousands	5	hundreds	7	tens	2	ones

And using hundreds, tens and ones:

6	9	5	hundreds	7	tens	2	ones

Only tens and ones:

6	9	5	7	tens	2	ones

Using only ones:

6	9	5	7	2	ones

69 572 has:
- 6 ten thousands
- 9 thousands
- 5 hundreds
- 7 tens
- 2 ones

Example 2: Complete the number expanders for 16 892.

1	ten thousands	6	thousands	8	hundreds	9	tens	2	ones

		thousands	8	hundreds	9	tens	2	ones

1	6	8	hundreds		tens		ones

				tens	2	ones

					ones

Check your answer on the video!

Your turn

Complete the number expanders for 53 247.

5	ten thousands	3	thousands	2	hundreds	4	tens	7	ones

		thousands		hundreds		tens		ones

			hundreds		tens		ones

				tens		ones

					ones

SELF CHECK Tick how you feel

Got it! ☐ Need help... ☐ I don't get it ☐

Check your answers

How many did you get correct? ☐

  ISBN: 9781925726169

PRACTICE

1 Write the numbers.

- 64 thousands, 3 hundreds, 4 tens, 1 ones 64 341

a 5 ten thousands, 6 tens, 5 ones, 1 hundred ________

b 8 hundreds, 7 ten thousands ________

c 472 hundreds, 4 ones, 9 tens ________

d 9642 tens, 8 ones ________

e 596 hundreds, 0 tens, 3 ones ________

f 84 thousands, 9 ones, 5 tens ________

2 Write the numbers on the number expanders.

• 63 249	6	ten thousands	3	thousands	2	hundreds	4	tens	9	ones
a 47 326		ten thousands		thousands		hundreds		tens		ones
b 81 016		ten thousands		thousands		hundreds		tens		ones
c 92 304		ten thousands		thousands		hundreds		tens		ones

3 Match the numbers to the number expanders.

- 47 230

a 27 349

b 91 462

c 82 954

d 90 365

e 57 952

2	7	3	4	tens	9	ones		
8	2	9	hundreds	5	tens	4	ones	
4	7	thousands	2	hundreds	3	tens	0	ones
9	0	3	6	tens	5	ones		
5	7	9	hundreds	5	tens	2	ones	
9	1	4	6	2	ones			

CATCH UP MATHS YEAR 5 BOOK A © PASCAL PRESS ISBN: 9781925726169

EXPANDED NUMBERS
FIVE-DIGIT NUMBERS

When a number is written to show the value of each digit, it is called an expanded number.

Example 1: Write 3629 in expanded form.

3629 = 3000 (3 thousands) + 600 (6 hundreds) + 20 (2 tens) + 9 (9 ones)

Example 2: Write 59 284 in expanded form.

59 284 = 50 000 (5 ten thousands) + 9000 (9 thousands) + 200 (2 hundreds) + 80 (8 tens) + 4 (4 ones)

Example 3: Write 85 409 in expanded form.

85 409 = ________ (8 ten thousands) + 5000 (__ thousands) + 400 (__ hundreds) + __ (9 ones)

Example 4: Write 16 265 in expanded form.

16 265 = ________ (1 ten thousand) + 6000 (__ thousands) + 200 (__ hundreds) + ___ (6 tens) + __ (5 ones)

Complete the expanded numbers.

Check your answer on the video!

- 56 324 = 50 000 + 6000 + 300 + 20 + 4

a 72 393 = 70 000 + 2000 + ____ + 90 + 3

b 1874 = ______ + 800 + 70 + 4

c 98 731 = 90 000 + 8000 + ____ + 30 + 1

d 5638 = 5000 + 600 + ___ + 8

Check your answers
How many did you get correct?

PRACTICE

1 Match the numbers to their expanded form.

	Number	Expanded form
●	24 937	100 + 90 + 3
a	15 471	5000 + 600 + 40 + 9
b	3584	50 + 7
c	29 762	20 000 + 9000 + 700 + 60 + 2
d	35 847	3000 + 500 + 80 + 4
e	57	10 000 + 5000 + 400 + 70 + 1
f	193	20 000 + 4000 + 900 + 30 + 7
g	5649	30 000 + 5000 + 800 + 40 + 7

2 Write these numbers in expanded form.

● 75 389 = 70 000 + 5000 + 300 + 80 + 9

a 22 387 = ______________________

b 91 476 = ______________________

c 89 295 = ______________________

d 73 049 = ______________________

3 Fill in the missing numbers.

● 62 583 = 60 000 + 2000 + 500 + 80 + 3

a 74 860 = ______ + 4000 + ____ + ___ + 0

b 81 060 = 80 000 + _____ + ___

c 16 036 = ______ + 6000 + ___ + __

d 97 398 = 90 000 + 7000 + ____ + 90 + __

e 23 529 = ______ + 3000 + ____ + 20 + __

f 54 207 = 50 000 + 4000 + ____ + __

CATCH UP MATHS YEAR 5 BOOK A © PASCAL PRESS ISBN: 9781925726169

MODELLING NUMBERS
FIVE-DIGIT NUMBERS

You can use Base 10 blocks and an abacus to model numbers.

Example 1: Model 24 385 with base 10 blocks and on the abacus.

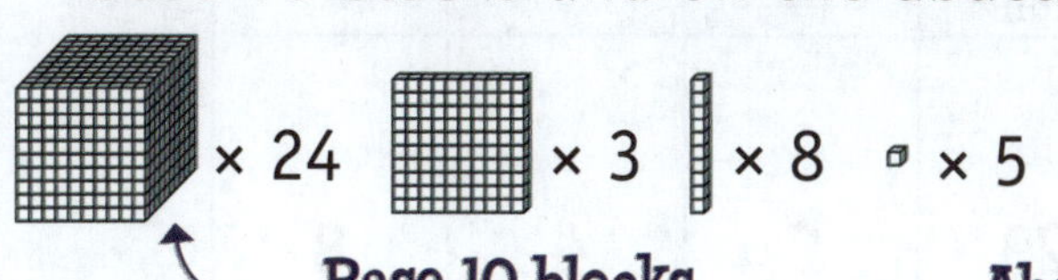

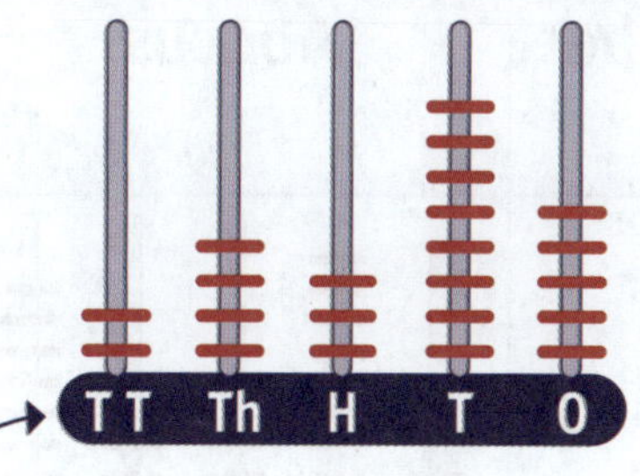

Example 2: Model 59 741.

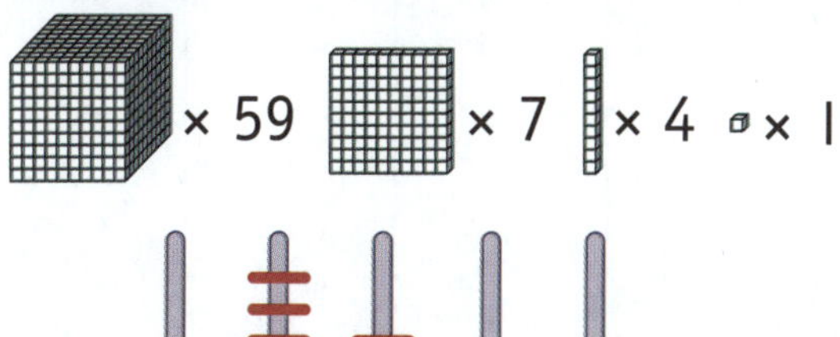

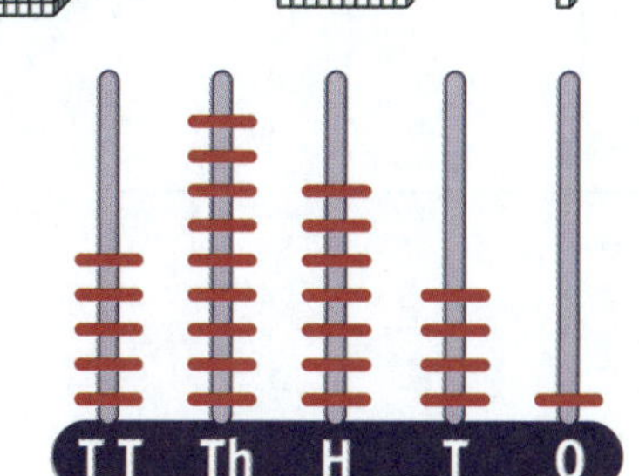

Example 3: Complete the number models for 63 812.

× 63 × ___ × ___ × 2

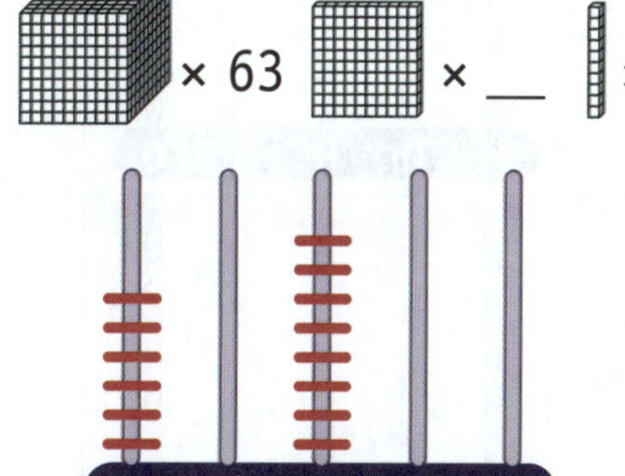

Your turn

1 Write the numbers that are modelled with base 10 blocks.

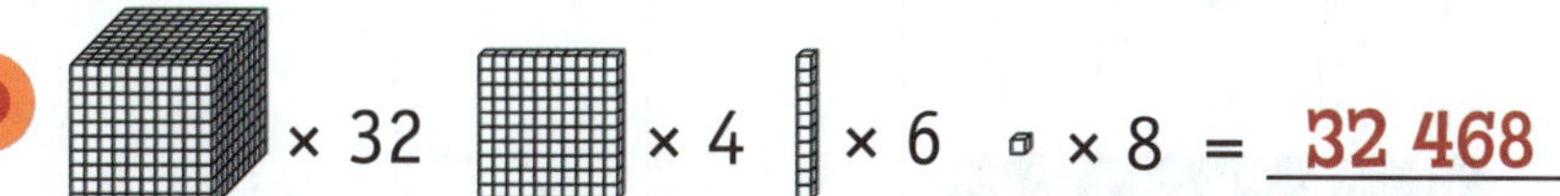

= 32 468

a × 16 × 2 × 0 × 9 = ________

2 Write the numbers that are modelled on the abacus.

23 616 a ________ b ________

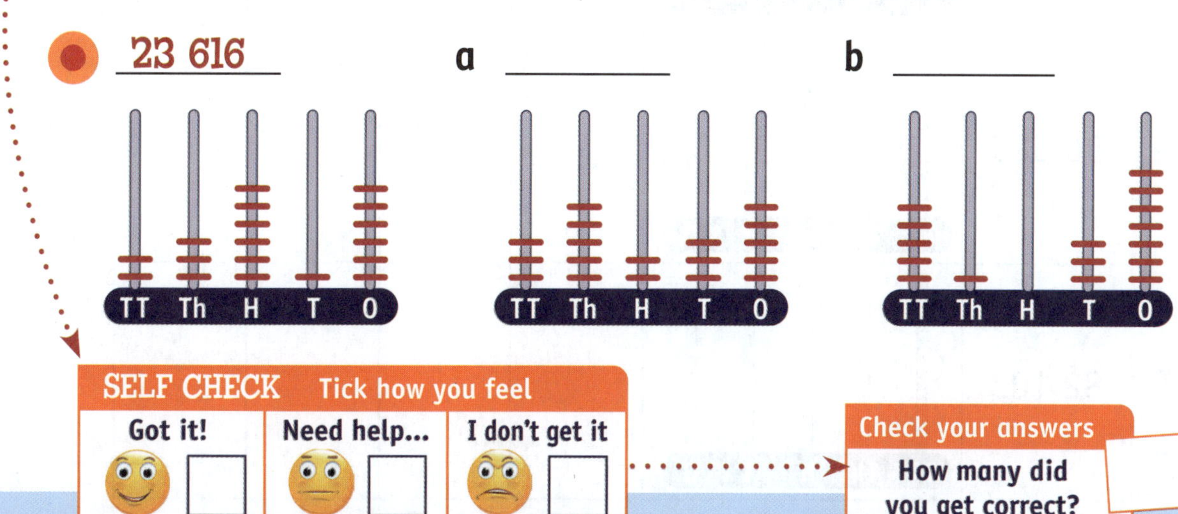

Check your answers
How many did you get correct?

PRACTICE

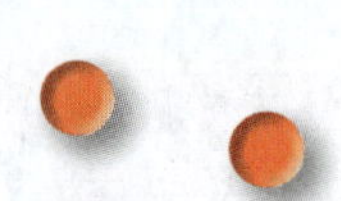

1 Model each number on the abacus and with Base 10 blocks.

	Number	Abacus	Base 10			
			thousands cube	hundreds flat	tens long	ones cube
●	29 348	TT Th H T O	29	3	4	8
a	37 647	TT Th H T O				
b	98 980	TT Th H T O				
c	74 342	TT Th H T O				
d	48 063	TT Th H T O				
e	96 386	TT Th H T O				
f	82 103	TT Th H T O				

CATCH UP MATHS YEAR 5 BOOK A © PASCAL PRESS ISBN: 9781925726169

ORDERING NUMBERS
FIVE-DIGIT NUMBERS

You can arrange numbers in ascending order or descending order.

Example 1: 36 342, 47 359, 53 128, 65 249, 73 142

These numbers are ordered from smallest to largest.
That's ascending order.

Ascending means going up. The smallest number is first.

Descending means going down. The largest number is first.

Example 2: 65 932, 54 638, 42 015, 37 025, 16 539

These numbers are ordered from largest to smallest.
That's descending order.

Example 3:
Write numbers that fit into this list of ascending numbers.

12 456, ________, 38 460, ________, 58 215

Example 4:
Write numbers that fit into this list of descending numbers.

89 995, 60 321, ________, 45 099, ________

Check your answer on the video!

Draw a ✓ beside the numbers in ascending order and a ✗ beside the numbers in descending order.

- 7352, 12 937, 25 864, 38 236 ✓

a 86 325, 87 526, 88 937, 89 473

b 43 249, 42 349, 40 389, 30 373

c 52 345, 50 553, 50 522, 50 001

d 62 523, 65 223, 66 278, 67 349

Check your answers
How many did you get correct?

Write each set of numbers in ascending and then descending order.

382, 6432, 1946, 1911, 2375

Ascending order **382, 1911, 1946, 2375, 6432**

Descending order **6432, 2375, 1946, 1911, 382**

a 3974, 1797, 9745, 4937, 2849

Ascending order ______________________

Descending order ______________________

b 59 263, 26 395, 69 352, 96 532, 29 356

Ascending order ______________________

Descending order ______________________

c 75 631, 13 657, 16 375, 57 136, 67 531

Ascending order ______________________

Descending order ______________________

d 23 468, 32 648, 86 423, 68 324, 48 632

Ascending order ______________________

Descending order ______________________

Order the prize money below from largest (1) to smallest (10). The first one has been done as an example.

Prize Money	Order
$57 293	
$75 216	
$52 999	
$10 195	
$12 639	

Prize Money	Order
$63 147	
$24 652	
$55 250	
$98 343	1
$60 249	

CATCH UP MATHS YEAR 5 BOOK A © PASCAL PRESS ISBN: 9781925726169

ODD AND EVEN NUMBERS

THREE-DIGIT, FOUR-DIGIT AND FIVE-DIGIT NUMBERS

Odd numbers end in 1, 3, 5, 7, 9.	Even numbers end in 2, 4, 6, 8, 0.

Example 1:
Write a 3-digit odd number.
371

Example 2:
Write a 4-digit odd number.
7933

Example 3:
Write a 5-digit odd number.
29 485

Example 4:
Write a 3-digit even number.
732

Example 5:
Write a 4-digit even number.
2374

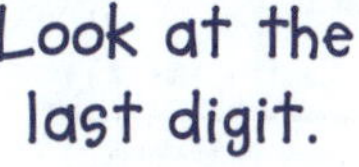

Example 6:
Write a 5-digit even number.
17 340

Example 7:
Write a last digit that makes this number odd:
14 43__

Example 8:
Write a last digit that makes this number even:
63__

Use red to circle the odd numbers and green to circle the even numbers.

2593 1367 389 13 247
8932 432 2468 415
25 386 146 3572 56 930
9009 71 529 133

SELF CHECK Tick how you feel

Got it!	Need help...	I don't get it

Check your answers
How many did you get correct?

PRACTICE

1 **Write the next three numbers in the series.**
Hint: Are the numbers odd or even, and ascending or descending?

- 2324, 2326, 2328, 2330, 2332

a 473, 475, ________, ________, ________

b 1820, 1822, ________, ________, ________

c 896, 898, ________, ________, ________

d 12 438, 12 440, ________, ________, ________

e 98 320, 98 322, ________, ________, ________

f 65 236, 65 234, ________, ________, ________

g 19 085, 19 083, ________, ________, ________

h 2402, 2400, ________, ________, ________

i 873, 871, ________, ________, ________

j 24 936, 24 938, ________, ________, ________

2 **Write the numbers in the correct box.**

- ~~4810~~ 235 767 57 975 1351 2492 42 436

148 646 13 571 25 920 4244 98 999 3593

Odd Numbers	Even Numbers
	4810

GREATER THAN, LESS THAN, EQUAL TO

THREE-DIGIT, FOUR-DIGIT AND FIVE-DIGIT NUMBERS

	=	
the symbol for greater than	the symbol for equal to	the symbol for less than (it looks like an L)

Example 1: 25 374 [>] 21 320

25 374 is greater than 21 320.

These symbols help you compare numbers.

Example 2: 12 347 [<] 42 135

12 347 is less than 42 135.

Example 3: 52 436 [=] 52 436

52 436 is equal to 52 436.

Example 4: 626 > 425

626 ________________ 425.

Example 5: 1345 = 1345

1345 ______________ 1345.

Example 6: 29 306 < 57 403

29 306 ______________ 57 403.

Write True or False.

- ● 32 357 < 33 295 True
- a 63 249 > 69 423 ________
- b 95 931 < 96 918 ________
- c 12 354 = 12 354 ________
- d 67 493 < 69 957 ________
- e 99 099 > 98 326 ________

Which number does the pointy end point to? That helps you tell the difference between < and >.

SELF CHECK Tick how you feel

Got it!	Need help...	I don't get it
 ☐	 ☐	 ☐

Check your answers

How many did you get correct? ☐

PRACTICE

1 Write <, > or = in the boxes and cross out the incorrect words.

- ● 53 824 [>] 52 852 because 53 824 is greater than / ~~less than~~ / ~~equal to~~ 52 852
- **a** 27 325 ☐ 53 725 because 27 325 is greater than / less than / equal to 53 725
- **b** 49 243 ☐ 35 935 because 49 243 is greater than / less than / equal to 35 935
- **c** 12 653 ☐ 12 653 because 12 653 is greater than / less than / equal to 12 653
- **d** 59 731 ☐ 15 463 because 59 731 is greater than / less than / equal to 15 463
- **e** 33 821 ☐ 97 328 because 33 821 is greater than / less than / equal to 97 328

2 Write >, < or = to make the statements true.

- ● 253 [<] 325
- **a** 1594 ☐ 1495
- **b** 6327 ☐ 7326
- **c** 149 ☐ 941
- **d** 1873 ☐ 1873
- **e** 14 385 ☐ 14 835
- **f** 25 849 ☐ 28 594
- **g** 38 542 ☐ 24 583
- **h** 98 990 ☐ 98 099
- **i** 347 ☐ 743
- **j** 842 ☐ 842
- **k** 15 385 ☐ 15 385
- **l** 5738 ☐ 8753
- **m** 9205 ☐ 9025
- **n** 47 382 ☐ 43 872

3 Tick the correct statements and cross the incorrect statements.

- ● 347 < 754 ✓
- **a** 1254 = 1254
- **b** 389 < 983
- **c** 7621 > 8723
- **d** 8491 < 8194
- **e** 8648 > 8846
- **f** 5347 < 5743
- **g** 1263 < 1362
- **h** 9801 > 9108
- **i** 12 534 > 13 435
- **j** 58 934 < 54 983
- **k** 72 105 > 75 107

CATCH UP MATHS YEAR 5 BOOK A © PASCAL PRESS ISBN: 9781925726169

ROUNDING TO THE NEAREST 100, 1000 AND 10 000

Rounding is useful when you need to estimate an answer.

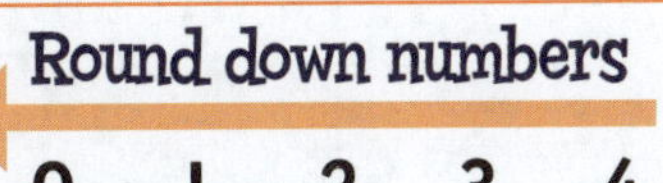

Round down numbers					Round up numbers				
0	1	2	3	4	5	6	7	8	9

Rounding to 100

1 Go to the hundreds column.

2 Write the round up number above the hundreds.

3 Circle the number in the tens column.

4 Is it a round up or round down number?

5 Round the number.

Example 1: Round to the nearest 100.

9 ← up
8(5)7

900

Rounding to 1000

1 Go to the thousands column.

2 Write the round up number above the thousands.

3 Circle the number in the hundreds column.

4 Is it a round up or round down number?

5 Round the number.

Example 2: Round to the nearest 1000.

6 ← up
45 (8)45

46 000

Rounding to 10 000

1 Go to the ten thousands column.

2 Write the round up number above the ten thousands.

3 Circle the number in the thousands column.

4 Is it a round up or round down number?

5 Round the number.

Example 3: Round to the nearest 10 000.

3
2(4) 556
down

20 000

Example 4:
Round 15 486 to the nearest 100.

5
15 4(8)6

______

Example 5:
Round 1742 to the nearest 1000.

2
1(7)42

______

Round to the nearest 100.

● 3 ← up
37 2(9)5
37 300

a 5834 ______

b 652 ______

SELF CHECK Tick how you feel

Got it!	Need help...	I don't get it
☐	☐	☐

Check your answers
How many did you get correct? ☐

PRACTICE

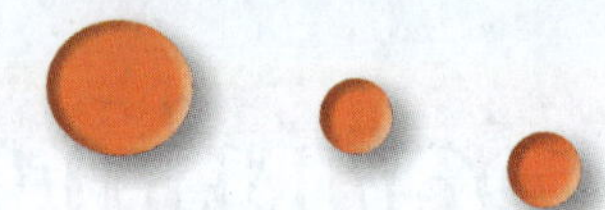

1 Round these numbers to the nearest 100.

- 6873 (9, up) — 6900
- c 9876 ______
- f 64 327 ______
- a 653 ______
- d 15 492 ______
- g 714 ______
- b 1524 ______
- e 59 821 ______
- h 8935 ______

2 Round these numbers to the nearest 1000.

- 74 238 (5, down) — 74 000
- b 2738 ______
- d 16 423 ______
- a 1543 ______
- c 6735 ______
- e 41 470 ______

3 Round these numbers to the nearest 10 000.

- 64 372 (7, down) — 60 000
- b 54 972 ______
- d 60 429 ______
- a 47 346 ______
- c 58 262 ______
- e 10 003 ______

4 Round these numbers.

	Nearest 100	Nearest 1000	Nearest 10 000
37 342	37 300	37 000	40 000
a 68 437			
b 40 305			
c 23 036			

CATCH UP MATHS YEAR 5 BOOK A © PASCAL PRESS ISBN: 9781925726169

LARGEST AND SMALLEST NUMBERS

THREE-DIGIT, FOUR-DIGIT AND FIVE-DIGIT NUMBERS

How would you move these numbers around to make the smallest number possible?

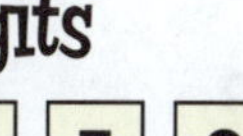

3 digits	4 digits	5 digits
5 8 1	4 3 2 5	6 5 3 7 8

To make the smallest possible number, start with the number that has the least value. Then write the next number with the least value. Keep going until you have used all the digits.

1 5 8	2 3 4 5	3 5 6 7 8

To make the largest possible number, start with the number that has the most value. Then write the next number with the most value. Keep going until you have used all the digits.

8 5 1	5 4 3 2	8 7 6 5 3

Example 1:
Write the missing digits to make the smallest possible number.

6 1 5 4

1 ☐ ☐ 6

Example 2:
Write the missing digits to make the largest possible number.

9 2 7 8 3

9 8 ☐ ☐ ☐

Your turn

1 Use the digits to write the smallest number you can.

● 3, 4, 2 234

a 5, 7, 3, 5 ________

b 4, 2, 6 ________

c 7, 1, 2, 0, 9 ________

2 Use the digits to write the largest number you can.

● 2, 5, 7 752

a 6, 3, 8, 4 ________

b 9, 2, 1 ________

c 8, 4, 9, 0, 9 ________

Check your answers
How many did you get correct? ☐

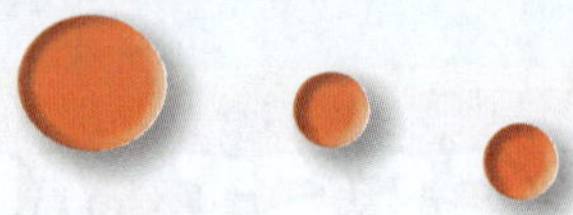

1 Use the numbers to write the smallest three-digit number you can.

- ● 3, 4, 9 349
- **a** 7, 0, 7 ______
- **b** 3, 9, 0 ______
- **c** 3, 6, 2 ______
- **d** 4, 1, 5 ______
- **e** 2, 8, 3 ______
- **f** 5, 5, 1 ______
- **g** 9, 0, 6 ______
- **h** 5, 3, 2 ______

2 Use the numbers to write the largest four-digit number you can.

- ● 2, 4, 6, 0 6420
- **a** 5, 8, 4, 3 ______
- **b** 9, 5, 2, 8 ______
- **c** 3, 3, 1, 3 ______
- **d** 6, 3, 0, 3 ______
- **e** 5, 0, 4, 0 ______
- **f** 1, 3, 5, 9 ______
- **g** 9, 6, 0, 8 ______
- **h** 7, 0, 0, 8 ______

3 Write the smallest and largest five-digit numbers using these digits.

	Digits	Smallest number	Largest number
●	3 1 6 4 5	13 456	65 431
a	2 4 5 3 2		
b	5 3 4 8 6		
c	7 6 4 3 9		
d	7 5 2 0 7		
e	8 6 1 8 1		
f	7 0 9 9 0		
g	8 0 9 1 2		

CATCH UP MATHS YEAR 5 BOOK A © PASCAL PRESS ISBN: 9781925726169

SIX-DIGIT NUMBERS

A six-digit number is a number that has 6 digits.
Numbers from 100 000 to 999 999 are all six-digit numbers.

125 374 **259 538** **643 875** **909 873**

Example 1:
The number 362 541 has:

3 hundred thousands
6 ten thousands
2 thousands
5 hundreds
4 tens
1 ones

How many 6-digit numbers are there?

Example 2: 980 431 has:

9 ______________
__ ten thousands
__ thousands
4 __________
3 ______
__ ones

Check your answer on the video!

Example 3:

Number	Hundred Thousands	Ten Thousands	Thousands	Hundreds	Tens	Ones
362 541	3	6	2	5	4	1
980 431		8	0			

Fill in the number chart.

	Number	Hundred Thousands	Ten Thousands	Thousands	Hundreds	Tens	Ones
●	362 592	3	6	2	5	9	2
a	137 397						
b	950 133						

SELF CHECK Tick how you feel

Got it!	Need help...	I don't get it
 ☐	 ☐	☐

Check your answers
How many did you get correct? ☐

PRACTICE

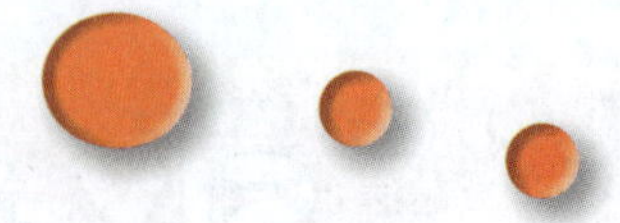

1 Write the numbers in words.

- 739 856 seven hundred and thirty-nine thousand, eight hundred and fifty-six

a 524 378 ____________________

b 954 048 ____________________

c 125 295 ____________________

d 180 032 ____________________

e 989 624 ____________________

2 Fill in the number chart.

	Number	Hundred Thousands	Ten Thousands	Thousands	Hundreds	Tens	Ones
•	625 437	6	2	5	4	3	7
a	102 252						
b		6	7	9	0	3	5
c		1	0	0	3	2	4
d	959 615						
e	980 711						
f	897 241						
g	948 111						
h		4	6	8	0	3	7
i		9	7	5	1	5	2

CATCH UP MATHS YEAR 5 BOOK A © PASCAL PRESS ISBN: 9781925726169

VALUE
SIX-DIGIT NUMBERS

The value of a digit depends on its place in the number.

SCAN to watch video

Example 1:

HTh TTh Th H T O

524 396

The value of the 5 is 500 000.

The value of the 2 is 20 000.

The value of the 4 is 4000.

The value of the 3 is 300.

The value of the 9 is 90.

The value of the 6 is 6.

5 has the greatest value.
It is worth 500 000 because it is in the hundred thousands place.

6 has the smallest value.
It is in the ones place and it is only worth 6.

Example 2:

HTh TTh Th H T O

724 861

The value of the 7 is ________.

The value of the 2 is ________.

The value of the 4 is ______.

The value of the 8 is _____.

The value of the 6 is ___.

The value of the 1 is __.

__ has the greatest value and __ has the smallest value.

Check your answer on the video!

Your turn

Fill in the missing values for 672 495.

a The value of 6 is ___________.

b The value of 7 is _________.

c The value of 2 is _______.

d The value of 4 is _____.

e The value of 9 is ___.

f The value of 5 is __.

SELF CHECK	Tick how you feel	
Got it! ☐	Need help... ☐	I don't get it ☐

Check your answers
How many did you get correct? ☐

PRACTICE

1 Write the value of the underlined digits.

- ● 437 851 30 000
- a 252 152 ______
- b 643 853 ______
- c 981 347 ______
- d 103 240 ______
- e 743 185 ______
- f 325 104 ______
- g 606 660 ______
- h 958 712 ______
- i 529 612 ______
- j 395 831 ______
- k 602 909 ______

2 What is the value of 2 in these numbers?

- ● 624 987 20 000
- a 178 342 ______
- b 238 596 ______
- c 42 375 ______
- d 60 302 ______
- e 53 826 ______
- f 97 253 ______
- g 72 438 ______
- h 293 851 ______
- i 624 381 ______
- j 35 825 ______
- k 739 023 ______

3 Write a six-digit number that has a 6 with the given value.

- ● 600 000 634 281
- a 60 000 ______
- b 60 ______
- c 6 ______
- d 600 ______
- e 6000 ______

4 What is the value of the 4?

- ● 64 259 4000
- a 49 327 ______
- b 12 543 ______
- c 837 420 ______
- d 13 504 ______
- e 495 272 ______

CATCH UP MATHS YEAR 5 BOOK A © PASCAL PRESS ISBN: 9781925726169

NUMBER EXPANDERS

SIX-DIGIT NUMBERS

This number expander shows 382 479.

3	hundred thousands	8	ten thousands	2	thousands	4	hundreds	7	tens	9	ones

SCAN to watch video

Example 1: 382 479 can also be shown using ten thousands, thousands, hundreds, tens and ones:

3	8	ten thousands	2	thousands	4	hundreds	7	tens	9	ones

Here 382 479 is shown using thousands, hundreds, tens and ones:

3	8	2	thousands	4	hundreds	7	tens	9	ones

Using hundreds, tens and ones:

3	8	2	4	hundreds	7	tens	9	ones

Using tens and ones:

3	8	2	4	7	tens	9	ones

Using ones:

3	8	2	4	7	9	ones

Example 2: Complete the number expanders for 856 243.

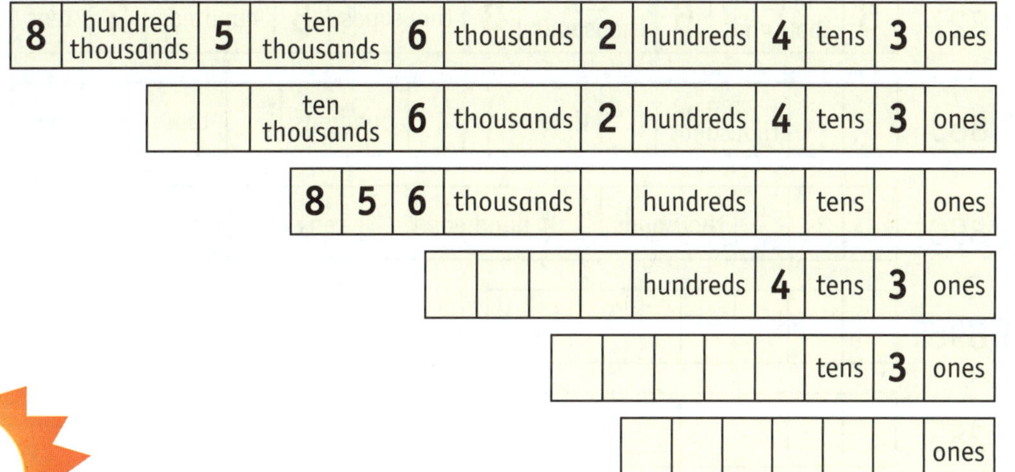

8	hundred thousands	5	ten thousands	6	thousands	2	hundreds	4	tens	3	ones

		ten thousands	6	thousands	2	hundreds	4	tens	3	ones

8	5	6	thousands		hundreds		tens		ones

				hundreds	4	tens	3	ones

					tens	3	ones

						ones

Check your answer on the video!

Your turn

Write the numbers shown on the number expanders.

● | 6 | 3 | 2 | 1 | hundreds | 4 | tens | 2 | ones | — 632 142

a | 2 | 7 | 9 | 1 | 1 | 3 | ones | — ____________

b | 7 | 1 | ten thousands | 3 | thousands | 4 | hundreds | 9 | tens | 0 | ones | — ____________

SELF CHECK Tick how you feel

Got it! | Need help... | I don't get it

Check your answers

How many did you get correct?

PRACTICE

Write the numbers.

- 67 ten thousands, 4 tens, 3 ones, 7 hundreds 670 743

a 438 thousands, 3 tens, 6 ones, 5 hundreds ____________

b 823 ones, 44 ten thousands, 4 thousands ____________

c 5 hundred thousands, 5 ones, 6 thousands, 4 ten thousands

____

d 55 342 tens, 9 ones ____________

e 873 290 ones ____________

f 437 thousands, 3 hundreds, 2 ones ____________

g 5381 hundreds, 7 tens, 8 ones ____________

2 Write the numbers onto the number expanders.

- 563 827

5	hundred thousands	6	ten thousands	3	thousands	8	hundreds	2	tens	7	ones

a 420 863

		ten thousands		thousands		hundreds		tens		ones

b 643 890

			thousands		hundreds		tens		ones

c 843 896

						ones

d 736 283

					tens		ones

3 Write the numbers shown on the number expanders.

-

6	hundred thousands	4	ten thousands	2	thousands	9	hundreds	0	tens	3	ones

642 903

a

8	6	9	thousands	3	hundreds	7	tens	5	ones

____

b

1	3	ten thousands	5	thousands	3	hundreds	5	tens	0	ones

____

c

2	4	0	5	hundreds	9	tens	3	ones

____

CATCH UP MATHS YEAR 5 BOOK A © PASCAL PRESS ISBN: 9781925726169

EXPANDED NUMBERS
SIX-DIGIT NUMBERS

When we write the value of each digit of a number, we are expanding the number.

Example 1:

172 = 100 + 70 + 2

hundreds, tens, ones

Numbers written like this are in 'expanded form'.

Example 2:

2659 = 2000 + 600 + 50 + 9

thousands

Example 3:

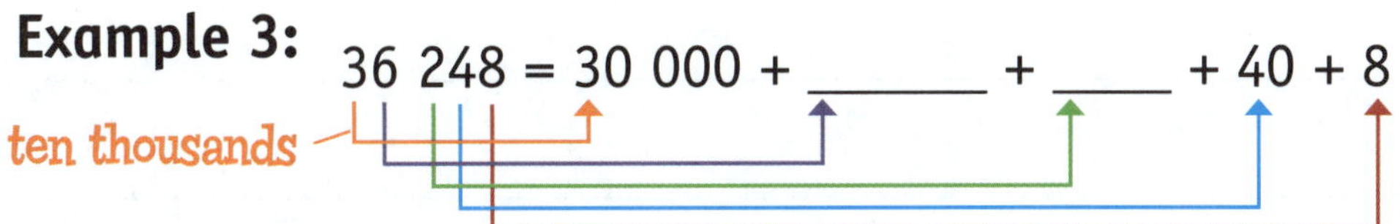

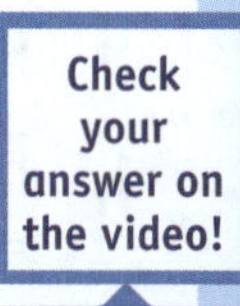

Example 4:

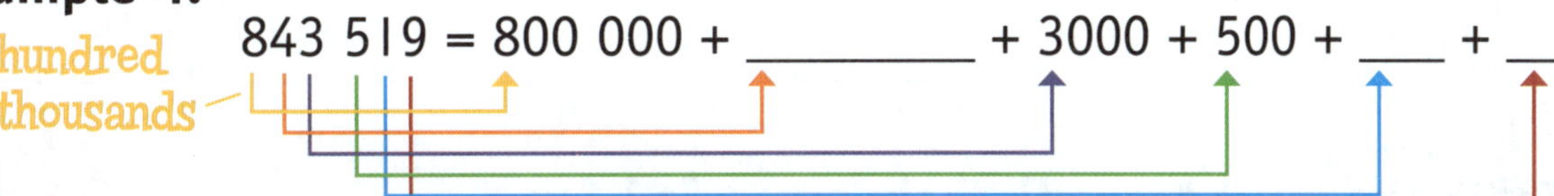

Your turn

Expand the numbers.

● 54 = 50 + 4

a 126 = ______________________

b 2593 = ______________________

c 32 846

= ______________________

d 743 291

= ______________________

Check your answers

How many did you get correct?

PRACTICE

1 Write each number in expanded form.

● 532 674 = 500 000 + 30 000 + 2000 + 600 + 70 + 4

a 29 = ______

b 6382 = ______

c 763 = ______

d 24 639 = ______

e 617 345 = ______

f 721 509 = ______

g 809 312 = ______

h 930 630 = ______

i 682 900 = ______

j 400 326 = ______

2 Match the numbers with their expanded form.

● 576 359	400 000 + 10 000 + 2000 + 500 + 10 + 6
a 249 205	300 000 + 20 000 + 6000 + 10
b 602 950	900 000 + 30 000 + 6000 + 5
c 870 323	600 000 + 2000 + 900 + 50
d 936 005	500 000 + 70 000 + 6000 + 300 + 50 + 9
e 938 505	200 000 + 40 000 + 9000 + 200 + 5
f 326 010	800 000 + 70 000 + 300 + 20 + 3
g 421 006	900 000 + 30 000 + 8000 + 500 + 5
h 725 400	700 000 + 20 000 + 5000 + 400
i 412 516	400 000 + 20 000 + 1000 + 6

ORDERING NUMBERS
SIX-DIGIT NUMBERS

These six-digit numbers are ordered in ascending order (smallest to largest):

326 581 523 486 695 254 723 982

smallest 6-digit number — largest 6-digit number

Now the numbers are ordered in descending order (largest to smallest):

723 982 695 254 523 486 326 581

Example 1: These numbers are arranged in ascending order:

112 371, 212 173, 221 317, 317 122, 722 113

Example 2: These numbers are arranged in descending order:

985 543, 854 953, 549 843, 453 895, 448 395

Example 3:
Write numbers that fit into this list of descending numbers.

245 923, __________, 219 654, __________, 196 026

Example 4:
Write numbers that fit into this list of ascending numbers.

__________, 186 249, 347 288, __________, 472 383

Circle the larger number and underline the smaller number in each pair of numbers.

●	623 587 (underlined)	653 827 (circled)	c	257 984	752 948
a	540 480	854 584	d	197 913	157 593
b	952 437	954 273	e	395 482	386 529

SELF CHECK Tick how you feel

Got it!	Need help...	I don't get it
☐	☐	☐

Check your answers
How many did you get correct? ☐

PRACTICE

1 Circle the largest number and underline the smallest number in each group.

●	<u>527 541</u>	(725 514)	554 721	c	382 563	335 274	438 293
a	637 985	589 736	985 673	d	245 896	568 924	294 865
b	239 147	129 734	473 921	e	586 920	508 290	908 250

2 Order these numbers from largest (1) to smallest (5).

●	248 283	382 842	832 842	883 242	488 422
	5	4	2	1	3
a	735 460	573 640	647 305	507 640	375 046
	☐	☐	☐	☐	☐
b	280 943	890 342	432 980	809 342	908 243
	☐	☐	☐	☐	☐
c	613 978	987 613	796 831	189 367	798 631
	☐	☐	☐	☐	☐

3 Order these numbers from smallest (1) to largest (5).

●	480 963	369 084	963 840	408 639	630 894
	3	1	5	2	4
a	273 651	165 372	235 617	756 213	156 273
	☐	☐	☐	☐	☐
b	524 378	857 423	875 432	542 738	427 853
	☐	☐	☐	☐	☐
c	671 054	104 576	765 410	567 104	410 675
	☐	☐	☐	☐	☐

CATCH UP MATHS YEAR 5 BOOK A © PASCAL PRESS ISBN: 9781925726169

ODD AND EVEN NUMBERS
SIX-DIGIT NUMBERS

SCAN to watch video

These six-digit numbers are even:

525 630   743 128   952 436   950 124   625 382

They end in 0, 2, 4, 6 or 8 and can be divided by 2.

These six-digit numbers are odd:

796 153   243 687   581 311   982 999   246 805

They end in 1, 3, 5, 7 or 9 and cannot be divided by 2.

Example 1: Is 259 328 odd or even?

Look at the last number. It is an 8.
Numbers that end in 2, 4, 6, 8 or 0 are even.
259 328 is an even number.

Example 2: Is 135 713 odd or even?

Look at the last number. It is a 3.
Numbers that end in 1, 3, 5, 7 or 9 are odd.
135 713 is an odd number.

It doesn't matter how big the number is – always look at the ones place to find out if it is odd or even.

Example 3:
Write a last digit that makes these numbers even.

159 24<u>6</u>

285 11__

Example 4:
Write a last digit that makes these numbers odd.

163 82<u>1</u>

839 29__

Use green to circle the even numbers and red to circle the odd numbers.

● 253 284   631 514   326 513   537 611

483 160   989 336   534 419

Check your answers
How many did you get correct?

PRACTICE

1 Write the numbers in the correct boxes.

- ~~427 359~~ 625 437 542 837 931 391 756 284 982 534 843 712 351 255 491 311 736 526 118 318 149 032 282 533

Odd Numbers	Even Numbers
427 359	

2 Cross out the odd number in each group.

- 765 432, 567 374, ~~765 347~~, 567 386

a 456 983, 634 598, 365 982, 659 834

b 724 982, 427 289, 982 724, 874 790

c 597 380, 803 592, 302 958, 203 859

d 632 784, 487 632, 874 326, 362 487

e 639 088, 508 963, 580 696, 396 802

3 Write the next three even numbers.

- 621 842, 621 844, 621 846, 621 848

a 273 510, ________, ________, ________

b 896 538, ________, ________, ________

c 911 086, ________, ________, ________

d 709 498, ________, ________, ________

e 431 504, ________, ________, ________

CATCH UP MATHS YEAR 5 BOOK A © PASCAL PRESS ISBN: 9781925726169

GREATER THAN, LESS THAN, EQUAL TO

SIX-DIGIT NUMBERS

> greater than	= equal to	< less than

SCAN to watch video

Example 1: 121 359 > 121 357
This statement is read as:
121 359 is greater than 121 357. It is true.

Using the symbols for greater than, less than and equal to is a way to compare two values.

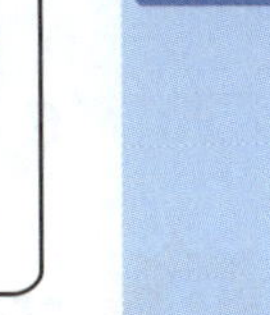

Example 2: 762 397 < 977 623
This statement is read as:
762 397 is less than 977 623. It is true.

Example 3: 529 612 = 529 612
This statement is read as:
529 612 is equal to 529 612. It is true.

Example 4: Write the words to complete the statement.

649 382 is greater than 545 743.

Example 5: Write the words to complete the statement.

125 093 ________________ 125 093.

Example 6: Write the words to complete the statement.

347 095 ________________ 865 237.

Your turn

Write T for true or F for false.

- 963 742 < 976 549 T
- **a** 264 173 > 246 371 ☐
- **b** 843 251 < 493 521 ☐
- **c** 764 258 > 747 285 ☐
- **d** 937 081 < 973 180 ☐
- **e** 843 420 = 843 420 ☐

SELF CHECK Tick how you feel

Got it!	Need help...	I don't get it
 ☐	 ☐	 ☐

Check your answers
How many did you get correct? ☐

PRACTICE

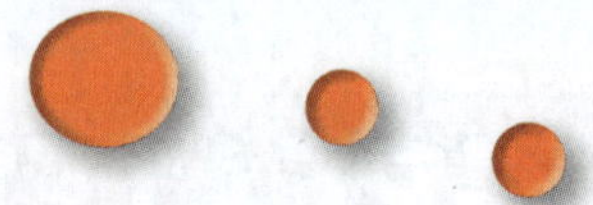

1 Complete by writing is greater than, is less than or is equal to.

- 563 928 is greater than ______ 539 826
- a 106 250 ______ 160 520
- b 392 533 ______ 392 533
- c 635 423 ______ 623 345
- d 742 411 ______ 744 211
- e 909 438 ______ 998 043

2 Use >, < or = to complete the following.

- 672 142 [<] 762 412
- a 246 848 [ ] 244 688
- b 124 852 [ ] 122 854
- c 259 630 [ ] 952 036
- d 371 142 [ ] 371 142
- e 480 963 [ ] 408 639
- f 950 133 [ ] 905 133
- g 137 397 [ ] 137 397
- h 524 949 [ ] 544 299
- i 180 913 [ ] 118 903
- j 723 456 [ ] 723 456
- k 642 955 [ ] 795 032

3 Tick the statements that are true and cross the ones that are not true.

- 374 158 < 347 851 ✗
- a 732 950 < 723 509
- b 963 795 > 997 635
- c 599 455 < 995 554
- d 523 690 = 523 690
- e 222 922 = 222 229
- f 302 241 < 220 314
- g 749 862 > 947 826
- h 593 452 < 953 524
- i 423 599 = 423 599
- j 172 054 > 172 095
- k 626 433 < 752 153
- l 849 386 < 993 399
- m 104 387 < 140 783
- n 291 359 > 219 935
- o 524 378 > 453 873

ROUNDING TO 10 000 AND 100 000

Rounding is useful when you need to estimate an answer.

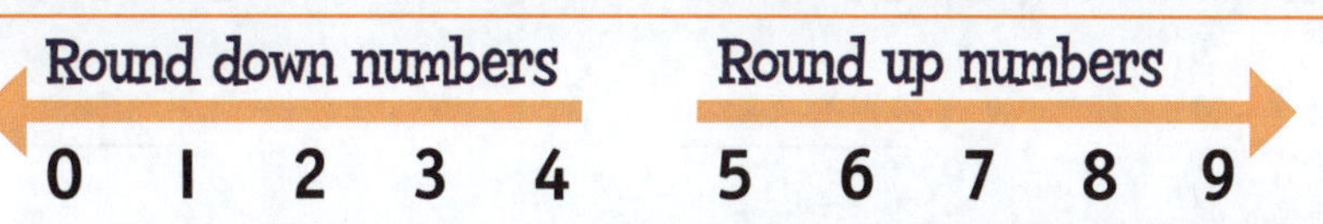

Rounding to 10 000

1 Go to the ten thousands column.
2 Write the round up number above the ten thousands.
3 Circle the number in the thousands column.
4 Is it a round up or round down number?
5 Round the number.

Example 1:
Round to the nearest 10 000.

3
24 936 down
20 000

Example 2:
Round to the nearest 10 000.

8
73 829
____

Rounding to 100 000

1 Go to the hundred thousands column.
2 Write the round up number above the hundred thousands.
3 Circle the number in the ten thousands column.
4 Is it a round up or round down number?
5 Round the number.

Example 3:
Round to the nearest 100 000.

7
682 459 up
700 000

Example 4:
Round to the nearest 100 000.

5
496 253
____

Your turn

Round these numbers.

● nearest 100 000
3
275 496 up
300 000

a nearest 10 000
32 854
____

b nearest 100 000
938 243
____

SELF CHECK Tick how you feel

Got it!	Need help...	I don't get it
 ☐	 ☐	☐

Check your answers
How many did you get correct? ☐

PRACTICE

1 Round these numbers to the nearest 10 000.

- 8 / 7(2) 854 down → 70 000
- b 59 545 ______
- d 43 298 ______
- a 36 592 ______
- c 21 325 ______
- e 84 297 ______

2 Round these numbers to the nearest 100 000.

- 8 / 7(2)4 836 down → 700 000
- b 375 011 ______
- d 752 359 ______
- a 837 453 ______
- c 439 256 ______
- e 674 219 ______

3 Round these numbers.

	Nearest 10 000	Nearest 100 000
365 102	370 000	400 000
a 658 317		
b 473 164		
c 518 901		
d 461 597		
e 174 126		
f 843 345		
g 252 238		
h 486 080		
i 727 723		
j 638 614		

CATCH UP MATHS YEAR 5 BOOK A © PASCAL PRESS ISBN: 9781925726169

LARGEST AND SMALLEST NUMBERS

SIX-DIGIT NUMBERS

SCAN to watch video

How would you move these six numbers around to make the smallest number possible? 7 2 3 5 8 9

2 3 5 7 8 9 — Put the 2 first as it has the least value. Next, put the 3, then the 5 then the 7, 8 and 9 to make the smallest 6-digit number possible.

9 8 7 5 3 2 — To make the largest 6-digit number possible, put the 9 first as it has the most value, then the 8, followed by the 7, 5, 3 and 2.

Example 1: Make the smallest possible number.

9 1 8 0 5 4

1 0 4 5 8 9

Example 2: Make the smallest possible number.

3 2 0 7 6 1

1 _ _ _ _ 7

Example 3: Make the largest possible number.

6 1 8 0 4 3

8 6 4 3 1 0

Example 4: Make the largest possible number.

2 1 2 5 7 9

9 _ _ _ _ 1

Check your answer on the video!

Your turn

Write the smallest and largest number you can using the given digits.

● 387435
Smallest: 334 578
Largest: 875 433

a 963827
Smallest: __________
Largest: __________

b 443726
Smallest: __________
Largest: __________

c 684097
Smallest: __________
Largest: __________

SELF CHECK Tick how you feel

Got it!	Need help...	I don't get it
☐	☐	☐

Check your answers
How many did you get correct? ☐

PRACTICE

Write the smallest and largest six-digit numbers using these digits.

	Digits	Smallest number	Largest number
●	6 3 7 8 5 2	235 678	876 532
a	7 5 4 2 2 3		
b	8 9 0 4 3 7		
c	9 2 5 3 6 2		
d	1 4 9 9 8 3		
e	8 8 7 4 9 5		
f	5 4 2 2 5 1		
g	4 7 9 3 2 6		
h	2 8 2 1 1 2		

Cross out the number that is NOT the largest number or smallest number that can be made with the digits.

● 7, 5, 9, 2, 3, 1: 123 579, ~~957 321~~, 975 321

a 8, 8, 3, 2, 5, 0: 205 388, 203 588, 885 320

b 6, 5, 7, 0, 3, 0: 765 300, 300 567, 300 576

c 7, 6, 2, 5, 3, 1: 765 321, 123 567, 132 567

d 2, 9, 0, 3, 7, 7: 230 977, 203 779, 977 320

e 2, 7, 0, 4, 7, 3: 203 477, 774 302, 774 320

f 1, 6, 0, 4, 0, 0: 164 000, 146 000, 641 000

g 7, 7, 0, 3, 1, 5: 103 577, 775 310, 775 301

h 0, 2, 0, 5, 0, 6: 652 000, 200 056, 526 000

i 6, 2, 3, 0, 5, 9: 203 569, 023 569, 965 320

THE ROLE OF ZERO

Zero (0) is a symbol used to describe a place holder for a value when no other number is in that place.

Example 1: 20
The zero is in the ones place.
It is holding the place of the ones.
Otherwise, the number would be 2.

Example 2: 508
The zero is holding the tens position.
Otherwise, the number would be 58.

Example 3: 3037
The zero is holding the hundreds place.
Otherwise, the number would be 337.

Example 4: 409 357
The zero is in the ten thousands place.
It is holding the place of the ________.
Otherwise, the number would be ________.

Example 5: 10 436
The zero is in the ________ place.
It is holding the place of the ________.
Otherwise, the number would be ________.

Write the place that zero is holding.

- 7302 tens
- **a** 603 ________
- **b** 10 438 ________

SELF CHECK Tick how you feel

Got it!	Need help...	I don't get it
☐	☐	☐

Check your answers
How many did you get correct? ☐

PRACTICE

1 What place is the zero holding?

- 320 436 thousands
- a 407 __________
- b 10 __________
- c 1035 __________
- d 2035 __________
- e 6230 __________
- f 504 362 __________
- g 115 308 __________
- h 7203 __________
- i 30 __________
- j 260 __________
- k 1063 __________
- l 52 160 __________
- m 60 354 __________
- n 107 623 __________
- o 350 249 __________
- p 623 502 __________
- q 524 309 __________

2 Write the numbers.

- Write three 3-digit numbers with a zero in the tens place.

  305, 407, 506

- a Write three 4-digit numbers with a zero in the ones place.

  ______, ______, ______

- b Write three 5-digit numbers with a zero in the hundreds place.

  ______, ______, ______

- c Write three 6-digit numbers with a zero in the ten thousands place.

  ______, ______, ______

- d Write three 2-digit numbers with a zero in the ones place.

  ___, ___, ___

- e Write three 6-digit numbers with a zero in the thousands place.

  ______, ______, ______

ABBREVIATIONS OF LARGE NUMBERS

You can use the letter K, the word 'thousand' or the abbreviation 'thous.' instead of writing three zeros at the end of a number.

Example 1:
Use the word 'thousand' to write these numbers.

5000 = 5 thousand

26 000 = 26 thousand

129 000 = ________________

Example 2:
Use 'thous.' to write these numbers.

350 000 = 350 thous.

19 000 = 19 thous.

562 000 = ________________

Example 3:
Use K to write these numbers.

350 000 = 350K

14 000 = 14K

3000 = ____

Example 4:
Use 3 zeroes to write these numbers.

129 thous. = 129 000

62K = 62 000

5 thousand = ______

1 Use K to write these numbers.

● 283 000 = 283K **a** 72 000 = ______ **b** 146 000 = ______

2 Use the word 'thousand' to write these numbers.

● 492 000 = 492 thousand **a** 138 000 = ______________

3 Use the abbreviation 'thous.' to write these numbers.

● 492 000 = 492 thous. **a** 916 000 = ____________ **b** 57 000 = ____________

SELF CHECK Tick how you feel

Got it!	Need help...	I don't get it
☐	 ☐	 ☐

Check your answers
How many did you get correct? ☐

PRACTICE

1 Match the numbers.

● 439 000	14K
a 246 000	951 thous.
b 810 000	43K
c 900 000	1K
d 5000	217 thousand
e 25 000	900 thous.
f 43 000	439K
g 217 000	5K
h 951 000	25 thousand
i 1000	246 thous.
j 14 000	810K

2 Use the abbreviation thous. to write these numbers.

● 42 000 = 42 thous.

a 315 000 = ____________

b 16 000 = ____________

c 400 000 = ____________

d 814 000 = ____________

e 2000 = ____________

f 8000 = ____________

g 97 000 = ____________

3 Use K to write these numbers.

● 743 000 = 743K

a 73 000 = ________

b 3000 = ________

c 52 000 = ________

d 15 000 = ________

e 376 000 = ________

f 988 000 = ________

g 1000 = ________

CATCH UP MATHS YEAR 5 BOOK A © PASCAL PRESS ISBN: 9781925726169

FACTORS

There are two ways to identify factors.

- A factor is a number that divides into a larger number without leaving a remainder.
  30 ÷ 6 = 5, so 5 and 6 are factors of 30.
- We know that 6 × 5 = 30, so any numbers multiplied to get the product are factors of that product.
  factor × factor = product

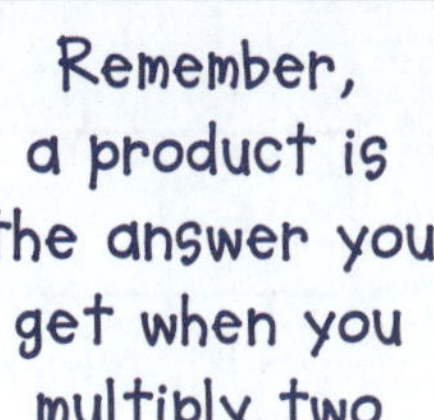

Example 1: What are the factors of 12?

We know 1 × 12 = 12, so 1 and 12 are factors.

2 × 6 = 12, so 2 and 6 are factors.

3 × 4 = 12, so 3 and 4 are factors.

We can write all the factors of 12 in order:

1, 2, 3, 4, 6, 12

Example 2: What are the factors of 21?

We know 1 × 21 = 21, so 1 and ___ are factors.

3 × 7 = 21, so 3 and __ are factors.

We can write all the factors of 21 in order:

__, __, __, ___

Circle the factors of each number.

● Factors of 8:
①②3④5 6⑧

b Factors of 9:
1 2 3 6 7 9

a Factors of 15:
1 2 3 5 7 8 9 15

c Factors of 7:
1 2 3 4 5 6 7

SELF CHECK Tick how you feel		
Got it!  ☐	Need help...  ☐	I don't get it ☐

Check your answers
How many did you get correct? ☐

PRACTICE

1 Write the factors of these numbers.

- 10 1, 2, 5, 10

a 14 ____________________

b 16 ____________________

c 30 ____________________

d 32 ____________________

e 36 ____________________

f 20 ____________________

g 22 ____________________

h 6 ____________________

2 Cross out the number that is not a factor of the number in bold.

- **21:** 1, ~~2~~, 3, 7, 21

a **27:** 1, 3, 6, 9, 27

b **40:** 1, 2, 3, 4, 5, 8, 10, 40

c **100:** 1, 2, 4, 5, 6, 10, 20, 25, 50, 100

d **18:** 1, 2, 3, 4, 6, 9, 18

e **24:** 1, 2, 3, 4, 6, 7, 8, 12, 24

f **25:** 1, 2, 5, 25

g **28:** 1, 2, 3, 4, 7, 14, 28

h **5:** 1, 2, 5

i **27:** 1, 3, 6, 9, 27

j **49:** 1, 4, 7, 49

k **33:** 1, 3, 10, 11, 33

CATCH UP MATHS YEAR 5 BOOK A © PASCAL PRESS ISBN: 9781925726169

HIGHEST COMMON FACTOR (HCF)

The Highest Common Factor (HCF) is the highest number that is a factor of two other numbers.

SCAN to watch video

Example 1:
What is the Highest Common Factor (HCF) of 8 and 12?

8: 1, 2, (4), 8
12: 1, 2, 3, (4), 6, 12

- Write all the factors of each number.
- Look for the highest number that is in both lists.

4 is the Highest Common Factor of 8 and 12.

Example 2:
What is the Highest Common Factor (HCF) of 15 and 45?

15: 1, 3, 5, (15)
45: 1, 3, 5, (15), 45

15 is the HCF of 15 and 45.

Example 3: What is the Highest Common Factor (HCF) of 3 and 9?

3: 1, (3)
9: 1, (3), 9

__ is the HCF of 3 and 9.

Example 4: What is the Highest Common Factor (HCF) of 4 and 12?

4: 1, 2, 4
12: 1, 2, 3, 4, 6, 12

__ is the HCF of 4 and 12.

Your turn

Circle the HCF for each pair of numbers.

● **8:** 1, (2), 4, 8
10: 1, (2), 5, 10

b **3:** 1, 3
9: 1, 3, 9

a **12:** 1, 2, 3, 4, 6, 12
16: 1, 2, 4, 8, 16

c **25:** 1, 5, 25
35: 1, 5, 7, 35

SELF CHECK Tick how you feel		
Got it! ☐	Need help... ☐	I don't get it ☐

Check your answers
How many did you get correct? ☐

PRACTICE

1 Work out the Highest Common Factor (HCF).

- 12: 1, 2, 3, 4, 6, 12
  15: 1, 3, 5, 15
  3 is the HCF of 12 and 15.

a 14: ____________
21: ____________
___ is the HCF of 14 and 21.

b 21: ____________
24: ____________
___ is the HCF of 21 and 24.

c 18: ____________
36: ____________
___ is the HCF of 18 and 36.

d 10: ____________
24: ____________
___ is the HCF of 10 and 24.

e 16: ____________
20: ____________
___ is the HCF of 16 and 20.

f 22: ____________
33: ____________
___ is the HCF of 22 and 33.

g 10: ____________
30: ____________
___ is the HCF of 10 and 30.

h 9: ____________
27: ____________
___ is the HCF of 9 and 27.

i 12: ____________
18: ____________
___ is the HCF of 12 and 18.

CATCH UP MATHS YEAR 5 BOOK A © PASCAL PRESS ISBN: 9781925726169

MULTIPLES

A multiple is the number you get when you multiply two factors together.

12 is a multiple of 3 and 4 because 3 × 4 = 12 (factor: 3, factor: 4, multiple: 12)

12 is also a multiple of 1 and 12 because 1 × 12 = 12 (factor: 1, factor: 12, multiple: 12)

It is also a multiple of 2 and 6 because 2 × 6 = 12 (factor: 2, factor: 6, multiple: 12)

Every number has both 1 and itself as factors.

Example 1: What are the multiples of 3?

The multiples of 3 are the numbers you get when you multiply 3 by another number.

3 × 1 = **3**, 3 × 2 = **6**, 3 × 3 = **9**, 3 × 4 = **12**, 3 × 5 = **15**, 3 × 6 = **18**

The first six multiples of 3 are **3**, **6**, **9**, **12**, **15** and **18**.

Example 2: What are the first five multiples of 5?

5 × 1 = ___, 5 × 2 = ___, 5 × 3 = ___, 5 × 4 = ___, 5 × 5 = ___

The first five multiples of 5 are ___, ___, ___, ___ and ___.

Write the first three multiples.

● 2: 2, 4, 6

a 3: ___, ___, ___

b 4: ___, ___, ___

c 12: ___, ___, ___

Check your answers
How many did you get correct?

PRACTICE

1 Write the first eight multiples.

- 1: 1, 2, 3, 4, 5, 6, 7, 8

a 2: ___, ___, ___, ___, ___, ___, ___, ___

b 4: ___, ___, ___, ___, ___, ___, ___, ___

c 5: ___, ___, ___, ___, ___, ___, ___, ___

d 7: ___, ___, ___, ___, ___, ___, ___, ___

e 9: ___, ___, ___, ___, ___, ___, ___, ___

f 10: ___, ___, ___, ___, ___, ___, ___, ___

g 11: ___, ___, ___, ___, ___, ___, ___, ___

2 Cross out the number that is not a multiple.

- 5: 55, 60, 65, 70, ~~73~~, 75

a 10: 90, 100, 110, 115, 120, 130

b 3: 21, 24, 27, 28, 30, 33

c 6: 6, 12, 18, 20, 24, 30

d 8: 8, 16, 24, 32, 40, 46

e 12: 12, 24, 28, 36, 48, 60

f 2: 18, 20, 22, 23, 24, 26

g 4: 36, 40, 44, 46, 48, 52

h 11: 22, 33, 35, 44, 55, 66

i 7: 21, 28, 35, 37, 42, 49

j 9: 54, 56, 63, 72, 81, 90

LOWEST COMMON MULTIPLES (LCM)

The Lowest Common Multiple (LCM) of two numbers is the smallest number that is a multiple of both numbers.

Example 1:
What is the Lowest Common Multiple of 4 and 8?

Multiples of **4**: 4, 8, 12, 16 ...

Multiples of **8**: 8, 16, 24 ...

The LCM of 4 and 8 is **8**.

Example 2:
What is the Lowest Common Multiple of 8 and 12?

Multiples of **8**: 8, 16, 24, 32 ...

Multiples of **12**: 12, 24, 36, 48 ...

The LCM of 8 and 12 is **24**.

Example 3:
What is the Lowest Common Multiple of 2 and 4?

Multiples of **2**: 2, 4, 6, 8 ...

Multiples of **4**: 4, 8, 12, 16 ...

The LCM of 2 and 4 is ___.

Circle the LCM for each pair of numbers.

● **2:** 2, 4, 6, 8 ...
8: 8, 16, 24, 32 ...

a **4:** 4, 8, 12, 16 ...
6: 6, 12, 18, 24 ...

b **2:** 2, 4, 6, 8 ...
3: 3, 6, 9, 12 ...

c **1:** 1, 2, 3, 4, 5, 6 ...
6: 6, 12, 18, 24, 30, 36 ...

Check your answers
How many did you get correct?

PRACTICE

1 What is the Lowest Common Multiple (LCM)?

2: 2, 4, 6, 8, (10)
5: 5, (10), 15, 20
The LCM is 10.

a 3: ____________
4: ____________
The LCM is ___.

b 5: ____________
10: ____________
The LCM is ___.

c 2: ____________
6: ____________
The LCM is ___.

d 3: ____________
6: ____________
The LCM is ___.

e 1: ____________
4: ____________
The LCM is ___.

f 4: ____________
10: ____________
The LCM is ___.

g 3: ____________
5: ____________
The LCM is ___.

h 2: ____________
10: ____________
The LCM is ___.

i 3: ____________
9: ____________
The LCM is ___.

j 4: ____________
8: ____________
The LCM is ___.

CATCH UP MATHS YEAR 5 BOOK A © PASCAL PRESS ISBN: 9781925726169

WHOLE NUMBERS REVIEW

1 Write the numbers in words.

a 62 ______________________

b 137 ______________________

c 563 ______________________

d 1250 ______________________

e 2030 ______________________

f 45 893 ______________________

____

g 86 724 ______________________

____

h 251 386 ______________________

____

2 Fill in the number chart.

	Number	Hundred Thousands	Ten Thousands	Thousands	Hundreds	Tens	Ones
a	56						
b	250						
c	1346						
d	8007						
e	32 430						
f	40 003						
g	100 200						
h	840 937						
i	647 300						
j	420 030						

REVIEW

What is the place value of the 7 in these numbers?

- a 473 ____________
- b 1257 ____________
- c 4732 ____________
- d 7890 ____________
- e 20 700 ____________
- f 762 381 ____________
- g 27 392 ____________
- h 173 258 ____________
- i 27 ____________
- j 387 ____________
- k 7321 ____________
- l 268 473 ____________

What is the value of the underlined digits?

- a 4$\underline{8}$27 ________
- b 15 3$\underline{2}$6 ________
- c $\underline{2}$9 ________
- d 16$\underline{4}$ ________
- e $\underline{2}$4 985 ________
- f 3$\underline{6}$ 438 ________
- g $\underline{1}$29 853 ________
- h 4$\underline{5}$7 906 ________
- i 695$\underline{7}$ ________
- j 47 $\underline{5}$90 ________
- k 531 2$\underline{8}$5 ________
- l $\underline{7}$36 150 ________
- m 1$\underline{6}$ ________
- n 4$\underline{9}$3 ________

Write the numbers.

- a 3 hundred thousands, 2 ten thousands, 8 thousands, 5 hundreds, 6 tens, 9 ones ________
- b 8 hundreds, 5 ones, 7 ten thousands, 6 hundred thousands, 4 thousands ________
- c 5 thousands, 3 ten thousands, 6 hundred thousands, 3 ones ________
- d 7 ones, 4 hundreds, 7 ten thousands, 2 hundred thousands ________

CATCH UP MATHS YEAR 5 BOOK A © PASCAL PRESS ISBN: 9781925726169

6 **Fill in the blank spaces.**

a 734 674

__ hundred thousands + _______

___ ten thousands + _______

_____ thousands + ______

______ hundreds + _____

_______ tens + ___

________ ones + __

b 583 652

__ hundred thousands + _______

___ ten thousands + _______

_____ thousands + ______

______ hundreds + _____

_______ tens + ___

________ ones + __

7 **Colour the digits in the hundred thousands yellow, ten thousands orange, thousands purple, hundreds green, tens blue and ones red.**

a 109 580 c 7620 e 89 g 346

b 961 d 87 314 f 73 554 h 406 789

8 **Write 6-digit numbers where the 7 is worth:**

a 70 ____________ d 700 000 ____________

b 7 ____________ e 70 000 ____________

c 700 ____________ f 7000 ____________

9 **Write 6-digit numbers where the 5 is worth:**

a 5000 ____________ d 50 000 ____________

b 50 ____________ e 500 000 ____________

c 5 ____________ f 500 ________

10 **Circle the numbers where the value of 4 is 400.**

236 400 4372 23 436 24 643 825 437 6840

14 593 983 402 127 483 1438 14 383 1534

REVIEW

11 Fill in the number expanders.

a 436 932

				hundreds		tens		ones

b 519 063

		ten thousands		thousands		hundreds		tens		ones

c 328 431

	hundred thousands		ten thousands		thousands		hundreds		tens		ones

d 754 902

						ones

e 147 389

					tens		ones

f 695 362

				hundreds		tens		ones

12 Complete these number expanders.

a 573 986

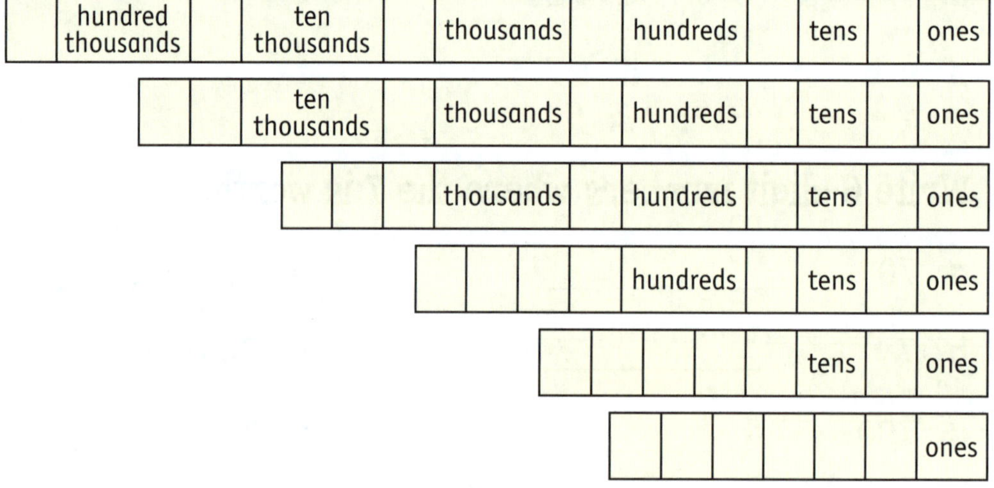

b 640 372

	hundred thousands		ten thousands		thousands		hundreds		tens		ones

		ten thousands		thousands		hundreds		tens		ones

			thousands		hundreds		tens		ones

				hundreds		tens		ones

					tens		ones

						ones

  ISBN: 9781925726169

13 Write these numbers in expanded form.

a 723 = ______________________________

b 1439 = ______________________________

c 533 = ______________________________

d 25 295 = ______________________________

e 384 629

= ______________________________

f 73 860 = ______________________________

g 509 350

= ______________________________

h 76 039

= ______________________________

i 630 284

= ______________________________

14 How many of each block do you need to make these numbers?

	Number	thousands block	hundreds flat	tens long	ones cube
a	536				
b	782				
c	1597				
d	7846				
e	39 873				
f	50 437				
g	142 583				
h	235 870				
i	643 209				

REVIEW

15 Show the numbers on the abacus.

a 7493

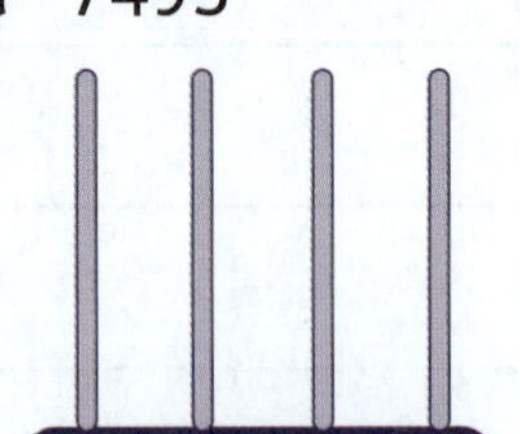

c 937 382

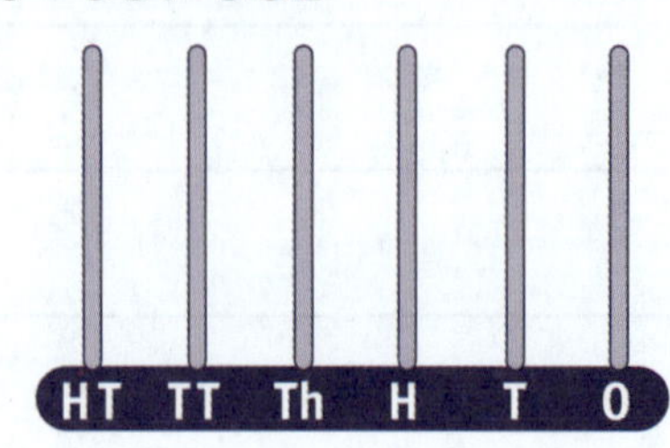

e 8498

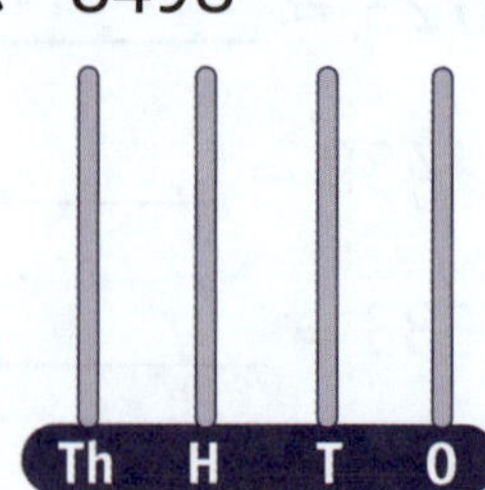

b 65 372

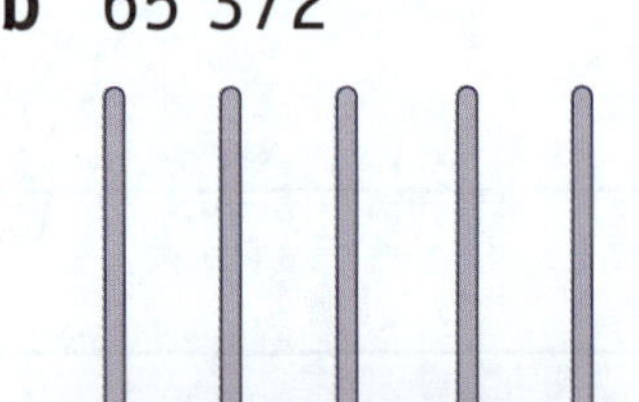

d 412 327

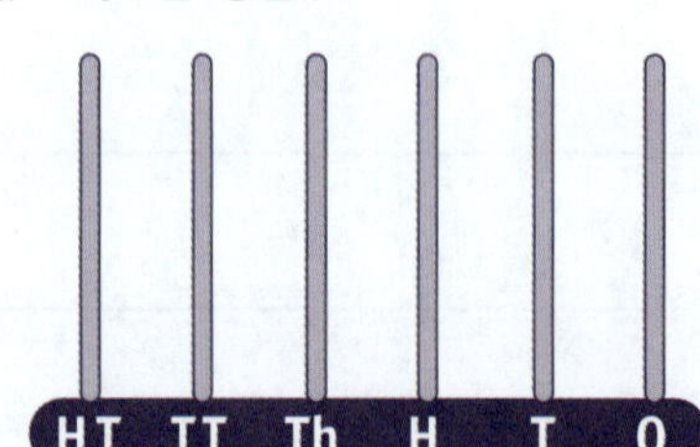

f 54 920

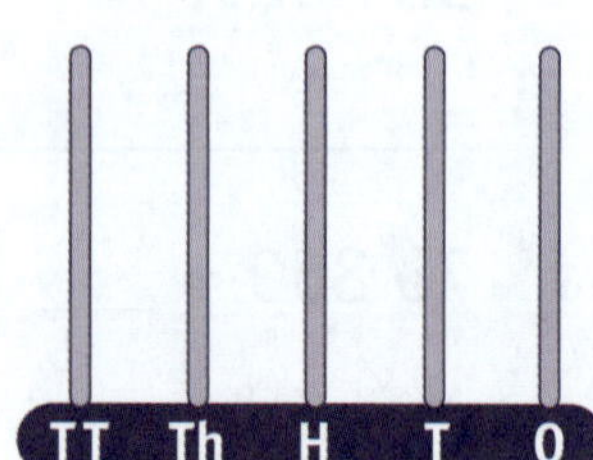

16 Write these numbers in ascending order.

a 6523, 3526, 5236, 5623, 5632

____

b 14 759, 41 579, 54 791, 91 547, 75 154

____

c 613 285, 582 316, 631 582, 531 682, 815 632

____

d 920 354, 453 029, 904 532, 290 543, 390 245

____

17 Write these numbers in descending order.

a 9524, 4529, 9254, 5429, 2945

____

b 37 390, 90 337, 97 330, 73 903, 97 033

____

CATCH UP MATHS YEAR 5 BOOK A © PASCAL PRESS ISBN: 9781925726169

c 61 379, 97 316, 69 137, 67 139, 79 613

____

d 459 352, 253 954, 495 234, 942 355, 395 542

____

e 864 023, 320 468, 682 403, 420 368, 403 862

____

18 **Use red to circle the odd numbers and green to circle the even numbers.**

16 138 27 53 847 849 342

3423 656 521 1526 16 430 439

19 **Write <, > or = to make the statements true.**

a 372 ☐ 273

b 249 ☐ 942

c 2835 ☐ 5382

d 7263 ☐ 3267

e 18 430 ☐ 10 834

f 29 372 ☐ 92 732

g 293 481 ☐ 934 812

h 76 439 ☐ 76 439

i 653 295 ☐ 625 395

j 843 252 ☐ 483 522

20 **Round the numbers.**

Round to	a 532 487	b 643 981	c 857 603
nearest 10			
nearest 100			
nearest 1000			
nearest 10 000			
nearest 100 000			

REVIEW

21 Write the smallest and largest numbers possible with these digits.

	Digits	Smallest number	Largest number
a	1 5 6		
b	3 5 7 9		
c	5 4 2 7 8		
d	9 3 6 4 1		
e	1 9 4 5 9 5		

22 Write the place that zero is holding.

a 307 ____________________

b 1230 ____________________

c 5079 ____________________

d 10 372 ____________________

e 51 095 ____________________

f 107 386 ____________________

g 134 302 ____________________

h 707 325 ____________________

23 Write five 4-digit numbers with a zero in the hundreds place.

__________, __________, __________, __________, __________

24 Write five 5-digit numbers with a zero in the tens place.

__________, __________, __________, __________, __________

25 Write five 6-digit numbers with a zero in the thousands place.

__________, __________, __________, __________, __________

26 Use K to write these numbers.

a 350 000 = __________

b 1000 = __________

c 291 000 = __________

d 739 000 = __________

e 400 000 = __________

f 395 000 = __________

CATCH UP MATHS YEAR 5 BOOK A © PASCAL PRESS ISBN: 9781925726169

27 **Use the abbreviation thous. to write these numbers.**

a 635 000 = ______________ d 849 000 = ______________

b 5000 = ______________ e 500 000 = ______________

c 173 000 = ______________ f 572 000 = ______________

28 **Write all the factors.**

a 12 ______________________ c 18 ______________________

b 16 ______________________ d 36 ______________________

29 **Write the Highest Common Factor (HCF).**

a 8: ______________________
12: ______________________
The HCF is ___.

b 20: ______________________
24: ______________________
The HCF is ___.

c 15: ______________________
18: ______________________
The HCF is ___.

30 **Write the first five multiples.**

a 2: ___, ___, ___, ___, ___ d 9: ___, ___, ___, ___, ___

b 8: ___, ___, ___, ___, ___ e 10: ___, ___, ___, ___, ___

c 5: ___, ___, ___, ___, ___ f 7: ___, ___, ___, ___, ___

31 **Find the Lowest Common Multiple (LCM).**

a 3: ______________________
4: ______________________
The LCM is ___.

b 2: ______________________
3: ______________________
The LCM is ___.

ADDING THREE OR MORE SINGLE-DIGIT NUMBERS

A quick and easy method of adding three numbers is to look for numbers that add to 10.

Example 1:

2 + 5 + 8 **2 + 8 = 10**

= 2 + 8 + 5

= 10 + 5

= 15

Example 2:

6 + 4 + 3

= 6 + 4 + 3

= ___ + ___

= ___

Example 3:

1 + 8 + 2 + 9

= 1 + 9 + 8 + 2

= 10 + 10

= 20

Example 4:

5 + 2 + 5 + 6

= ___ + ___ + ___ + ___

= ___ + ___

= ___

Look for numbers that add to 10 and then solve.

● 6 + 5 + 4

= **6 + 4 + 5**

= **10 + 5**

= **15**

b 3 + 5 + 2 + 7

= ___

= ___

= ___

a 2 + 5 + 8

= ___

= ___

= ___

c 4 + 7 + 6 + 2

= ___

= ___

= ___

SELF CHECK Tick how you feel

Got it!	Need help...	I don't get it
  ☐	 ☐	 ☐

Check your answers

How many did you get correct? ☐

CATCH UP MATHS YEAR 5 BOOK A © PASCAL PRESS ISBN: 9781925726169

PRACTICE

1 Look for numbers that add to 10 and then solve.

6 + 2 + 4
= 6 + 4 + 2
= 10 + 2
= 12

a 3 + 8 + 7
= ______
= ______
= ______

b 5 + 5 + 9
= ______
= ______
= ______

c 1 + 7 + 9
= ______
= ______
= ______

d 6 + 5 + 4
= ______
= ______
= ______

e 2 + 1 + 8
= ______
= ______
= ______

2 Look for numbers that add to 10 and then solve.

8 + 1 + 7 + 2
= 8 + 2 + 1 + 7
= 10 + 8
= 18

a 6 + 4 + 3 + 4
= ______
= ______
= ______

b 2 + 9 + 3 + 8
= ______
= ______
= ______

c 5 + 6 + 5 + 7
= ______
= ______
= ______

d 4 + 3 + 6 + 9
= ______
= ______
= ______

e 5 + 4 + 6 + 1
= ______
= ______
= ______

SUM

The sum is the total you get when numbers are added together.

Example 1:

3 + 5 + 10 = 18

18 is the sum of 3, 5 and 10.

Example 2:

8 + 3 + 9 = ___

___ is the sum of 8, 3 and 9.

Example 3:

What is the sum of 6, 8 and 12?

6 + 8 + 12 = 26

26 is the sum of 6, 8 and 12.

Example 4:

What is the sum of 4, 7 and 16?

4 + 7 + 16 = ___

___ is the sum of 4, 7 and 16.

What is the sum?

8, 4, 10

8 + 4 + 10

= 22

The sum of 8, 4 and 10 is 22.

a 6, 3, 1

____

= _____

The sum of 6, 3 and 1 is _____.

b 5, 9, 2

____

= _____

The sum of 5, 9 and 2 is _____.

c 24 + 10 + 26

= _____

The sum of 24, 10 and 26 is _____.

Got it!	Need help...	I don't get it

Check your answers

How many did you get correct?

CATCH UP MATHS YEAR 5 BOOK A © PASCAL PRESS ISBN: 9781925726169

PRACTICE

1 Match the additions with the correct sum.

● 3 + 8 + 12	The sum is 67.
a 5 + 4 + 15	The sum is 82.
b 19 + 29 + 52	The sum is 23.
c 2 + 20 + 45	The sum is 130.
d 37 + 25 + 26	The sum is 198.
e 21 + 42 + 19	The sum is 88.
f 77 + 16 + 37	The sum is 24.
g 41 + 69 + 88	The sum is 100.

2 What is the sum?

● 14, 13, 52

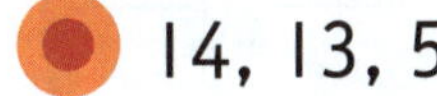

14 + 13 + 52 = 79

a 21, 31, 41

____

b 62, 14, 13

____

c 120, 110, 90

____

d 51, 65, 77

____

e 32, 8, 50

____

f 127, 107, 117

____

g 103, 157, 209

____

h 410, 350, 420

____

i 152, 410, 310

____

RELATING ADDITION AND SUBTRACTION

Addition and subtraction are related.
They are opposite actions.

We can use the three numbers 7, 13 and 20 to find out two addition facts and two subtraction facts.

Fact 1	7 + 13 = 20	Addition facts
Fact 2	13 + 7 = 20	
Fact 3	20 − 7 = 13	Subtraction facts
Fact 4	20 − 13 = 7	

The four facts you find out are a fact family.

Example 1:
Write the fact family.

(13) (23) (36)

Fact 1 13 + 23 = 36

Fact 2 23 + 13 = 36

Fact 3 36 − 13 = 23

Fact 4 36 − 23 = 13

Example 2:

(24) (15) (39)

15 + ___ = 39

___ + 15 = 39

39 − 15 = ___

39 − ___ = 15

Check your answer on the video!

Example 3:

(44) (28) (16)

___ + ___ = ___

___ + ___ = ___

___ − ___ = ___

___ − ___ = ___

Write the fact family for each set of three numbers.

● (52) (16) (68)

16 + 52 = 68

52 + 16 = 68

68 − 16 = 52

68 − 52 = 16

a (15) (73) (88)

___ + ___ = ___

___ + ___ = ___

___ − ___ = ___

___ − ___ = ___

b (47) (15) (32)

___ + ___ = ___

___ + ___ = ___

___ − ___ = ___

___ − ___ = ___

SELF CHECK Tick how you feel

Got it!	Need help...	I don't get it
☐	☐	☐

Check your answers
How many did you get correct? ☐

CATCH UP MATHS YEAR 5 BOOK A © PASCAL PRESS ISBN: 9781925726169

1 Fill in the missing numbers in each fact family.

● 24 + 9 = 33
9 + 24 = 33
33 − 9 = 24
33 − 24 = 9

a 51 + 25 = 76
25 + ___ = 76
76 − 25 = ___
76 − ___ = 25

b 83 + 15 = 98
15 + ___ = 98
___ − 15 = 83
98 − ___ = 15

2 Write the fact family for each set of numbers.

● (14) (17) (31)
14 + 17 = 31
17 + 14 = 31
31 − 17 = 14
31 − 14 = 17

b (17) (25) (42)
◯ + ◯ = ◯
◯ + ◯ = ◯
◯ − ◯ = ◯
◯ − ◯ = ◯

d (15) (34) (19)
◯ + ◯ = ◯
◯ + ◯ = ◯
◯ − ◯ = ◯
◯ − ◯ = ◯

a (24) (11) (35)
◯ + ◯ = ◯
◯ + ◯ = ◯
◯ − ◯ = ◯
◯ − ◯ = ◯

c (34) (42) (76)
◯ + ◯ = ◯
◯ + ◯ = ◯
◯ − ◯ = ◯
◯ − ◯ = ◯

e (67) (23) (44)
◯ + ◯ = ◯
◯ + ◯ = ◯
◯ − ◯ = ◯
◯ − ◯ = ◯

3 Write two addition and two subtraction facts using each set of numbers.

● [12] [14] [26]
12 + 14 = 26
14 + 12 = 26
26 − 12 = 14
26 − 14 = 12

a [45] [77] [32]
☐ + ☐ = ☐
☐ + ☐ = ☐
☐ − ☐ = ☐
☐ − ☐ = ☐

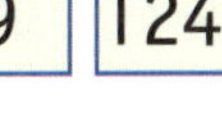

b [133] [9] [124]
☐ + ☐ = ☐
☐ + ☐ = ☐
☐ − ☐ = ☐
☐ − ☐ = ☐

ADDITION WITHOUT TRADING

TWO-DIGIT AND THREE-DIGIT NUMBERS

SCAN to watch video

Addition is when we combine two or more numbers to make one bigger number.

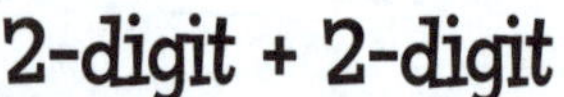

Example 1:

	Tens	Ones
	7	2
+	1	5
	8	7

Example 2:

	Tens	Ones
	3	4
+	2	5
	5	9

Example 3:

	Tens	Ones
	3	2
+	4	1

2-digit + 3-digit

Example 4:

	H	T	O
	4	3	4
+		5	2
	4	8	6

Example 5:

	H	T	O
	6	0	6
+		7	3
	6	7	9

Example 6:

	H	T	O
	7	5	6
+		3	2

3-digit + 3-digit

Example 7:

	H	T	O
	4	7	1
+	1	2	8
	5	9	9

Example 8:

	H	T	O
	3	0	5
+	2	9	3
	5	9	8

Example 9:

	H	T	O
	6	2	0
+	2	5	9

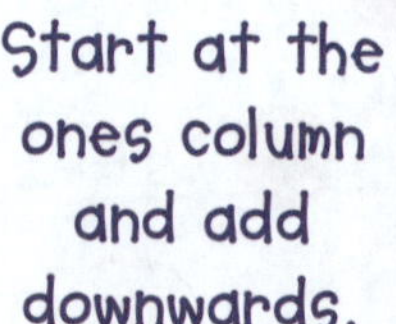

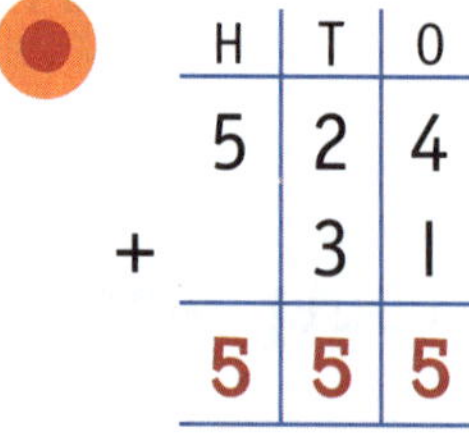

Complete these additions.

●

	H	T	O
	5	2	4
+		3	1
	5	5	5

a

	H	T	O
		4	3
+		1	4

b

	H	T	O
	1	3	2
+		4	5

c

	H	T	O
	2	5	1
+	1	2	8

d

	H	T	O
	3	5	7
+	5	1	2

e

	H	T	O
	8	6	2
+	1	3	5

SELF CHECK Tick how you feel

Got it!	Need help...	I don't get it
 ☐	 ☐	☐

Check your answers

How many did you get correct? ☐

CATCH UP MATHS YEAR 5 BOOK A © PASCAL PRESS ISBN: 9781925726169

1 Solve these additions.

Example:

	Tens	Ones
	4	3
+	2	6
	6	9

a

	Tens	Ones
	5	2
+	1	4

b

	Tens	Ones
	7	1
+	3	3

c

	Tens	Ones
	4	4
+	1	5

2 Complete these additions.

Example: $62 + 32 = 94$

a $81 + 24 =$ ____

b $80 + 34 =$ ____

c $16 + 31 =$ ____

3 Solve.

Example:

	H	T	O
	1	4	3
+		2	4
	1	6	7

a

	H	T	O
	2	3	4
+		2	5

b

	H	T	O
	4	3	1
+		1	6

c

	H	T	O
		7	2
+	3	1	5

4 Add.

Example: $46 + 152 = 198$

a $514 + 82 =$ ____

b $348 + 51 =$ ____

c $703 + 92 =$ ____

5 Complete.

Example: $425 + 134 = 559$

a $324 + 153 =$ ____

b $106 + 193 =$ ____

c $426 + 352 =$ ____

d $519 + 430 =$ ____

e $243 + 445 =$ ____

f $730 + 300 =$ ____

g $431 + 504 =$ ____

ADDING WITH TRADING

TWO-DIGIT AND THREE-DIGIT NUMBERS

Sometimes when we add, the sum of the two digits in a place value column is more than 9. Then we trade over ten ones for one ten.

SCAN to watch video

2-digit + 2-digit	2-digit + 3-digit	3-digit + 3-digit
Example 1: $^{1}5\ 3 + 6\ 9 = 12\ 2$	**Example 4:** $^{1}3\ ^{1}6\ 7 + 4\ 8 = 4\ 1\ 5$	**Example 7:** $^{1}6\ ^{1}3\ 4 + 5\ 6\ 7 = 12\ 0\ 1$
Example 2: $^{1}7\ 5 + 4\ 8 = 12\ 3$	**Example 5:** $5\ ^{1}3\ 8 + 5\ 6 = 5\ 9\ 4$	**Example 8:** $^{1}3\ ^{1}8\ 5 + 2\ 4\ 6 = 6\ 3\ 1$
Example 3: $8\ 7 + 3\ 5 =$ ____	**Example 6:** $6\ 2\ 4 + 8\ 9 =$ ____	**Example 9:** $7\ 4\ 8 + 5\ 8\ 6 =$ ____

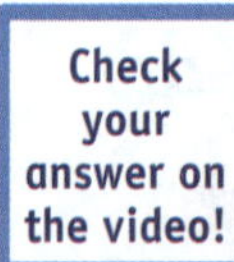

Check your answer on the video!

Your turn

Solve these additions.

- ● $^{1}7\ ^{1}4\ 3 + 5\ 8 = 8\ 0\ 1$
- **a** $9\ 9 + 4\ 6 =$ ____
- **b** $5\ 2 + 1\ 9 =$ ____
- **c** $5\ 0\ 7 + 3\ 5 =$ ____
- **d** $6\ 3\ 8 + 2\ 5 =$ ____
- **e** $8\ 0\ 3 + 3\ 4\ 8 =$ ____

SELF CHECK Tick how you feel

Got it!	Need help...	I don't get it
 ☐	 ☐	 ☐

Check your answers

How many did you get correct? ☐

CATCH UP MATHS YEAR 5 BOOK A © PASCAL PRESS ISBN: 9781925726169

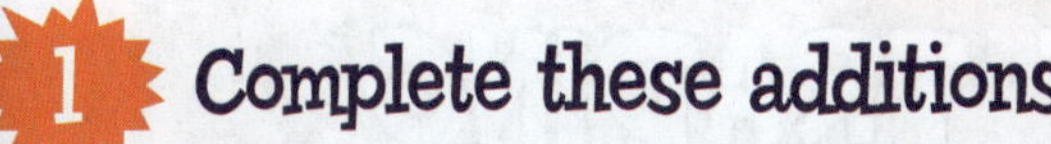

PRACTICE

1 Complete these additions.

Example:

	T	O
	[1]7	3
+	5	8
	13	1

a

	T	O
	3	6
+	4	9

b

	T	O
	7	8
+	1	4

c

	T	O
	8	3
+	2	7

2 Answer these additions.

Example: $^{1}19 + 68 = 87$

a $24 + 66 =$ ____

b $38 + 35 =$ ____

c $54 + 57 =$ ____

3 Solve.

Example:

	H	T	O
	[1]3	[1]5	7
+		4	8
	4	0	5

a

	H	T	O
	4	9	5
+		4	7

b

	H	T	O
	3	8	2
+		7	5

c

	H	T	O
	5	0	9
+		4	9

4 Add.

Example: $^{1}\ ^{1}68 + 257 = 325$

a $93 + 249 =$ ____

b $42 + 875 =$ ____

c $817 + 85 =$ ____

5 Complete.

Example: $3\,^{1}29 + 748 = 1077$

a $436 + 154 =$ ____

b $259 + 645 =$ ____

c $110 + 899 =$ ____

d $629 + 780 =$ ____

e $533 + 874 =$ ____

f $429 + 275 =$ ____

g $570 + 299 =$ ____

ADDING WITH AND WITHOUT TRADING

FOUR-DIGIT NUMBERS

Here are examples of adding 4-digit numbers with 2-digit, 3-digit and 4-digit numbers, with and without trading.

SCAN to watch video

	4-digit + 2-digit	4-digit + 3-digit	4-digit + 4-digit
Without trading	**Example 1:** 3042 + 53 3095	**Example 3:** 6381 + 608 6989	**Example 5:** 4573 + 5321 9894
With trading	**Example 2:** 56 ¹3 9 + 45 5684	**Example 4:** 5 ¹4 ¹8 2 + 159 5641	**Example 6:** ¹5 ¹9 ¹4 7 + 2655 8602

Example 7:
2537
+ 98
______

Example 8:
9008
+ 990
______

Example 9:
6837
+ 1895
______

Solve these additions.

● ¹3 ¹8 ¹5 4
+ 156
4010

a 2152
+ 215
______

b 5354
+ 1345
______

c 6824
+ 5495
______

d 1579
+ 43
______

e 8097
+ 49
______

SELF CHECK Tick how you feel

Got it!	Need help...	I don't get it
 ☐	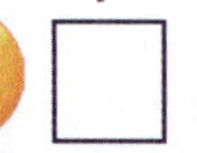 ☐	 ☐

Check your answers
How many did you get correct? ☐

CATCH UP MATHS YEAR 5 BOOK A © PASCAL PRESS ISBN: 9781925726169

PRACTICE

1 Answer these.

● (worked example)

	Th	H	T	O
	4	[1]3	[1]5	7
+			5	3
	4	4	1	0

a

	Th	H	T	O
	5	9	8	1
+			3	8

b

	Th	H	T	O
	5	6	2	5
+			4	3

2 Add these.

● 7 [1]0 [1]9 5 + 5 6 = 7 1 5 1

a 8 9 7 3 + 2 4 = ____

b 6 2 6 1 + 4 9 = ____

c 8 7 8 9 + 3 9 = ____

3 Solve.

● 7 5 3 1 + 2 4 6 = 7 7 7 7

b 5 8 1 6 + 3 4 7 = ____

d 4 3 8 2 + 3 4 1 = ____

f 6 9 2 0 + 1 8 9 = ____

a 8 9 9 4 + 5 3 8 = ____

c 6 2 4 3 + 1 8 5 = ____

e 5 7 1 1 + 3 9 9 = ____

g 4 9 2 0 + 2 3 9 = ____

4 Add.

● 4 9 2 5 + 1 0 3 1 = 5 9 5 6

b 6 7 8 9 + 2 3 4 5 = ____

d 8 5 7 3 + 5 4 8 0 = ____

f 7 2 6 7 + 5 2 4 3 = ____

a 5 4 3 1 + 1 5 4 2 = ____

c 9 9 9 0 + 7 3 4 8 = ____

e 9 3 2 4 + 1 5 8 8 = ____

g 9 3 7 3 + 2 7 7 9 = ____

ROUNDING TO ESTIMATE ADDITION ANSWERS

We can use rounding to get an answer that is close to the actual answer.

An estimate is near the answer, but it isn't perfectly accurate.

Round each number to the nearest 10 and then add to estimate the answer.

Example 1:

138 + 21 is about 160

140 20

Example 2:

142 + 57 is about ___

___ ___

Round to the nearest 100 to estimate the answer.

Example 3:

346 + 287 is about 600

300 300

Example 4:

415 + 695 is about ___

___ ___

Check your answer on the video!

Your turn

Estimate the answer.

Round to the nearest 10.

514 + 32 is about 540

510 30

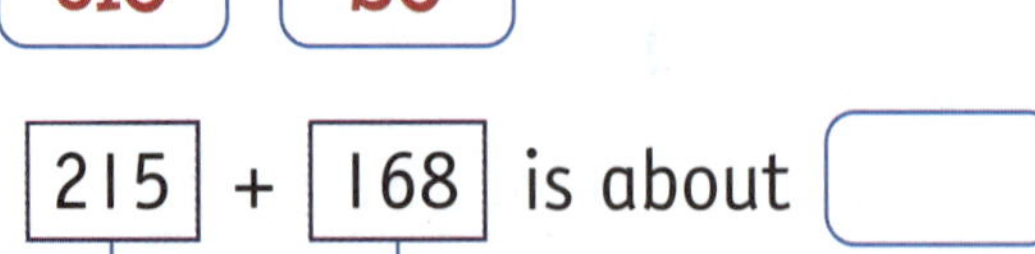

a Round to the nearest 100.

215 + 168 is about ___

___ ___

b Round to the nearest 10.

547 + 98 is about ___

___ ___

SELF CHECK Tick how you feel

Check your answers

How many did you get correct?

CATCH UP MATHS YEAR 5 BOOK A © PASCAL PRESS ISBN: 9781925726169

1 Estimate the answer by rounding each number to the nearest 10.

- 154 + 38 is about **190**
  - 150 40
- c 281 + 36 is about ___
  - ___ ___
- a 63 + 45 is about ___
  - ___ ___
- d 26 + 584 is about ___
  - ___ ___
- b 68 + 14 is about ___
  - ___ ___
- e 33 + 436 is about ___
  - ___ ___

2 Round each number to the nearest 100 to estimate the answer.

- 158 + 273 is about **500**
  - 200 300
- a 519 + 647 is about ___
  - ___ ___
- b 595 + 235 is about ___
  - ___ ___
- c 909 + 367 is about ___
  - ___ ___
- d 435 + 815 is about ___
  - ___ ___

USING THE JUMP STRATEGY TO SOLVE ADDITION

TWO-DIGIT AND THREE-DIGIT NUMBERS

We can use a number line and the jump strategy to solve addition problems.

SCAN to watch video

Example 1: 58 + 26 = 84

2-digit + 2-digit

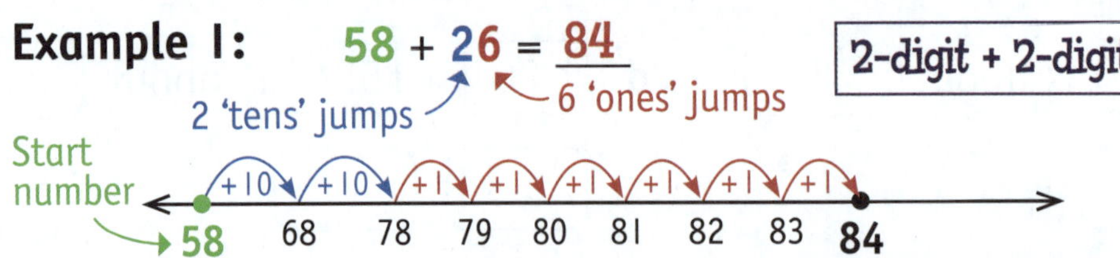

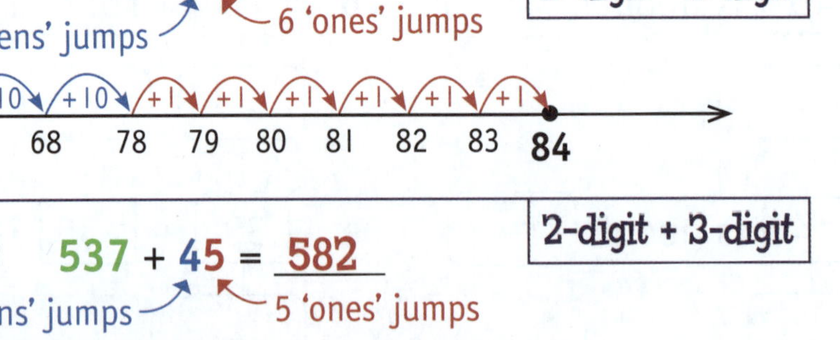

Example 2: 537 + 45 = 582

2-digit + 3-digit

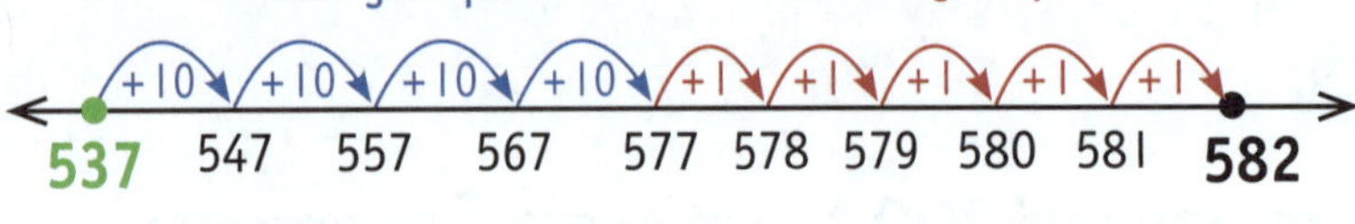

Example 3: 65 + 23 = ___

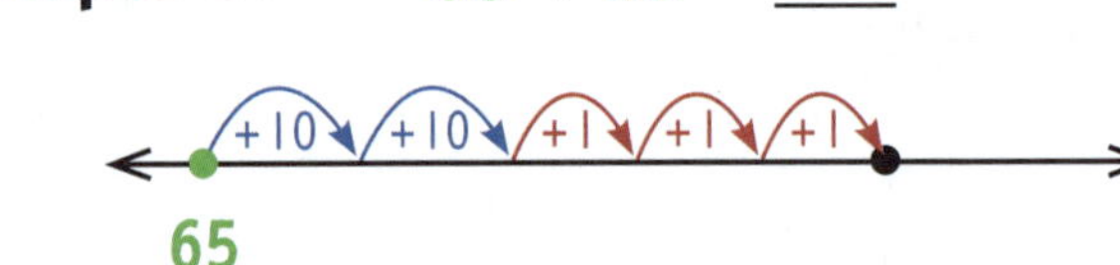

Use the number lines to solve the additions.

426 + 35 = 461

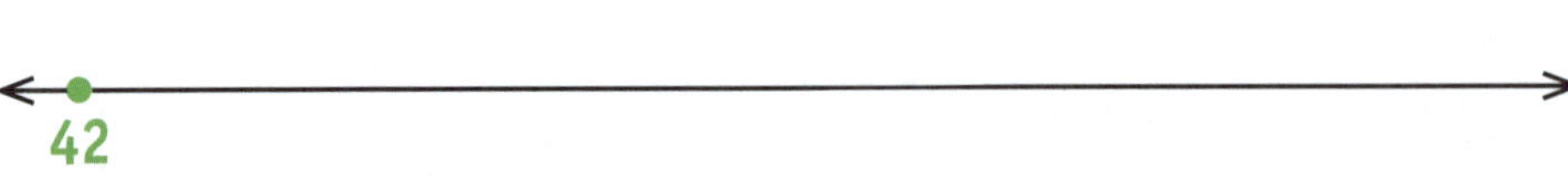

a 543 + 27 = ______

543

b 42 + 35 = ______

42

Check your answers
How many did you get correct?

CATCH UP MATHS YEAR 5 BOOK A © PASCAL PRESS ISBN: 9781925726169

PRACTICE

1 Use the number lines and the jump strategy to solve these additions.

● 43 + 67 = 110

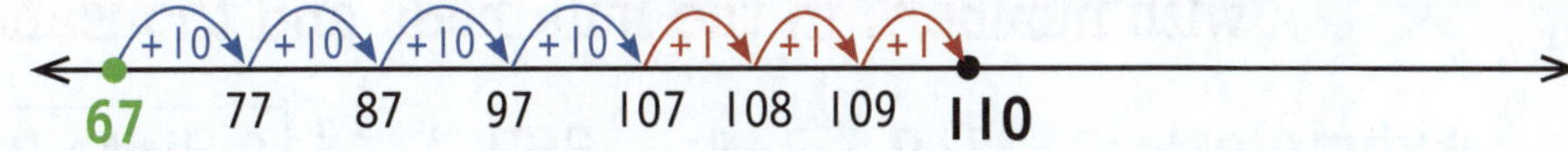

a 32 + 46 = ____

b 53 + 24 = ____

c 378 + 91 = ____

d 243 + 32 = ____

e 819 + 36 = ____

f 702 + 43 = ____

g 515 + 21 = ____

h 240 + 40 = ____

USING THE JUMP STRATEGY

THREE-DIGIT AND FOUR-DIGIT NUMBERS

SCAN to watch video

We can use the jump strategy to solve addition problems with numbers in the hundreds and thousands.

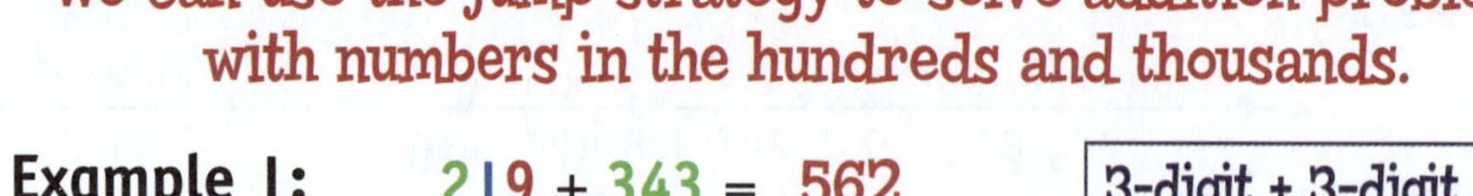

Example 1: 219 + 343 = 562

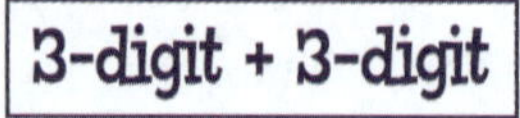

2 'hundreds' jumps, 1 'tens' jump, 9 'ones' jumps

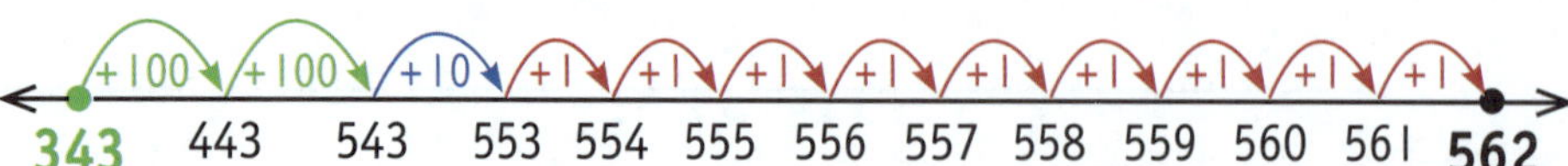
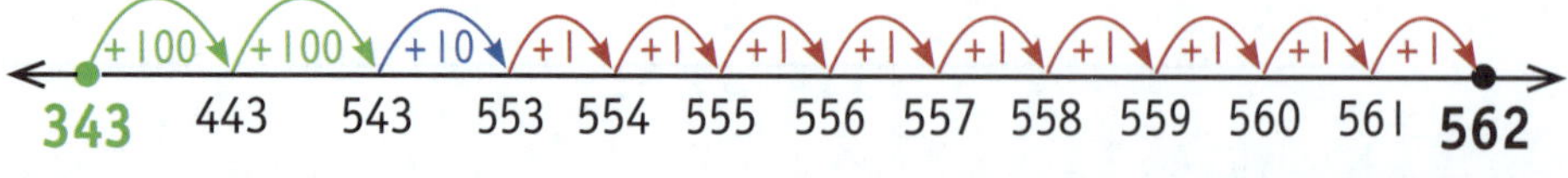

Start with the bigger number.

Example 2: 5382 + 132 = 5514

4-digit + 3-digit

Example 3: 4385 + 1521 = 5906

4-digit + 4-digit

1 'thousands' jump, 5 'hundreds' jumps, 2 'tens' jumps, 1 'ones' jump

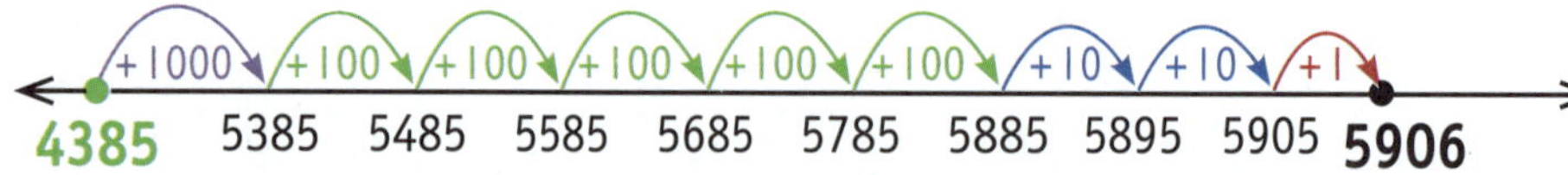

Example 4: 6243 + 412 = ______

____ ____ ____ ____ ____ ____ ____

Check your answer on the video!

Your turn

Use the number lines to solve the additions.

● 3265 + 1231 = 4496

+1000 +100 +100 +10 +10 +10 +1

3265 4265 4365 4465 4475 4485 4495 4496

a 1432 + 123 = ______

1432

SELF CHECK Tick how you feel

Got it!	Need help...	I don't get it
☐	☐	 ☐

Check your answers

How many did you get correct? ☐

CATCH UP MATHS YEAR 5 BOOK A © PASCAL PRESS ISBN: 9781925726169

1 Solve using the jump strategy and the number lines.

326 + 437 = 763

a 435 + 213 = ______

b 152 + 275 = ______

c 378 + 314 = ______

d 425 + 105 = ______

e 515 + 1243 = ______

f 1707 + 345 = ______

g 1432 + 2436 = ______

h 8724 + 2153 = ______

USING THE SPLIT STRATEGY TO SOLVE ADDITION

TWO-DIGIT AND THREE-DIGIT NUMBERS

With the split strategy, you split the numbers to show their value, and then add.

SCAN to watch video

Example 1:

2-digit + 2-digit

34 + 23

30 + 20 = 50

4 + 3 = 7

50 + 7

= 57

Example 2:

3-digit + 3-digit

424 + 315

400 + 300 = 700

20 + 10 = 30

4 + 5 = 9

700 + 30 + 9

= 739

Example 3:

2-digit + 3-digit

523 + 16

500

20 + ___ = 30

___ + 6 = 9

500 + ___ + ___

= 

Your turn

Use the split strategy to solve these additions.

● 54 + 33

50 + 30 = 80

4 + 3 = 7

80 + 7 = 87

a 462 + 35

___

___ + ___ = ___

___ + ___ = ___

___ + ___ + ___ = ___

b 524 + 133

___ + ___ = ___

___ + ___ = ___

___ + ___ = ___

___ + ___ + ___ = ___

c 642 + 336

___ + ___ = ___

___ + ___ = ___

___ + ___ = ___

___ + ___ + ___ = ___

SELF CHECK Tick how you feel

Got it!	Need help...	I don't get it
 ☐	 ☐	  ☐

Check your answers

How many did you get correct? ☐

CATCH UP MATHS YEAR 5 BOOK A © PASCAL PRESS ISBN: 9781925726169

PRACTICE

1 Use the split strategy to solve these additions.

325 + 243

300 + 200 = 500

20 + 40 = 60

5 + 3 = 8

500 + 60 + 8 = 568

a 36 + 23

___ + ___ = ___

___ + ___ = ___

___ + ___ = ___

b 534 + 235

___ + ___ = ___

___ + ___ = ___

___ + ___ = ___

___ + ___ + ___ = ___

c 132 + 43

___ + ___ = ___

___ + ___ = ___

___ + ___ = ___

___ + ___ + ___ = ___

d 742 + 835

___ + ___ = ___

___ + ___ = ___

___ + ___ = ___

___ + ___ + ___ = ___

e 858 + 41

___ + ___ = ___

___ + ___ = ___

___ + ___ = ___

___ + ___ + ___ = ___

f 552 + 425

___ + ___ = ___

___ + ___ = ___

___ + ___ = ___

___ + ___ + ___ = ___

g 283 + 14

___ + ___ = ___

___ + ___ = ___

___ + ___ = ___

___ + ___ + ___ = ___

h 607 + 254

___ + ___ = ___

___ + ___ = ___

___ + ___ = ___

___ + ___ + ___ = ___

USING THE SPLIT STRATEGY

THREE-DIGIT AND FOUR-DIGIT NUMBERS

You can use the split strategy to solve addition with larger numbers, in the hundreds and thousands.

Example 1:

3-digit + 4-digit

1253 + 431

1000

200 + 400 = 600

50 + 30 = 80

3 + 1 = 4

1000 + 600 + 80 + 4

= 1684

Example 2:

4-digit + 4-digit

2381 + 1315

2000 + 1000 = 3000

300 + 300 = 600

80 + 10 = 90

1 + 5 = 6

3000 + 600 + 90 + 6

= 3696

Example 3:

3-digit + 3-digit

536 + 251

500 + _____ = _____

30 + ____ = ____

___ + 1 = ___

_____ + ____ + ___

= ______

You can use this strategy to add numbers with any number of digits – 2, 3, 4, 5 or more!

Your turn

Solve using the split strategy.

● 1324 + 351

1000 + 0 = 1000

300 + 300 = 600

20 + 50 = 70

4 + 1 = 5

1000 + 600 + 70 + 5

= 1675

a 2525 + 4132

______ + ______ = ______

_____ + _____ = _____

____ + ____ = ____

___ + ___ = ___

______ + _____ + ____ + ___

= ______

SELF CHECK Tick how you feel

Got it!	Need help...	I don't get it
 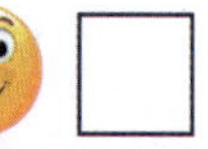 ☐	 ☐	  ☐

Check your answers

How many did you get correct? ☐

CATCH UP MATHS YEAR 5 BOOK A © PASCAL PRESS ISBN: 9781925726169

PRACTICE

1 **Use the split strategy to solve these additions.**

● 4341 + 1536

4000 + 1000 = 5000

300 + 500 = 800

40 + 30 = 70

1 + 6 = 7

5000 + 800 + 70 + 7

= 5877

a 561 + 215

____ + ____ = ____

___ + ___ = ___

__ + __ = __

____ + ___ + __ = ____

b 3251 + 4323

_____ + _____ = _____

____ + ____ = ____

___ + ___ = ___

__ + __ = __

____ + ____ + ___ + __

= _____

c 8732 + 146

_____ + _____ = _____

____ + ____ = ____

___ + ___ = ___

__ + __ = __

____ + ____ + ___ + __

= _____

d 757 + 121

____ + ____ = ____

___ + ___ = ___

__ + __ = __

____ + ___ + __ = ____

e 7267 + 1321

_____ + _____ = _____

____ + ____ = ____

___ + ___ = ___

__ + __ = __

____ + ____ + ___ + __

= _____

ROUNDING MONEY

When we shop, the totals often end in numbers that are not 0 or 5, like \$10.24, \$3.58 or \$22.51.

We can't use Australian coins to make 1 cent, 2 cents, 3 cents, 4 cents, 6 cents, 7 cents, 8 cents or 9 cents, so we round the cents to 0 or 5.

Example 1: Round \$10.36 to the nearest 5 cents.

Look at the last number.

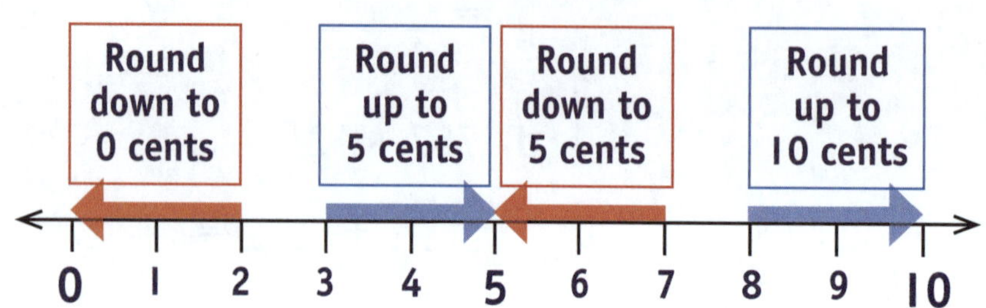

6 rounds down to 5 cents because 6 is closer to 5 than 0.

\$10.36 rounded to the nearest 5 cents is \$10.35.

Example 2: Round \$7.32 to the nearest 5 cents.

Look at the last number.

The number 2 rounds _____ to 0 cents because 2 is closer to 0 than 5.

\$7.32 rounded to the nearest 5 cents is _______.

Round to the nearest 5 cents.

● \$9.88 \$9.90

a \$8.57 _______

b \$14.33 _______

c \$15.24 _______

d \$20.66 _______

e \$84.97 _______

SELF CHECK Tick how you feel

Got it!	Need help...	I don't get it
☐	☐	☐

Check your answers
How many did you get correct? ☐

CATCH UP MATHS YEAR 5 BOOK A © PASCAL PRESS ISBN: 9781925726169

PRACTICE

Round these amounts to the nearest 5 cents.

- e.g. $2.43 — $2.45
- g $23.93 ______
- n $9.86 ______
- a $1.58 ______
- h $27.78 ______
- o $101.94 ______
- b $0.97 ______
- i $7.28 ______
- p $35.48 ______
- c $3.42 ______
- j $41.87 ______
- q $123.53 ______
- d $5.54 ______
- k $0.12 ______
- r $25.67 ______
- e $11.31 ______
- l $33.91 ______
- s $15.67 ______
- f $13.66 ______
- m $31.03 ______
- t $0.83 ______

2 Circle the correct answer to round each amount to the nearest 5 cents.

e.g. **$1.52**

($1.50 circled) $1.55 $1.60

a **$2.46**

$2.45 $2.40 $2.50

b **$3.73**

$3.70 $3.80 $3.75

c **$24.14**

$24.10 $24.15 $24.20

d **$41.58**

$41.60 $41.55 $41.50

e **$157.53**

$157.50 $157.60 $157.55

f **$275.98**

$275.90 $275.95 $276.00

g **$199.99**

$199.95 $199.90 $200.00

WORKING OUT CHANGE

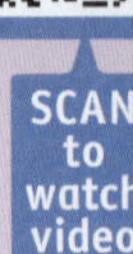
SCAN to watch video

If you use cash to buy something, you can work out the change by counting on.

Example 1:
If you spend $3.60, what is your change from $5.00?

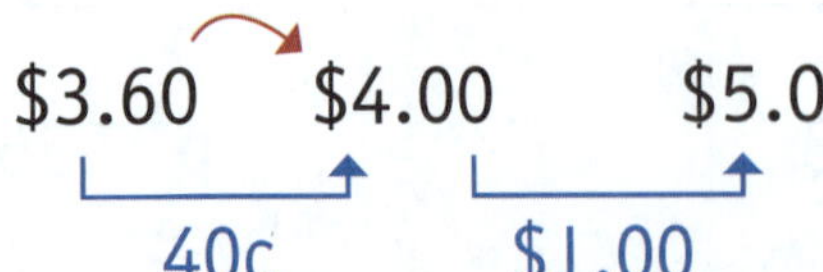

up to the next dollar

$3.60 → $4.00 → $5.00

40c $1.00 The change is $1.40.

Example 2: If you spend $5.25, what is your change from $10?

up to the next dollar

$5.25 → $6.00 → $10.00

75c $4.00 The change is $4.75.

Example 3: If you spend $5.80, what is your change from $20?

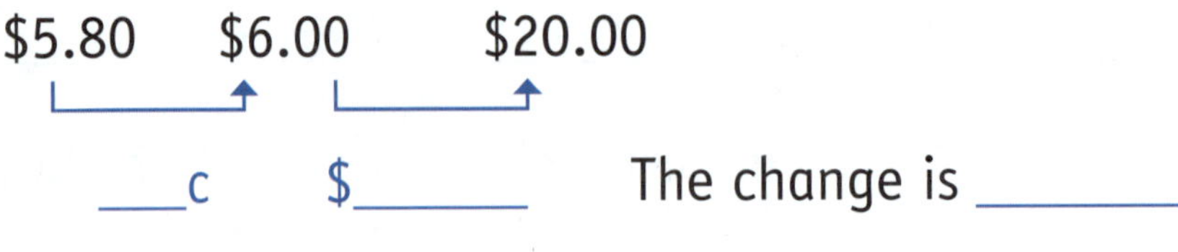

$5.80 → $6.00 → $20.00

____c $______ The change is ________.

Check your answer on the video!

Your turn

Work out the change.

	Spend		Amount given	Change
●	$1.25	$2.00	$10.00	$8.75
	75c	$8.00		
a	$2.75	$3.00	$5.00	
b	$11.55	$12.00	$20.00	

SELF CHECK Tick how you feel

Got it!	Need help...	I don't get it
☐	☐	☐

Check your answers
How many did you get correct?

CATCH UP MATHS YEAR 5 BOOK A © PASCAL PRESS ISBN: 9781925726169

PRACTICE

1 Work out the change by counting on to the nearest dollar and then counting on the dollars to the amount given.

	Spend		Amount given	Change
●	25c	$1.00	$2.00	$1.75
		75c	$1.00	

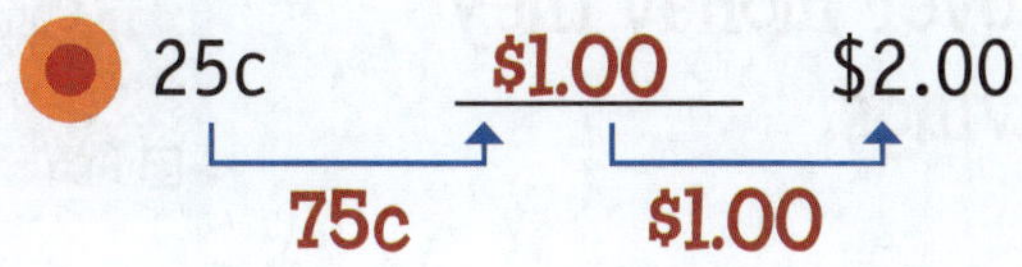

	Spend		Amount given	Change
a	$1.35	______	$3.00	
b	$2.10	______	$5.00	
c	$3.95	______	$5.00	
d	$1.80	______	$5.00	
e	$2.55	______	$10.00	
f	$5.20	______	$10.00	
g	$8.30	______	$10.00	
h	$12.50	______	$20.00	
i	$8.75	______	$20.00	
j	$23.55	______	$50.00	

MONEY PROBLEMS SPENDING AND SAVING

People spend money on lots of different things. These are called expenses. Any leftover money they do not spend is called savings.

Example 1:
Antonia gets paid $95.70 per week. Here are her expenses:

Food $22.50 | Drum lesson $15.00 | Clothes $20.00 | Gift $18.75

Antonia spends $76.25:

	$ 22.50	Food
	$ 15.00	Drum lesson
	$ 20.00	Clothes
+	$ 18.75	Gift
	$ 76.25	

Antonia saves $19.45:

	$ 95.70
–	$ 76.25
	$ 19.45

Example 2: Add up Paul's expenses.

Food $10.50 | Guitar lesson $45.00 | Football $20.00 | Shoes $18.75

______ + ______ + ______ + ______ = ______

If Paul earns $395.00 per week, how much can he save?

$395 – ______ = ______

Your turn

Stelios spends $25.20 on food and $15.00 on clothes. He earned $77.50.

● How much did Stelios spend altogether?

	$ 25.20
+	$ 15.00
	$ 40.20

Stelios spent $40.20.

a How much does Stelios save?

Stelios saves ______.

SELF CHECK Tick how you feel

Got it!	Need help...	I don't get it
☐	☐	☐

Check your answers
How many did you get correct? ☐

CATCH UP MATHS YEAR 5 BOOK A © PASCAL PRESS ISBN: 9781925726169

PRACTICE

1 Danny and Kai are working out how much they spend (expenses) and how much they can save this week.

Danny

Payslip	$594.30	
Expenses	$28.75	Lunches
	$30.00	Movies
	$162.30	Groceries
	$25.00	Gift

Kai

Payslip	$553.75	
Expenses	$24.35	Lunches
	$15.00	Parking fee
	$155.70	Groceries
	$42.30	Restaurant

a Work out Danny's and Kai's total expenses.

Danny: ________ Kai: ________

b How much can Danny save? __________

c How much can Kai deposit? __________

d Who earns more money? __________

e Who saves more money? __________

2 If you put money into a bank account, you deposit the money. If you take money out of a bank account, it is called a withdrawal.

Here, Deborah deposits $230.20 and withdraws $20:

Deposit	Withdraw	Balance
$230.20		$230.20
	$20.00	$210.20

Fill in the tables below to show Danny and Kai depositing their savings and then withdrawing $20.

a Danny

Deposit	Withdraw	Balance
		$ 0.00
$		$
	$	$

b Kai

Deposit	Withdraw	Balance
		$ 0.00
$		$
	$	$

3 Look at Syd's bank statement.

Date	Deposit	Withdrawal	Balance
15/3	$150.00		$150.00
17/3		$20.00	$130.00
18/3	$100.00		$230.00
21/3	$157.00		$287.00
23/3	$80.00		$367.00
27/3		$72.00	$295.00
29/3	$520.00		$815.00
1/4		$100.00	$715.00

What were the total withdrawals made by Syd?

```
  $  20.00
  $  72.00
+ $ 100.00
  $ 192.00
```

a How much money was deposited on 18/3? ____________

b How much money was taken out on 27/3? ____________

c What was the total amount deposited? ____________

d What was the balance on 15/3? ____________

e What was the balance on 21/3? ____________

f What was the balance on 29/3? ____________

g What was the balance on 1/4? ____________

h When was $80 deposited? ____________

i When was $100 withdrawn? ____________

j What was the highest balance? ____________

k What was the largest deposit? ____________

l What was the smallest withdrawal? ____________

m What date was the balance $367.00? ____________

n What date was the balance $150.00? ____________

CATCH UP MATHS YEAR 5 BOOK A © PASCAL PRESS ISBN: 9781925726169

Write the deposits and withdrawals in the correct columns on Jo's bank statement. Don't forget to fill out the balance!

- 5/5 Deposit $20.00
- 7/5 Deposit $30.00
- 9/5 Deposit $120.00
- 15/5 Withdraw $50
- 21/5 Withdraw $20
- 30/5 Deposit $60
- 1/6 Withdraw $10
- 6/6 Deposit $250

Date	Deposit	Withdrawal	Balance
5/5	$20.00		$20.00
7/5			
9/5			
15/5			
21/5			
30/5			
1/6			
6/6			

a What was the balance on 7/5? ____________

b What was the balance on 15/5? ____________

c What was the balance on 30/5? ____________

d What was the balance on 6/6? ____________

e What was the largest deposit made? ____________

f How many deposits were made? ____________

g How many withdrawals were made? ____________

h What was the total amount of deposits? ____________

i What was the total amount of withdrawals? ____________

j Which date was $60 deposited? ____________

k How much money was deposited up to and including 30/5?

____

l How much money was withdrawn up to and including 30/5?

____

Write these items in the correct part of Han's earnings and expenses chart.

Pay $979.80	Guitar lesson $55.00	Pay for chores $65.50	Food $127.50
Soccer fees $120.00	Pay for walking dogs $75.00	Clothes $135.50	Movies $55.00

Han's Account

Earnings		Expenses	
Item	**Amount**	**Item**	**Amount**
Total		Total	

a Han's expenses total $__________.

b How much did Han get paid for doing chores and walking dogs?

____

c What were Han's total earnings? __________

d Han can deposit $__________ into his bank account.

e What is Han's biggest expense? ____________________

f If Han started with $150.50 in his bank account, how much would he have after depositing his savings?

____

CATCH UP MATHS YEAR 5 BOOK A © PASCAL PRESS ISBN: 9781925726169

6 Necessities are things that are necessary to live. Luxuries are things that are nice to have, but they are not necessary.

Sort the items below as necessities or luxuries and then calculate the total costs.

Necessity	Amount
Total	

Luxury	Amount
Total	

Soft drink $4.20

Toy car $15.20

Take away $42.00

Fruit and vegetables $29.50

Flowers $22.50

Clothes $55.00

Milk $4.20

Medicine $16.90

Bread $5.00

Chocolate $5.50

a What two items would you choose not to buy if you were trying to save money?

____

b How much money was spent on food and drink? ____________

c If you only had $60 to spend, what would you buy, and why?

____

____

ADDITION REVIEW

1 Look for combinations that make 10 to help you add these.

a 6 + 7 + 4

= ____________

= ________

= ___

c 3 + 7 + 8

= ____________

= ________

= ___

e 1 + 8 + 2 + 9

= ____________

= ________

= ___

b 5 + 5 + 3

= ____________

= ________

= ___

d 8 + 1 + 2 + 5

= ____________

= ________

= ___

f 7 + 4 + 6 + 3

= ____________

= ________

= ___

2 What is the sum?

a 8, 4, 11

____

c 7 + 3 + 5 + 4

____

e 16, 2, 33, 8

____

b 2 + 9 + 8

____

d 13, 7, 15, 5

____

f 9, 9, 45, 16

____

3 Write a fact family using these numbers.

a 5, 15, 20

___ + ___ = ___

___ + ___ = ___

___ − ___ = ___

___ − ___ = ___

b 60, 48, 12

___ + ___ = ___

___ + ___ = ___

___ − ___ = ___

___ − ___ = ___

c 121, 20, 101

___ + ___ = ___

___ + ___ = ___

___ − ___ = ___

___ − ___ = ___

CATCH UP MATHS YEAR 5 BOOK A © PASCAL PRESS ISBN: 9781925726169

4 Write two addition and two subtraction facts.

a 24 35 59

○ + ○ = ○
○ + ○ = ○
○ − ○ = ○
○ − ○ = ○

b 47 62 15

○ + ○ = ○
○ + ○ = ○
○ − ○ = ○
○ − ○ = ○

c 84 53 31

○ + ○ = ○
○ + ○ = ○
○ − ○ = ○
○ − ○ = ○

5 Complete these additions.

a 36 + 52 = ____

b 57 + 22 = ____

c 72 + 15 = ____

d 43 + 25 = ____

e 17 + 91 = ____

6 Solve.

a 735 + 121 = ____

b 424 + 61 = ____

c 532 + 415 = ____

d 732 + 142 = ____

e 1382 + 1411 = ____

f 2461 + 1327 = ____

g 3253 + 1512 = ____

h 4210 + 1037 = ____

7 Complete these.

a 74 + 39 = ____

b 26 + 89 = ____

c 45 + 76 = ____

d 35 + 58 = ____

e 76 + 89 = ____

f 135 + 429 = ____

g 576 + 836 = ____

h 897 + 123 = ____

REVIEW

i
```
  7 0 9
+ 8 9 9
```

k
```
  6 2 8
+ 9 4 6
```

m
```
  9 5 0 1
+ 3 3 8 9
```

o
```
  7 3 8 5
+ 7 4 9 3
```

j
```
  4 5 7
+ 7 5 3
```

l
```
  2 4 3 6
+ 5 8 2 0
```

n
```
  6 2 0 6
+ 8 5 4 9
```

p
```
  7 8 2 5
+ 3 0 9 3
```

8 Round these numbers to the nearest 10 to estimate the answers.

a 27 + 32 is about ____

c 143 + 58 is about ____

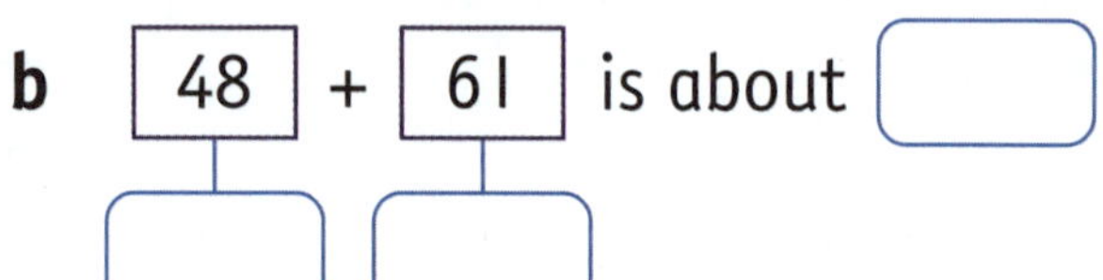

b 48 + 61 is about ____

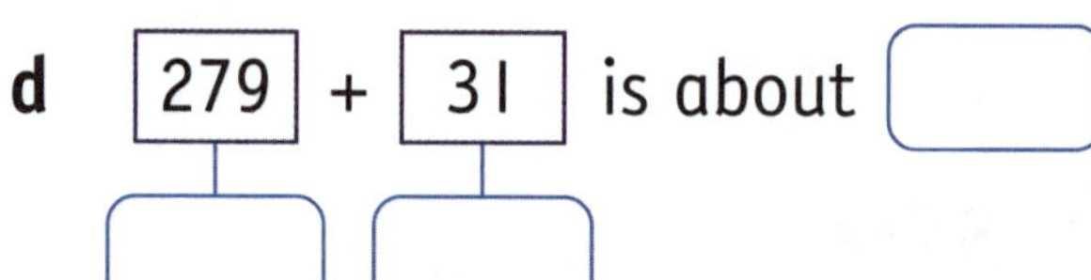

d 279 + 31 is about ____

9 Round these numbers to the nearest 100 to estimate the answers.

a 327 + 141 is about ____

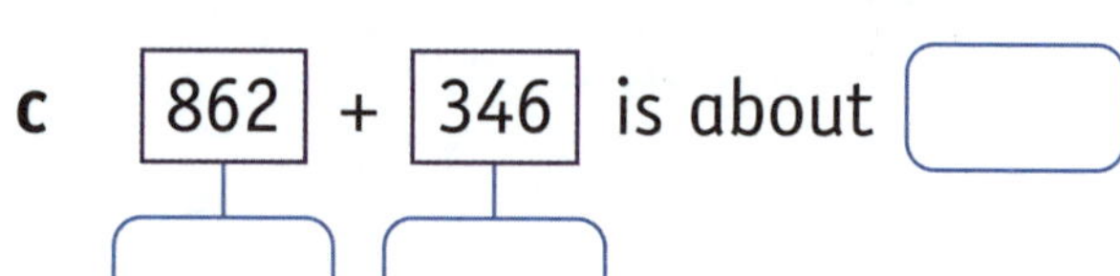

c 862 + 346 is about ____

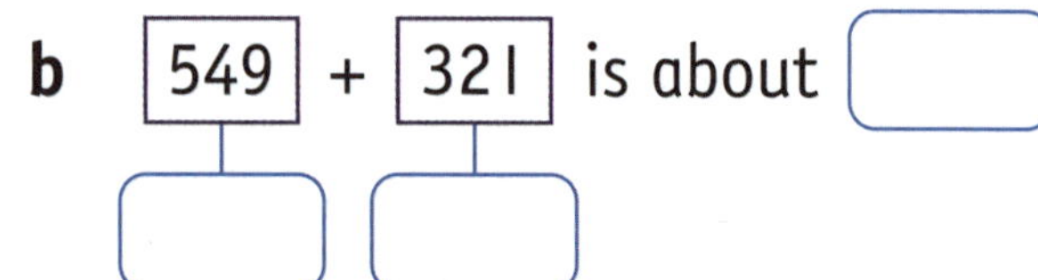

b 549 + 321 is about ____

d 416 + 711 is about ____

10 Use the jump strategy and the number lines to solve these additions.

a 37 + 25 = ______

CATCH UP MATHS YEAR 5 BOOK A © PASCAL PRESS ISBN: 9781925726169

b 42 + 68 = ______

c 32 + 453 = ______

d 72 + 181 = ______

e 5812 + 311 = ________

f 6057 + 211 = ________

g 3341 + 1234 = ________

h 5312 + 6341 = ________

Use the split strategy to solve these additions.

a 35 + 52

___ + ___ = ___

__ + __ = __

___ + __ = ___

b 62 + 71

___ + ___ = ____

__ + __ = __

___ + __ = ____

REVIEW

c 542 + 55

____ + ____ = ____

___ + ___ = ___

__ + __ = __

____ + ___ + __ = ____

d 734 + 23

____ + ____ = ____

___ + ___ = ___

__ + __ = __

____ + ___ + __ = ____

e 535 + 234

____ + ____ = ____

___ + ___ = ___

__ + __ = __

____ + ___ + __ = ____

f 652 + 143

____ + ____ = ____

___ + ___ = ___

__ + __ = __

____ + ___ + __ = ____

g 251 + 347

____ + ____ = ____

___ + ___ = ___

__ + __ = __

____ + ___ + __ = ____

h 5784 + 1213

_____ + _____ = _____

____ + ____ = ____

___ + ___ = ___

__ + __ = __

_____ + ____ + ___ + __

= _____

i 2613 + 5311

_____ + _____ = _____

____ + ____ = ____

___ + ___ = ___

__ + __ = __

_____ + ____ + ___ + __

= _____

12 Round these amounts to the nearest 5 cents.

a $1.26 ________

b $2.57 ________

c $8.49 ________

d $14.13 ________

e $25.41 ________

f $72.56 ________

g $102.52 ________

h $251.38 ________

CATCH UP MATHS YEAR 5 BOOK A © PASCAL PRESS ISBN: 9781925726169

13 **Work out the change.**

	Spend	Amount given	Change
a	65c	$2.00	
b	$1.55	$5.00	
c	$2.35	$5.00	
d	$3.70	$10.00	
e	$6.45	$10.00	

	Spend	Amount given	Change
f	$15.20	$20.00	
g	$7.65	$20.00	
h	$3.40	$50.00	
i	$28.10	$50.00	
j	$14.75	$50.00	

14 **Annie gets paid $495.15 per week. She spends:**

How much can Annie deposit into her savings account after paying her expenses?

____

Item	Amount
Food	$123.45
Movie	$16.50
Basketball	$10.00
Gift	$30.00

15 **Lina and Ali are working out their expenses to see how much money they can save this week.**

Lina

Payslip	**$637.20**	
Expenses	$231.00	Food
	$55.00	Clothes
	$50.00	Gift

Ali

Payslip	**$589.30**	
Expenses	$150.00	Food
	$40.00	Petrol
	$30.00	Gift
	$10.00	Movies

a Work out Lina's and Ali's expenses.

Lina ___________ Ali ___________

b Who saved more? _________

c Who had the larger payslip? ___________

d Who spent more on expenses? ___________

e Who spent less on food? ___________

f How much in savings do Lina and Ali have altogether? __________

USING THE JUMP STRATEGY TO SOLVE SUBTRACTION

TWO-DIGIT AND THREE-DIGIT NUMBERS

You can use the jump strategy on a number line to solve subtraction.

Example 1: Solve 67 − 32.

67 − 32 ← 2 'ones' jumps
3 'tens' jumps

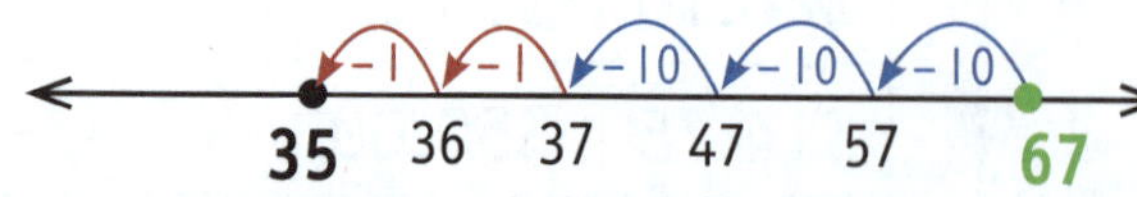

So, 67 − 32 = 35

Example 2: Solve 658 − 241.

658 − 241 ← 1 'ones' jump
2 'hundreds' jumps
4 'tens' jumps

−1 −10 −10 −10 −10 −100 −100

417 418 428 438 448 458 558 658

So, 658 − 241 = 417

Example 3: Solve 739 − 14.

739 − 14

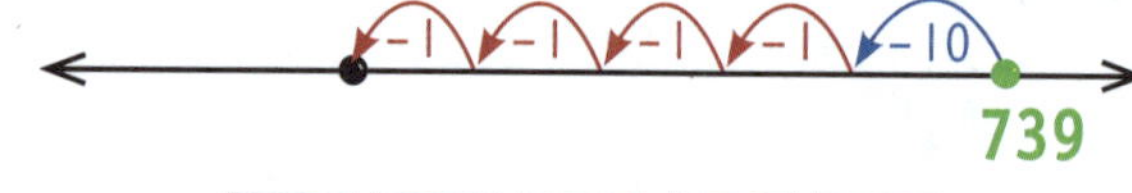

___ ___ ___ ___ ___

So, 739 − 14 = ______

Your turn

Solve using the jump strategy and the number line.

983 − 241 = 742

−1 −10 −10 −10 −10 −100 −100

742 743 753 763 773 783 883 983

a 342 − 35 = ______

−1 −1 −1 −1 −1 −10 −10 −10

342

b 51 − 23 = ______

−1 −1 −1 −10 −10

51

c 536 − 252 = ______

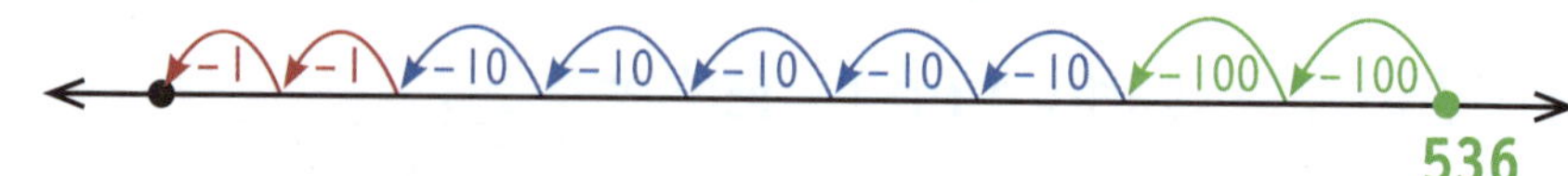

SELF CHECK Tick how you feel

Got it!	Need help...	I don't get it
 ☐	 ☐	 ☐

Check your answers
How many did you get correct? ☐

CATCH UP MATHS YEAR 5 BOOK A © PASCAL PRESS ISBN: 9781925726169

PRACTICE

1 Use the jump strategy and the number lines to solve these subtractions.

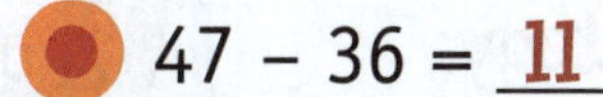

47 – 36 = 11

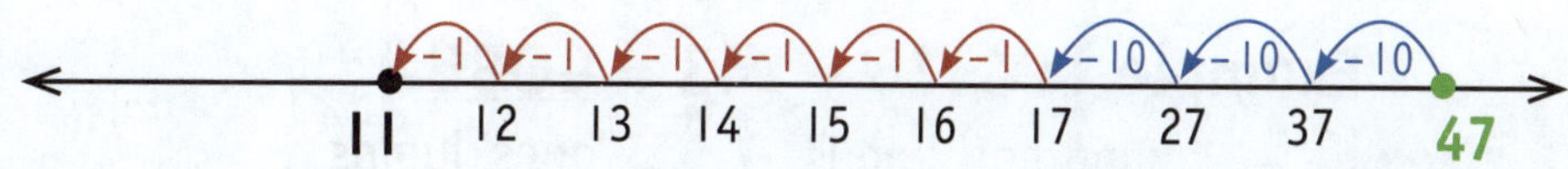

a 58 – 44 = ___

b 72 – 35 = ___

c 81 – 53 = ___

d 524 – 33 = ___

e 923 – 431 = ___

f 611 – 203 = ___

g 802 – 130 = ___

h 689 – 400 = ___

USING THE JUMP STRATEGY

THREE-DIGIT AND FOUR-DIGIT NUMBERS

These subtraction problems have three-digit and four-digit numbers. You can solve them using the jump strategy.

Example 1: 4328 − 243 = 4085

2 'hundreds' jumps, 4 'tens' jumps, 3 'ones' jumps

−1 −1 −1 −10 −10 −10 −10 −100 −100

4085 4086 4087 4088 4098 4108 4118 4128 4228 4328

Example 2: 6597 − 2341 = 4256

2 'thousands' jumps, 3 'hundreds' jumps, 4 'tens' jumps, 1 'ones' jump

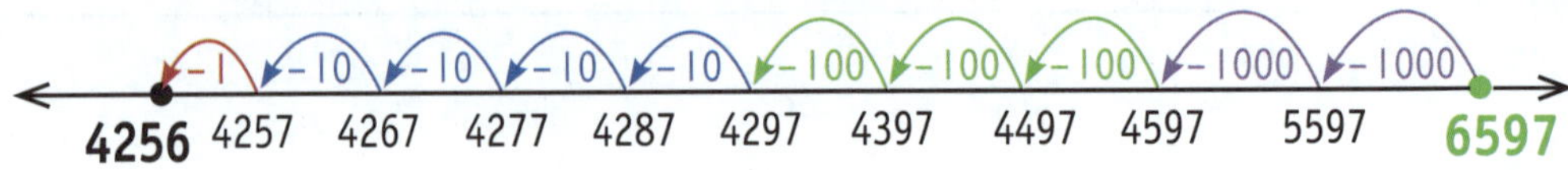

4256 4257 4267 4277 4287 4297 4397 4497 4597 5597 6597

Example 3: 8739 − 1213 = _____

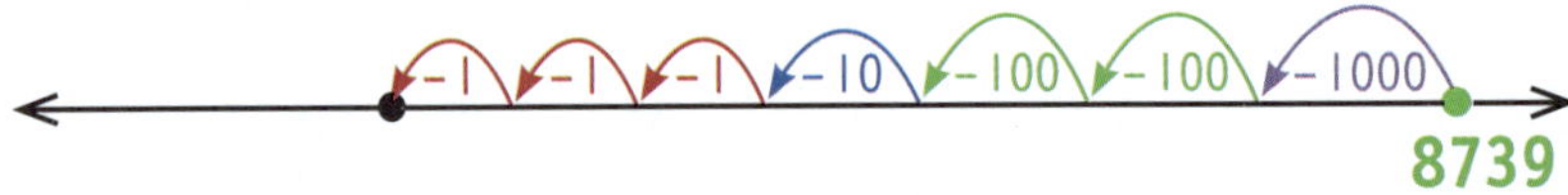

8739

____ ____ ____ ____ ____ ____ ____

Your turn

Solve using the jump strategy and the number line.

● 7529 − 1036 = 6493

−1 −1 −1 −1 −1 −1 −10 −10 −10 −1000

6493 6494 6495 6496 6497 6498 6499 6509 6519 6529 7529

a 5267 − 341 = _____

b 8421 − 2324 = _____

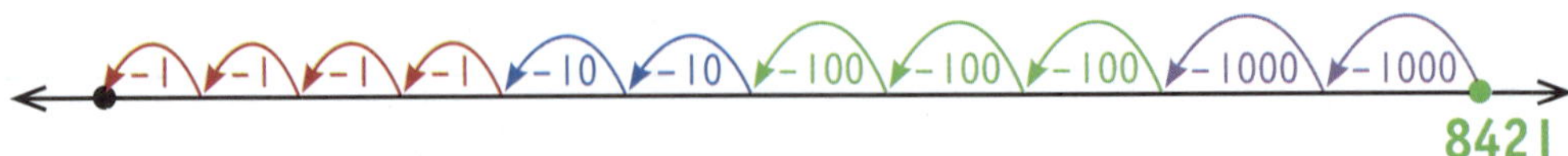

SELF CHECK Tick how you feel

Got it!	Need help...	I don't get it
☐	☐	☐

Check your answers

How many did you get correct?

CATCH UP MATHS YEAR 5 BOOK A © PASCAL PRESS ISBN: 9781925726169

PRACTICE

Work out the answers to these subtraction problems using the number lines and the jump strategy.

7438 – 3421 = 4017

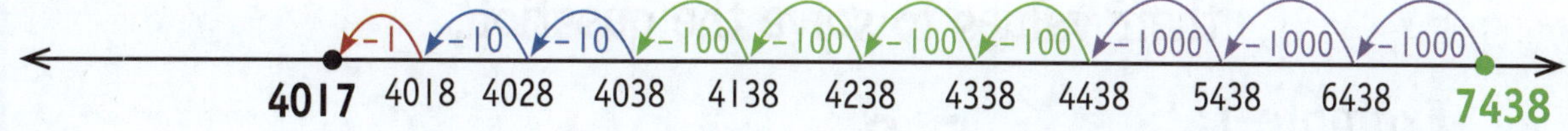

a 5981 – 136 = ______

b 2482 – 531 = ______

c 9834 – 102 = ______

d 5324 – 3140 = ______

e 8240 – 3213 = ______

f 6829 – 2432 = ______

g 1832 – 1431 = ______

h 3891 – 2132 = ______

USING THE SPLIT STRATEGY TO SOLVE SUBTRACTION

TWO-DIGIT AND THREE-DIGIT NUMBERS

With the split strategy, we 'split' the numbers into their values to solve the question.

Example 1:

5 | 2 − 3 | 1

tens | ones   tens | ones

5 has a value of 50

2 has a value of 2

3 has a value of 30

1 has a value of 1

52 − 31

50 − 30 = 20

2 − 1 = 1

We add here → 20 + 1 = 21

Split the numbers into tens and ones, and then subtract.

Example 2:

Solve 195 − 42

100

90 − 40 = 50

5 − 2 = 3

100 + 50 + 3 = 153

Example 3:

Solve 385 − 23

300

80 − ___ = ___

__ − 3 = __

300 + ___ + __ = _____

Your turn

Solve using the split strategy.

● 249 − 123

200 − 100 = 100

40 − 20 = 20

9 − 3 = 6

100 + 20 + 6 = 126

a 362 − 51

_____ − _____ = _____

____ − ____ = ____

__ − __ = __

_____ + ____ + __ = _____

SELF CHECK Tick how you feel

Got it!	Need help...	I don't get it
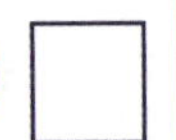 ☐	 ☐	☐

Check your answers

How many did you get correct? ☐

CATCH UP MATHS YEAR 5 BOOK A © PASCAL PRESS ISBN: 9781925726169

1 Solve these subtraction problems using the split strategy.

525 – 213

500 – 200 = 300

20 – 10 = 10

5 – 3 = 2

300 + 10 + 2 = 312

a 589 – 378

___ – ___ = ___

___ – ___ = ___

___ – ___ = ___

___ + ___ + ___ = ___

b 82 – 41

___ – ___ = ___

___ – ___ = ___

___ + ___ = ___

c 99 – 27

___ – ___ = ___

___ – ___ = ___

___ + ___ = ___

d 373 – 22

___ – ___ = ___

___ – ___ = ___

___ – ___ = ___

___ + ___ + ___ = ___

e 834 – 21

___ – ___ = ___

___ – ___ = ___

___ – ___ = ___

___ + ___ + ___ = ___

f 854 – 213

___ – ___ = ___

___ – ___ = ___

___ – ___ = ___

___ + ___ + ___ = ___

g 789 – 526

___ – ___ = ___

___ – ___ = ___

___ – ___ = ___

___ + ___ + ___ = ___

h 842 – 430

___ – ___ = ___

___ – ___ = ___

___ – ___ = ___

___ + ___ + ___ = ___

USING THE SPLIT STRATEGY

THREE-DIGIT AND FOUR-DIGIT NUMBERS

Here are examples of how the split strategy can be used to solve subtraction problems with larger numbers.

Example 1:

3842 – 321

3 has a value of 3000
8 has a value of 800
4 has a value of 40
2 has a value of 2
3 has a value of 300
2 has a value of 20
1 has a value of 1

3842 – 321

3000

800 – 300 = 500

40 – 20 = 20

2 – 1 = 1

3000 + 500 + 20 + 1

= 3521

Split the numbers and solve by using the value of each number.

Example 2: Solve 5392 – 1231

5000 – ______ = ______

300 – _____ = _____

90 – ____ = ____

2 – ___ = ___

______ + _____ + ____ + ___

= ______

Your turn Solve using the split strategy.

● 2435 – 214

2000 – 0 = 2000

400 – 200 = 200

30 – 10 = 20

5 – 4 = 1

2000 + 200 + 20 + 1

= 2221

a 4732 – 2132

______ – ______ = ______

_____ – _____ = _____

____ – ____ = ____

___ – ___ = ___

______ + _____ + ____ + ___

= ______

SELF CHECK Tick how you feel

Got it!	Need help...	I don't get it
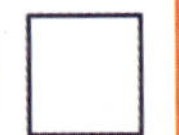 ☐	 ☐	☐

Check your answers

How many did you get correct? ☐

CATCH UP MATHS YEAR 5 BOOK A © PASCAL PRESS ISBN: 9781925726169

PRACTICE

1 **Use the split strategy to solve these subtraction problems.**

● 5342 – 3120 = 2222

5000 – 3000 = 2000

300 – 100 = 200

40 – 20 = 20

2 – 0 = 2

2000 + 200 + 20 + 2

= 2222

a 7238 – 2103

_____ – _____ = _____

____ – ____ = ____

___ – ___ = ___

__ – __ = __

_____ + ____ + ___ + __

= _____

b 6948 – 543

_____ – _____ = _____

____ – ____ = ____

___ – ___ = ___

__ – __ = __

_____ + ____ + ___ + __

= _____

c 4752 – 611

_____ – _____ = _____

____ – ____ = ____

___ – ___ = ___

__ – __ = __

_____ + ____ + ___ + __

= _____

d 8939 – 5213

_____ – _____ = _____

____ – ____ = ____

___ – ___ = ___

__ – __ = __

_____ + ____ + ___ + __

= _____

e 3856 – 724

_____ – _____ = _____

____ – ____ = ____

___ – ___ = ___

__ – __ = __

_____ + ____ + ___ + __

= _____

SUBTRACTION WITHOUT TRADING

TWO-DIGIT AND THREE-DIGIT NUMBERS

We can solve subtraction problems using only numbers and symbols.

Example 1:

	T	O
	6	4
–	2	3
	4	**1**

Start at the ones column:
4 – 3 = 1
Then work out 6 – 2.
Always subtract downwards.

Example 2:

	H	T	O
	5	9	3
–	2	4	1
	3	**5**	**2**

Start at the ones column.
Always subtract downwards.

Example 3:

	H	T	O
	4	9	7
–		2	4
	4	**7**	**3**

Example 4:

	H	T	O
	6	8	4
–	3	7	3

Complete these subtractions.

● 503 – 401 = **102**

a 562 – 31 = ____

b 936 – 124 = ____

c 59 – 23 = ____

d 795 – 24 = ____

e 648 – 443 = ____

Check your answers
How many did you get correct?

CATCH UP MATHS YEAR 5 BOOK A © PASCAL PRESS ISBN: 9781925726169

PRACTICE

1 Complete these subtractions.

Example:

	H	T	O
	7	3	0
−	2	2	0
	5	**1**	**0**

a

	H	T	O
	5	9	4
−		5	3

b

	T	O
	5	4
−	2	1

c

	H	T	O
	6	3	4
−		3	1

2 Now complete these subtractions.

Example: 69 − 37 = **32**

a 297 − 134 = ____

b 849 − 18 = ____

c 767 − 146 = ____

d 795 − 274 = ____

e 68 − 38 = ____

f 495 − 293 = ____

g 710 − 610 = ____

h 27 − 13 = ____

i 693 − 22 = ____

j 587 − 53 = ____

k 43 − 31 = ____

3 Fill in the missing digits.

Example:
```
  5 3 4
− 1 2 2
  4 1 2
```
(4, 2 and 4 filled in)

a
```
  _ 3
− 4 _
  2 2
```

b
```
  _ 9 2
− 3 _ 2
  4 5 _
```

c
```
  5 3 _
− _ 2 4
  4 _ 1
```

d
```
  7 6 8
− _ 2 _
  4 _ 8
```

e
```
  9 9 6
− 7 _ 2
  _ 2 _
```

f
```
  8 3 _
− _ 2 7
  3 _ 1
```

g
```
  _ 9 2
− 3 6 _
  1 _ 0
```

SUBTRACTION WITHOUT TRADING

THREE-DIGIT AND FOUR-DIGIT NUMBERS

Here are subtraction algorithms with three-digit and four-digit numbers.

Example 1:

	Th	H	T	O
	4	3	8	2
−		2	6	1
	4	**1**	**2**	**1**

Start at the ones column.
Then move to the tens column.
Next, the hundreds.
Finally, the thousands.

Example 2:

	Th	H	T	O
	5	8	5	4
−	1	7	4	0
	4	**1**	**1**	**4**

Always subtract downwards.

Example 3:

	Th	H	T	O
	3	8	4	9
−	3	7	4	5
		1		**4**

Example 4:

	Th	H	T	O
	8	7	4	4
−		6	4	2
		1	**0**	

Your turn

Solve these subtraction algorithms.

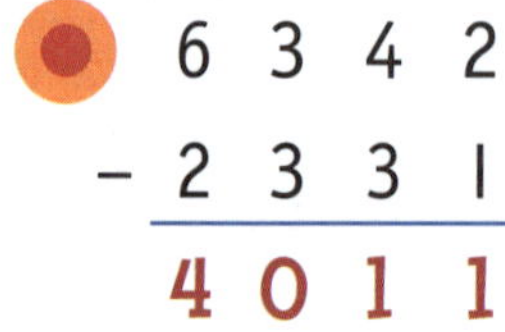

●
```
  6 3 4 2
- 2 3 3 1
  4 0 1 1
```

a
```
  5 9 4 5
- 1 3 2 4
```

b
```
  7 4 9 0
-   2 3 0
```

c
```
  8 4 9 3
-   3 1 2
```

d
```
  8 7 4 2
- 7 4 3 2
```

e
```
  2 6 0 4
- 2 4 0 3
```

SELF CHECK Tick how you feel

Got it!	Need help...	I don't get it
 ☐	  ☐	 ☐

Check your answers

How many did you get correct?

CATCH UP MATHS YEAR 5 BOOK A © PASCAL PRESS ISBN: 9781925726169

PRACTICE

1 Solve these subtraction algorithms.

Example:

	Th	H	T	O
	8	2	3	6
−		1	0	3
	8	1	3	3

a

	Th	H	T	O
	9	2	3	8
−	1	2	0	3

b

	Th	H	T	O
	5	3	4	5
−		2	0	4

c

	Th	H	T	O
	7	5	8	4
−	1	3	8	3

2 Complete these subtractions.

Example: 9836 − 4203 = 5633

a 7549 − 428 = ____

b 6381 − 4030 = ____

c 5395 − 284 = ____

3 Solve and then check your answer using addition.

Example: 8493 − 4371 = 4122; 4122 + 4371 = 8493 ✓

a 7845 − 3414 = ____; ____ + 3414 = ____

b 9498 − 236 = ____; ____ + 236 = ____

c 5982 − 1872 = ____; ____ + 1872 = ____

d 6653 − 4233 = ____; ____ + 4233 = ____

e 4998 − 3986 = ____; ____ + 3986 = ____

f 8324 − 103 = ____; ____ + 103 = ____

g 9372 − 161 = ____; ____ + 161 = ____

SUBTRACTION WITHOUT TRADING

FOUR-DIGIT AND FIVE-DIGIT NUMBERS

Here are examples of four-digit and five-digit subtraction algorithms.

Example 1: Solve.

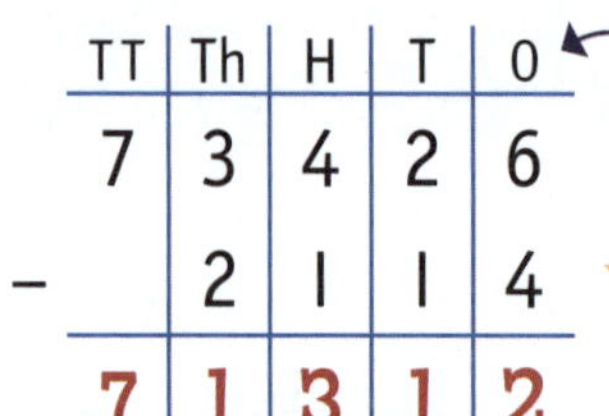

	TT	Th	H	T	O
	7	3	4	2	6
–		2	1	1	4
	7	**1**	**3**	**1**	**2**

Start at the ones column.

Always subtract downwards.

Example 2: Solve, then check using addition.

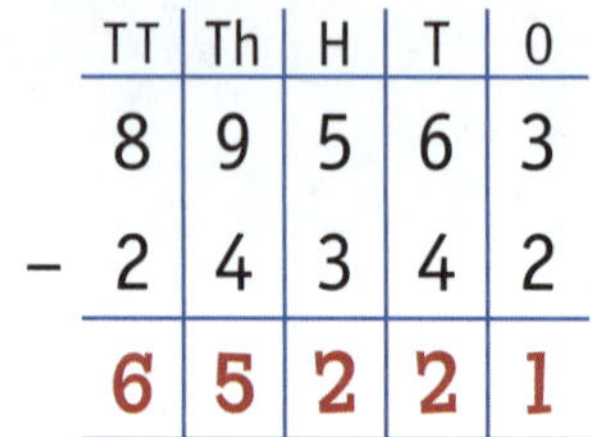

	TT	Th	H	T	O
	8	9	5	6	3
–	2	4	3	4	2
	6	**5**	**2**	**2**	**1**

	TT	Th	H	T	O
	6	5	2	2	1
+	2	4	3	4	2
	8	**9**	**5**	**6**	**3**

✓

If the added total equals the top number you subtracted from, you are correct!

Example 3: Solve, then check using addition.

	TT	Th	H	T	O
	9	2	4	7	3
–	6	2	3	4	1
	3	**0**	**1**	**3**	**2**

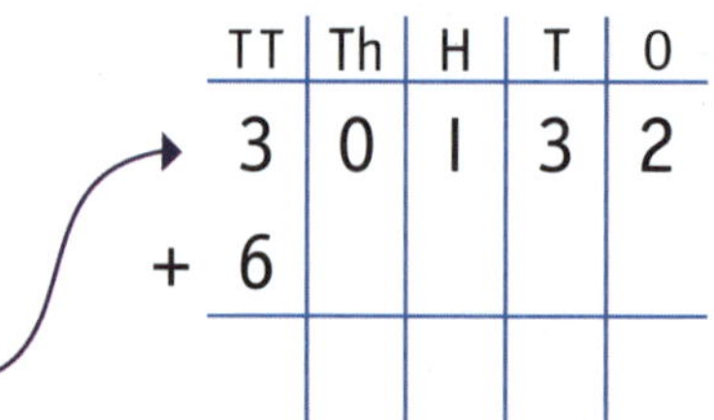

	TT	Th	H	T	O
	3	0	1	3	2
+	6				

Your turn

Solve, then check using addition.

●
```
  5 2 4 3 7
– 1 2 3 2 4
  4 0 1 1 3 ✓

  4 0 1 1 3
+ 1 2 3 2 4
  5 2 4 3 7
```

a
```
  8 3 9 2 5
– 4 2 3 1 4
  _________

+ 4 2 3 1 4
  _________
```

b
```
  8 7 8 8 9
– 1 5 7 3 4
  _________

+ 1 5 7 3 4
  _________
```

SELF CHECK Tick how you feel

Got it!	Need help...	I don't get it
 ☐	 ☐	☐

Check your answers

How many did you get correct? ☐

CATCH UP MATHS YEAR 5 BOOK A © PASCAL PRESS ISBN: 9781925726169

PRACTICE

1 Solve these subtraction algorithms.

Example:

	TT	Th	H	T	O
	8	5	9	3	4
−		3	4	2	3
	8	2	5	1	1

a

	TT	Th	H	T	O
	6	5	9	3	2
−	5	4	3	2	1

b

	TT	Th	H	T	O
	7	5	8	4	9
−		3	2	1	9

2 Solve these subtraction algorithms.

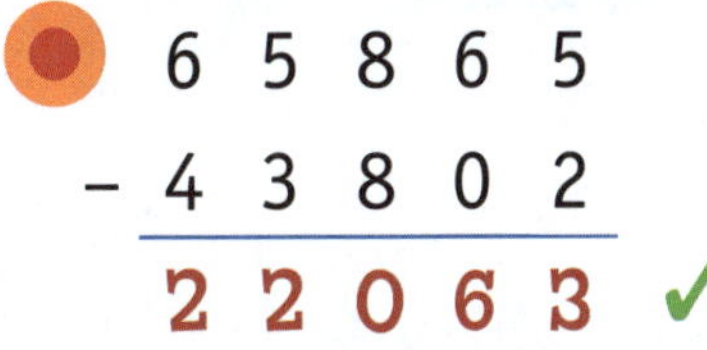

Example: 68398 − 13254 = 55144

a 55864 − 1354 = ______

b 42037 − 10025 = ______

c 85937 − 14325 = ______

d 38473 − 2351 = ______

e 76593 − 43293 = ______

3 Solve, then check using addition.

Example: 65865 − 43802 = 22063 ✓

Check: 22063 + 43802 = 65865

a 96394 − 26103 = ______ Check: ______ + 26103 = ______

b 97436 − 82304 = ______ Check: ______ + 82304 = ______

c 43829 − 22718 = ______ Check: ______ + 22718 = ______

d 56574 − 24253 = ______ Check: ______ + 24253 = ______

e 68490 − 53480 = ______ Check: ______ + 53480 = ______

  ISBN: 9781925726169

SUBTRACTION WITH TRADING

TWO-DIGIT AND THREE-DIGIT NUMBERS

SCAN to watch video

Trading can help us solve subtraction problems.

Example 1:

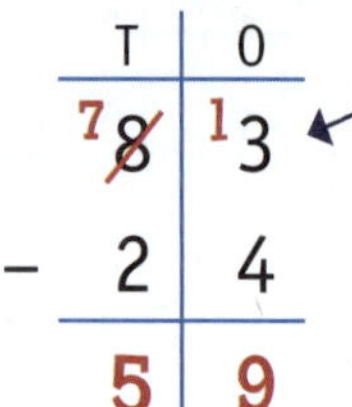

	T	O
	$^{7}\not{8}$	$^{1}3$
−	2	4
	5	**9**

We cannot take 4 from 3, so we trade 1 ten for 10 ones to make 13. Then we take 4 away from 13.

Always subtract downwards.

Example 2:

	H	T	O
	5	$^{2}\not{3}$	$^{1}4$
−	2	2	5
	3	**0**	**9**

Example 3:

	T	O
	$^{7}\not{8}$	$^{1}8$
−	4	9
	3	**9**

Example 4:

	H	T	O
	$^{5}\not{6}$	$^{9}\not{0}$	$^{1}2$
−	2	4	8

Example 5:

	H	T	O
	$^{3}\not{4}$	$^{12}\not{3}$	$^{1}0$
−	1	4	5

When you cannot take away a number in a subtraction, you must trade from the next place value.

Solve these algorithms.

● $5\ {}^{3}\not{4}\ {}^{1}2 - 439 = $ **103**

a $307 - 243 = $ ____

b $741 - 202 = $ ____

c $63 - 44 = $ ____

d $673 - 457 = $ ____

e $67 - 39 = $ ____

SELF CHECK Tick how you feel

Got it!	Need help...	I don't get it
☐	☐	☐

Check your answers

How many did you get correct? ____

CATCH UP MATHS YEAR 5 BOOK A © PASCAL PRESS ISBN: 9781925726169

PRACTICE

1 Complete these algorithms.

●

	T	O
	[5]~~6~~	[1]5
−	2	6
	3	9

a

	H	T	O
	3	4	7
−	1	3	8

b

	T	O
	8	1
−	1	9

c

	H	T	O
	9	1	4
−	2	6	0

2 Now complete these algorithms.

● 5 [3]~~4~~ [1]2 − 4 3 9 = 1 0 3

a 93 − 47 = ____

b 891 − 247 = ____

c 608 − 542 = ____

d 92 − 48 = ____

e 730 − 253 = ____

f 946 − 278 = ____

g 77 − 48 = ____

3 Fill in the missing digits.

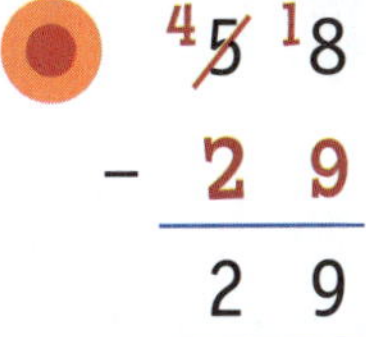

● [4]~~5~~ [1]8 − 2 9 = 2 9

a		4	**e**		3	**i**	5 2 4	
−	2	5	−	4	8	−	3	
	4			1			2 5	

a ◻4 − 25 = 4◻

b 5◻ − 36 = ◻9

c 88 − 6◻ = ◻9

d 9◻ − 78 = ◻3

e ◻3 − 48 = 1◻

f 42 − ◻7 = 2◻

g ◻2 − 1◻ = 23

h 32◻ − 133 = ◻◻3

i 524 − ◻3◻ = 2◻5

j 837 − ◻◻3 = 59◻

k ◻03 − 34◻ = 5◻1

l 741 − ◻3◻ = 5◻5

m 847 − 2◻8 = ◻39

n 42◻ − ◻36 = 2◻8

o 625 − ◻8◻ = 3◻0

SUBTRACTION WITH TRADING

THREE-DIGIT AND FOUR-DIGIT NUMBERS

Here is an example of a four-digit subtraction.
We have to trade 1 ten for 10 ones.

Example 1:

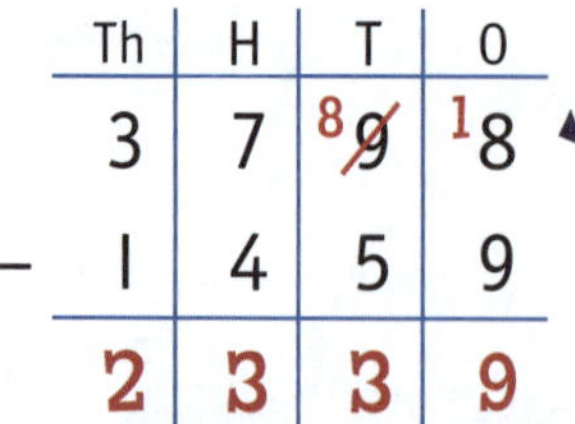

	Th	H	T	O
	3	7	^{8}~~9~~	^{1}8
−	1	4	5	9
	2	**3**	**3**	**9**

We cannot take 9 from 8, so we trade 1 ten for 10 ones. Now we can subtract 9 from 18.

Example 2:

	Th	H	T	O
	^{3}~~4~~	^{11}~~2~~	^{9}~~0~~	^{1}3
−	2	5	8	6
	1	**6**	**1**	**7**

Example 3:

	Th	H	T	O
	^{6}~~7~~	^{11}~~2~~	^{14}~~5~~	^{1}3
−	5	8	9	7

Example 4:

	Th	H	T	O
	3	7	^{0}~~1~~	^{1}1
−		2	0	6

Start at the ones column.
Then move to the tens column.
Next, the hundreds.
Finally, the thousands.

Solve these subtraction algorithms.

● ^{8}~~9~~ ^{9}~~0~~ ^{1}2 6
− 3 5 6 3
5 4 6 3

a 8 2 4 9
− 2 3 6 8
____

b 7 1 3 0
− 4 2 1 9
____

c 6 5 0 2
− 2 3 6 1
____

Check your answers

How many did you get correct?

CATCH UP MATHS YEAR 5 BOOK A © PASCAL PRESS ISBN: 9781925726169

PRACTICE

1 Solve these algorithms.

Example:

	Th	H	T	O
	2	8	8~~9~~	13
−	1	4	0	7
	1	4	8	6

a

	Th	H	T	O
	4	9	3	4
−	2	4	1	5

b

	Th	H	T	O
	7	0	3	2
−		1	5	3

2 Now solve these algorithms.

Example:
$$\begin{array}{r} 8\ {}^{2}\not{3}\ {}^{1}5\ 4 \\ -\ 3\ 1\ 9\ 2 \\ \hline 5\ 1\ 6\ 2 \end{array}$$

b
$$\begin{array}{r} 8034 \\ -\ 5093 \\ \hline \end{array}$$

d
$$\begin{array}{r} 6506 \\ -\ 1207 \\ \hline \end{array}$$

a
$$\begin{array}{r} 7267 \\ -\ 349 \\ \hline \end{array}$$

c
$$\begin{array}{r} 5721 \\ -\ 800 \\ \hline \end{array}$$

e
$$\begin{array}{r} 5862 \\ -\ 1374 \\ \hline \end{array}$$

3 Solve and then check your answer with addition.

Example:
$$\begin{array}{r} 4\ {}^{7}\not{8}\ {}^{9}\not{0}\ {}^{1}3 \\ -\ 1\ 3\ 5\ 4 \\ \hline 3\ 4\ 4\ 9 \end{array} \checkmark$$

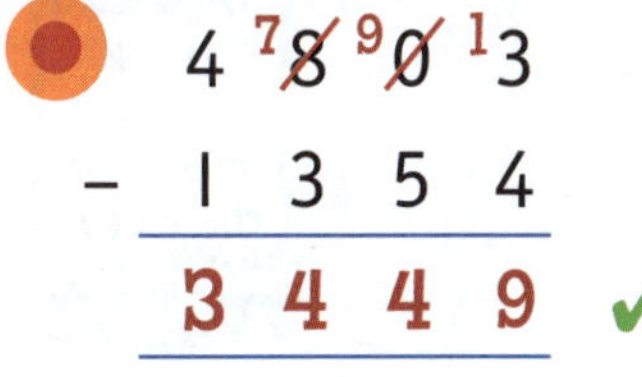

$$\begin{array}{r} 3\ {}^{1}4\ {}^{1}4\ 9 \\ +\ 1\ 3\ 5\ 4 \\ \hline 4\ 8\ 0\ 3 \end{array}$$

b
$$\begin{array}{r} 8541 \\ -\ 432 \\ \hline \end{array}$$

$$\begin{array}{r} \\ +\ 432 \\ \hline \end{array}$$

d
$$\begin{array}{r} 8322 \\ -\ 1703 \\ \hline \end{array}$$

$$\begin{array}{r} \\ +\ 1703 \\ \hline \end{array}$$

a
$$\begin{array}{r} 5204 \\ -\ 1373 \\ \hline \end{array}$$

$$\begin{array}{r} \\ +\ 1373 \\ \hline \end{array}$$

c
$$\begin{array}{r} 9307 \\ -\ 4358 \\ \hline \end{array}$$

$$\begin{array}{r} \\ +\ 4358 \\ \hline \end{array}$$

e
$$\begin{array}{r} 4836 \\ -\ 409 \\ \hline \end{array}$$

$$\begin{array}{r} \\ +\ 409 \\ \hline \end{array}$$

SUBTRACTION WITH TRADING

FIVE-DIGIT NUMBERS

SCAN to watch video

This is a subtraction algorithm where we need to trade 1 ten for 10 ones and then 1 hundred for 10 tens.

Example 1:

Here we traded 1 hundred for 10 tens.

	TT	Th	H	T	O
	8	6	^{2}~~3~~	^{13}~~4~~	^{1}3
–	3	2	1	7	8
	5	**4**	**1**	**6**	**5**

We couldn't take 8 from 3, so we traded 1 ten for 10 ones.

Example 2:

	TT	Th	H	T	O
	^{6}~~7~~	^{1}2	5	^{2}~~3~~	^{1}6
–	1	5	4	2	8
	5	**7**	**1**	**0**	**8**

Example 3:

	TT	Th	H	T	O
	4	^{7}~~8~~	^{12}~~3~~	^{9}~~0~~	^{1}2
–	1	0	5	6	3
	3	**7**	**7**	**3**	**9**

Example 4:

	TT	Th	H	T	O
	5	3	^{5}~~6~~	^{1}4	5
–		2	3	7	4
					1

Check your answer on the video!

Solve these subtraction algorithms.

● 3 ^{3}~~4~~ ^{10}~~1~~ ^{1}0 6
– 3 4 2 5
3 0 6 8 1

b 5 3 8 4 7
– 3 2 9 8 5

a 7 3 8 2 4
– 5 3 2 8

c 6 5 1 4 3
– 1 9 0 3 7

SELF CHECK Tick how you feel		
Got it! ☐	Need help... ☐	I don't get it ☐

Check your answers
How many did you get correct? ☐

CATCH UP MATHS YEAR 5 BOOK A © PASCAL PRESS ISBN: 9781925726169

PRACTICE

1 Solve these subtraction algorithms.

Example:

	TT	Th	H	T	O
	4	⁶~~7~~	⁹~~0~~	¹3	5
−	1	4	2	4	3
	3	2	7	9	2

a

	TT	Th	H	T	O
	7	2	4	3	1
−		1	4	3	5

b

	TT	Th	H	T	O
	9	2	0	7	4
−	4	3	2	5	7

2 Now solve these subtraction algorithms.

Example:

```
  5 ⁵6̸ ¹2 ³4̸ ¹0
−    5  7  3  5
  5  0  5  0  5
```

a
```
  1 6 0 3 0
−   4 9 5 2
```

b
```
  5 4 3 7 3
−   5 4 2 5
```

c
```
  4 0 3 2 4
− 2 1 4 0 8
```

d
```
  8 3 4 2 0
−   4 8 3 5
```

e
```
  6 7 0 8 5
− 3 2 9 8 6
```

3 Solve and then check using addition.

Example:

```
  ¹2̸ ¹³4̸ ¹3 ⁷8̸ ¹4
−  1   4  7  2  9
       9  6  5  5  ✓

  ¹   ¹9  6 ¹5  5
+  1   4  7  2  9
   2   4  3  8  4
```

a
```
  4 9 3 4 0
− 1 5 2 4 3

+ 1 5 2 4 3
```

b
```
  9 2 7 3 1
− 2 4 3 5 2

+ 2 4 3 5 2
```

4 Fill in the missing digits.

Example:

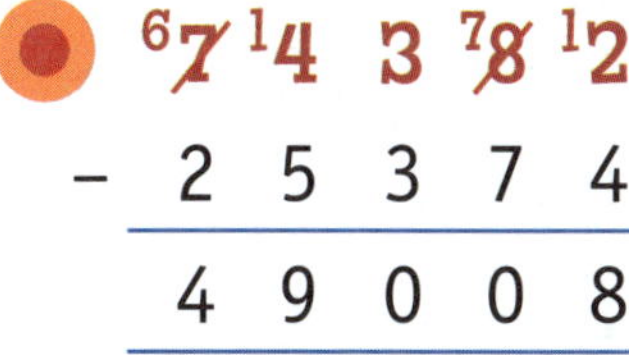

```
  ⁶7̸ ¹4 3 ⁷8̸ ¹2
−  2  5 3  7  4
   4  9 0  0  8
```

a
```
  5 8 2 7 3
−
  1 6 8 9 1
```

b
```

− 1 4 2 7 5
  3 8 2 1 8
```

c
```

− 4 4 2 8 4
  1 9 1 4 1
```

d
```
  8 2 3 1 0
−
  3 8 8 9 7
```

e
```
  9 1 4 7 2
−
  4 7 6 1 2
```

TRADING FROM LARGER PLACE VALUES

When you are subtracting, sometimes you can't trade from the next place value.

Example 1:

```
 [5]~~6~~ [9]~~0~~ [1]0
-         2     7
-----------------
   5      7     3
```

Example 2:

```
 [6]~~7~~ [9]~~0~~ [9]~~0~~ [1]0
-         2        4        3
--------------------------------
   6      7        5        7
```

Example 3:

```
 [7]~~8~~ [9]~~0~~ [9]~~0~~ [9]~~0~~ [1]0
-         7        3        2        8
-----------------------------------------
   7      2        6        7        2
```

Example 4:

```
 [8]~~9~~ [9]~~0~~ [9]~~0~~ [9]~~0~~ [9]~~0~~ [1]0
-         1        4        3        2        3
--------------------------------------------------
   8      8        5        6        7        7
```

Example 5:

```
 [5]~~6~~ [9]~~0~~ [9]~~0~~ [9]~~0~~ [1]0
-  3      4        5        9        5
-----------------------------------------

```

Solve these subtractions.

● (worked example)

```
  ~~1~~ [9]~~0~~ [9]~~0~~ [9]~~0~~ [1]0  0
-       4        3        2        4     0
------------------------------------------
        5        6        7        6     0
```

a

```
  3 0 0 0 0
- 1 3 4 2 0
-----------

```

b

```
  4 0 0 0
- 3 1 3 4
---------

```

c

```
  5 0 0
- 2 3 2
-------

```

SELF CHECK Tick how you feel

Got it!	Need help...	I don't get it
 ☐	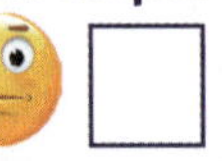 ☐	 ☐

Check your answers

How many did you get correct? ☐

CATCH UP MATHS YEAR 5 BOOK A © PASCAL PRESS ISBN: 9781925726169

PRACTICE

Solve these subtractions.

Example: $500000 - 12345 = 487655$ (working: ⁴5 ⁹0 ⁹0 ⁹0 ⁹0 ¹0)

a $70000 - 2573$

b $5000 - 147$

c $10000 - 3823$

d $40000 - 20302$

e $200000 - 74003$

f $28000 - 3436$

g $80000 - 35213$

h $6000 - 253$

i $400 - 132$

j $62000 - 13472$

k $33000 - 14004$

2 Solve these money subtractions.

Example: $100.00 - $43.65 = $56.35 (working: ¹0 ⁹0 ⁹0 . ⁹0 ¹0)

a $50.00 − $13.62

b $20.00 − $13.50

c $80.00 − $35.17

d $90.00 − $42.50

e $70.00 − $25.25

f $360.00 − $154.95

g $75.00 − $62.46

h $290.00 − $187.57

WHAT'S THE DIFFERENCE?

The difference is the answer you get when you subtract one number from another. The difference is the amount one number is bigger or smaller than another number.

SCAN to watch video

Example 1: What is the difference between 216 and 428?

	4	2	8
−	2	1	6
	2	**1**	**2**

The difference between 428 and 216 is 212.

Example 2: Is 256 112 the difference between 598 243 and 342 131?

	5	9	8	2	4	3
−	3	4	2	1	3	1
	2	**5**	**6**	**1**	**1**	**2**

Yes, 256 112 is the difference between 598 243 and 342 131.

Example 3: Find the difference between 5932 and 1341.

	5	~~9~~ 8	13	2
−	1	3	4	1
			9	**1**

Difference = __________

Your turn

Find the difference.

● 523 and 210

	5	2	3
−	2	1	0
	3	**1**	**3**

Difference = 313

a 4328 and 342

	4	3	2	8
−		3	4	2

Difference = ________

b 598 425 and 137 911

	5	9	8	4	2	5
−	1	3	7	9	1	1

Difference = __________

SELF CHECK Tick how you feel

Got it!	Need help...	I don't get it
 ☐	 ☐	☐

Check your answers
How many did you get correct? ☐

CATCH UP MATHS YEAR 5 BOOK A © PASCAL PRESS ISBN: 9781925726169

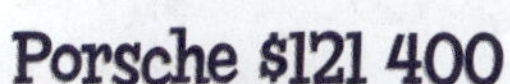

PRACTICE

Porsche $121 400

Ferrari $475 999

Lamborghini $325 643

Maserati $256 920

Mercedes $153 242

Alfa Romeo $143 800

1 Find the difference in price between each pair of cars.

● Lamborghini and Maserati

```
  ²3̶ ¹¹2̶ ¹⁴5̶ ¹6 4 3
–  2   5   6  9 2 0
       6   8  7 2 3
```

The difference is $68 723.

b Alfa Romeo and Mercedes

–

The difference is ________.

a Porsche and Ferrari

–

The difference is ________.

c Lamborghini and Porsche

–

The difference is ________.

2 What is the change from $500 000 when you buy this car?

● Alfa Romeo

```
  ⁴5̶ ⁹0̶ ⁹0̶ ¹0 0 0
–  1  4  3  8 0 0
   3  5  6  2 0 0
```

The change is $356 200.

b Maserati

–

The change is ________.

a Mercedes

–

The change is ________.

c Ferrari

–

The change is ________.

ROUNDING TO ESTIMATE SUBTRACTION ANSWERS

You can use rounding to get an approximate answer that is close to the actual answer.

Example 1: Round to the nearest 10 to estimate the answer.

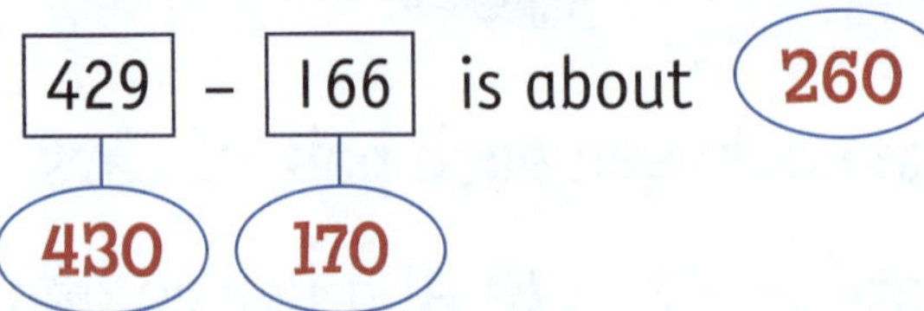

Example 2: Round to the nearest 100 to estimate the answer.

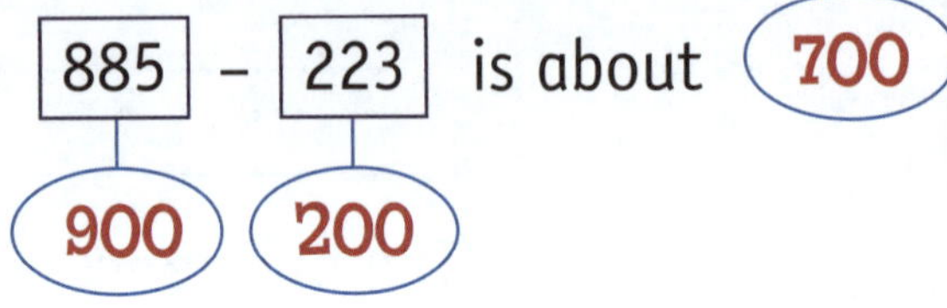

Example 3: Round to the nearest 100 to estimate the answer.

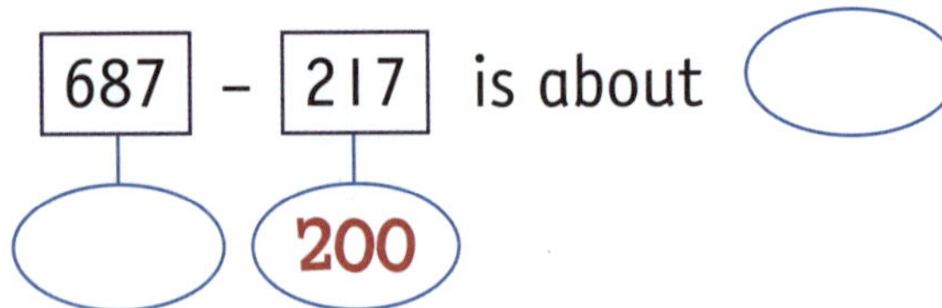

Your turn

1 Round to the nearest 10 to estimate the answer.

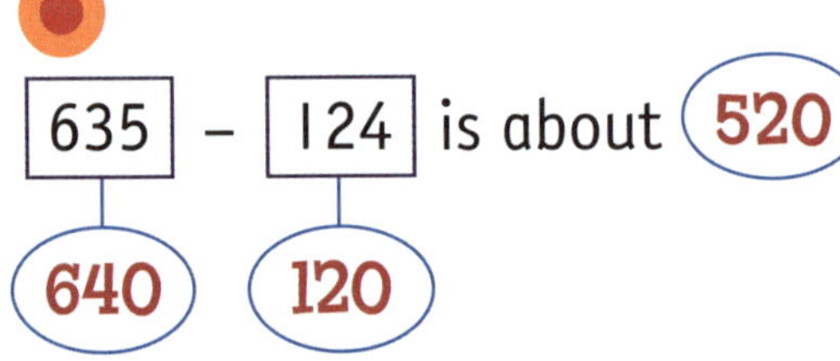

a

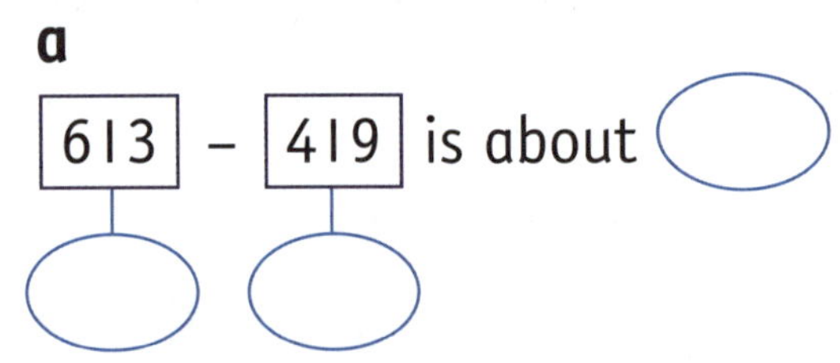

2 Round to the nearest 100 to estimate the answer.

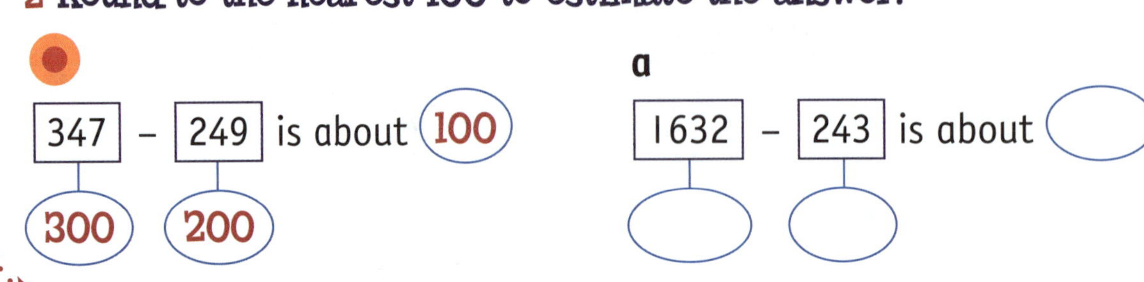

Check your answers

How many did you get correct?

CATCH UP MATHS YEAR 5 BOOK A © PASCAL PRESS ISBN: 9781925726169

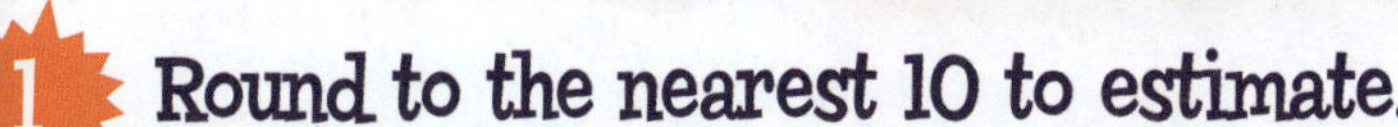
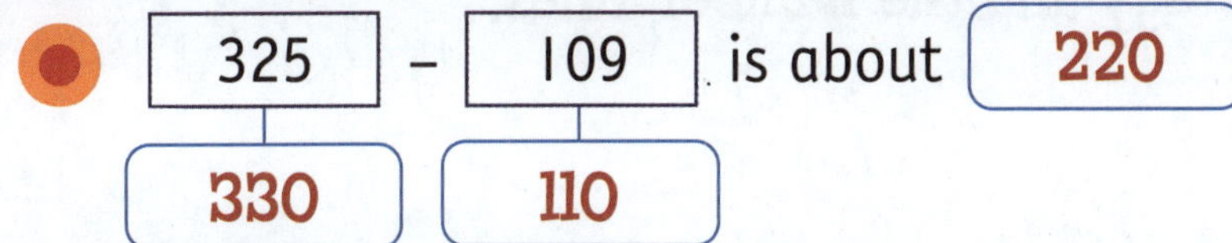

PRACTICE

1 Round to the nearest 10 to estimate.

- 325 – 109 is about 220 (330, 110)

a 742 – 257 is about ☐ (☐, ☐)

b 1357 – 243 is about ☐ (☐, ☐)

c 43 523 – 36 419 is about ☐ (☐, ☐)

d 73 425 – 34 561 is about ☐ (☐, ☐)

2 Round to the nearest 100 to estimate.

- 38 427 – 26 498 is about 11 900 (38 400, 26 500)

a 73 582 – 21 530 is about ☐ (☐, ☐)

b 343 – 257 is about ☐ (☐, ☐)

c 5984 – 3349 is about ☐ (☐, ☐)

SUBTRACTION REVIEW

1 **Solve using the jump strategy and the number lines.**

a 68 – 37 = ___

b 84 – 46 = ___

c 285 – 82 = _____

d 423 – 51 = _____

e 542 – 310 = _____

f 1326 – 243 = ______

g 8437 – 362 = ______

h 5982 – 1432 = ______

i 6384 – 2034 = ______

  ISBN: 9781925726169

2 Use the split strategy to solve these problems.

a 58 – 42

___ – ___ = ___

___ – ___ = ___

___ + ___ = ___

b 73 – 50

___ – ___ = ___

___ – ___ = ___

___ + ___ = ___

c 563 – 41

___ – ___ = ___

___ – ___ = ___

___ – ___ = ___

___ + ___ + ___ = ___

d 862 – 42

___ – ___ = ___

___ – ___ = ___

___ – ___ = ___

___ + ___ + ___ = ___

e 427 – 102

___ – ___ = ___

___ – ___ = ___

___ – ___ = ___

___ + ___ + ___ = ___

f 593 – 241

___ – ___ = ___

___ – ___ = ___

___ – ___ = ___

___ + ___ + ___ = ___

g 5943 – 231

___ – ___ = ___

___ – ___ = ___

___ – ___ = ___

___ – ___ = ___

___ + ___ + ___ + ___
= ___

h 9342 – 212

___ – ___ = ___

___ – ___ = ___

___ – ___ = ___

___ – ___ = ___

___ + ___ + ___ + ___
= ___

REVIEW

i 7536 − 4312

_____ − _____ = _____

____ − ____ = ____

___ − ___ = ___

__ − __ = __

_____ + ____ + ___ + __

= _____

j 9342 − 3021

_____ − _____ = _____

____ − ____ = ____

___ − ___ = ___

__ − __ = __

_____ + ____ + ___ + __

= _____

3 Solve these subtractions.

a $\begin{array}{r} 53 \\ -\ 21 \\ \hline \end{array}$

b $\begin{array}{r} 74 \\ -\ 53 \\ \hline \end{array}$

c $\begin{array}{r} 88 \\ -\ 42 \\ \hline \end{array}$

d $\begin{array}{r} 72 \\ -\ 35 \\ \hline \end{array}$

e $\begin{array}{r} 81 \\ -\ 43 \\ \hline \end{array}$

f $\begin{array}{r} 52 \\ -\ 35 \\ \hline \end{array}$

g $\begin{array}{r} 589 \\ -\ 37 \\ \hline \end{array}$

h $\begin{array}{r} 684 \\ -\ 142 \\ \hline \end{array}$

i $\begin{array}{r} 980 \\ -\ 450 \\ \hline \end{array}$

j $\begin{array}{r} 846 \\ -\ 305 \\ \hline \end{array}$

k $\begin{array}{r} 840 \\ -\ 235 \\ \hline \end{array}$

l $\begin{array}{r} 903 \\ -\ 42 \\ \hline \end{array}$

4 Now solve these subtractions.

a $\begin{array}{r} 5242 \\ -\ 131 \\ \hline \end{array}$

b $\begin{array}{r} 4953 \\ -\ 1742 \\ \hline \end{array}$

c $\begin{array}{r} 8972 \\ -\ 340 \\ \hline \end{array}$

d $\begin{array}{r} 3854 \\ -\ 2413 \\ \hline \end{array}$

e $\begin{array}{r} 6930 \\ -\ 543 \\ \hline \end{array}$

f $\begin{array}{r} 7529 \\ -\ 3453 \\ \hline \end{array}$

CATCH UP MATHS YEAR 5 BOOK A © PASCAL PRESS ISBN: 9781925726169

g
```
  54287
- 41152
```

i
```
  82630
-  1410
```

k
```
  62152
- 35983
```

h
```
  39857
- 21245
```

j
```
  72465
- 51323
```

l
```
  73154
- 42565
```

5 Solve and then use addition to check your answers.

a
```
  57
- 43
```
\+

d
```
  1532
-  415
```
\+

g
```
  6304
- 1251
```
\+

b
```
  143
-  28
```
\+

e
```
  5983
- 2741
```
\+

h
```
  58937
-  3977
```
\+

c
```
  342
- 152
```
\+

f
```
  5843
- 2709
```
\+

i
```
  95256
-  3039
```
\+

j	k	l
48111 − 25370	72531 − 39409	93007 − 24056
+	+	+

6 Complete.

a 400 − 32	**g** 3000 − 2375	**m** 70000 − 5352
b 600 − 143	**h** 9000 − 4030	**n** 60000 − 7053
c 800 − 768	**i** 4000 − 1307	**o** 50000 − 8321
d 5000 − 153	**j** 30000 − 3524	**p** 30000 − 21350
e 7000 − 583	**k** 50000 − 152	**q** 40000 − 15290
f 6000 − 421	**l** 20000 − 73	**r** 76000 − 31304

CATCH UP MATHS YEAR 5 BOOK A © PASCAL PRESS ISBN: 9781925726169

7 Work out the difference.

a 59 and 23

Difference = ______

b 512 and 37

Difference = ______

c 658 and 109

Difference = ______

d 8909 and 225

Difference = ______

e 7324 and 5231

Difference = ______

f 39 872 and 6384

Difference = ______

8 Round to the nearest 10 to estimate.

a 342 – 173 is about ☐

☐ ☐

b 3581 – 432 is about ☐

☐ ☐

c 17 437 – 1241 is about ☐

☐ ☐

9 Round to the nearest 100 to estimate.

a 436 – 132 is about ☐

☐ ☐

b 53 421 – 5368 is about ☐

☐ ☐

SKIP COUNTING

Skip counting means to skip the same amount of numbers every time. You can skip count forwards or backwards.

When you skip count forwards by 10, count every tenth number: 10, 20, 30, 40, 50.

Example 1:
Skip count forwards by 4, starting at 8.

8, 12, 16, 20, 24, 28

Example 2:
Skip count forwards by 7, starting at 21.

21, 28, ___, ___, 49, 56

Example 3:
Skip count backwards by 3, starting at 21.

21, 18, 15, 12, 9, 6

Example 4:
Skip count backwards by 5, starting at 60.

60, 55, ___, ___, 40, 35

1 Skip count forwards by:

- 2: 2, 4, 6, 8, 10, 12
- **a** 5: 5, ___, ___, ___, ___, ___
- **b** 4: 4, ___, ___, ___, ___, ___

When you practise counting in equal groups, it helps you learn to multiply.

2 Skip count backwards by:

- 4: 48, 44, 40, 36, 32, 28
- **a** 6: 36, ___, ___, ___, ___, ___
- **b** 8: 88, ___, ___, ___, ___, ___

SELF CHECK Tick how you feel

Got it!	Need help...	I don't get it
 ☐	 ☐	 ☐

Check your answers
How many did you get correct? ☐

CATCH UP MATHS YEAR 5 BOOK A © PASCAL PRESS ISBN: 9781925726169

PRACTICE

1 Follow the skip counting instructions.

●	Count up by 3s	3	6	9	12	15	18	21	24
a	Count up by 8s	8							
b	Count up by 9s	9							
c	Count up by 5s	5							
d	Count up by 1s	1							
e	Count up by 12s	12							
f	Count back by 2s	24							
g	Count back by 7s	84							
h	Count back by 6s	66							
i	Count back by 4s	40							
j	Count back by 10s	70							

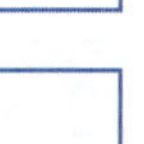

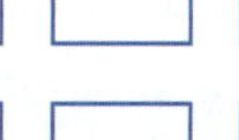

2 Fill in the missing numbers.

● 72, 64, 56, 48, 40

a 10, 9, 8, ___, ___

b 24, ___, ___, 60, 72

c 21, 28, ___, 42, ___

d 55, ___, 45, 40, ___

e 18, ___, 30, ___, 42

f ___, ___, 24, 28, 32

g 93, ___, 89, ___, 85

h 24, 25, 26, ___, ___

i 9, ___, ___, 36, 45

j 97, 107, ___, ___, 137

k ___, 66, 55, ___, 33

l ___, 70, ___, 50, 40

m 89, 98, ___, 116, ___

PRODUCT, FACTORS & MULTIPLES

When you multiply numbers, the answer is called the product.

A **factor** is a number that can multiply with another to give a multiple.	A **multiple** is the answer you get when you multiply two numbers together.

6 and 4 are factors of 24. 24 is a multiple of 6 and 4.

$6 \times 4 = 24$ ← multiple

(6 and 4: factors)

Example 1: What are the factors of 12?

1 × 12 = 12
2 × 6 = 12
3 × 4 = 12

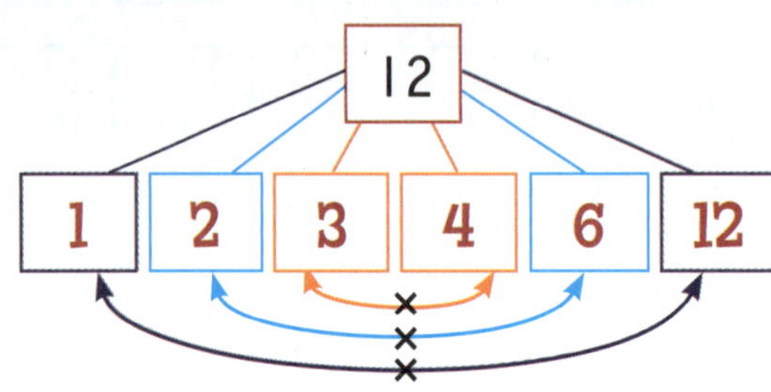

1, 2, 3, 4, 6 and 12 are the factors of 12.

12 is a multiple of 1, 2, 3, 4, 6 and 12.

Example 2: What are the factors of 10?

1 × ___ = 10

___ × ___ = 10

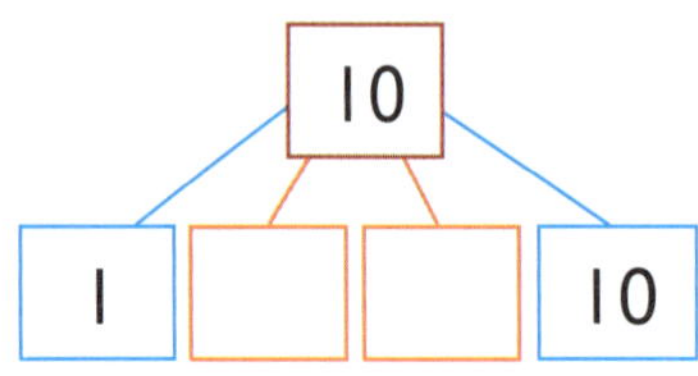

1, ___, ___ and 10 are the factors of 10.

10 is a multiple of 1, ___, ___ and 10.

Your turn

1 Cross out the number that is NOT a factor.

- 10: 1, 2, ~~3~~, 5, 10

a 15: 1, 2, 3, 5, 15

b 20: 1, 2, 3, 4, 5, 10

2 Circle the number that is NOT a multiple.

- 10: 10, 50, 60, (25), 20

a 5: 5, 15, 30, 42, 50

b 6: 12, 18, 23, 24, 30

c 2: 1, 2, 4, 10, 20, 8

SELF CHECK Tick how you feel

Got it!	Need help...	I don't get it
 ☐	 ☐	 ☐

Check your answers
How many did you get correct? ☐

CATCH UP MATHS YEAR 5 BOOK A © PASCAL PRESS ISBN: 9781925726169

PRACTICE

1 Write the factors.

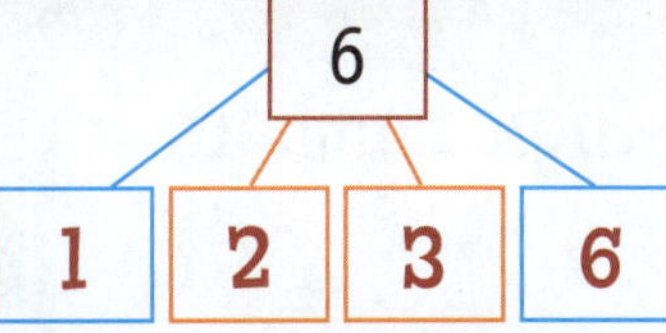

6: 1, 2, 3, 6

1 × 6 = 6

2 × 3 = 6

a 18

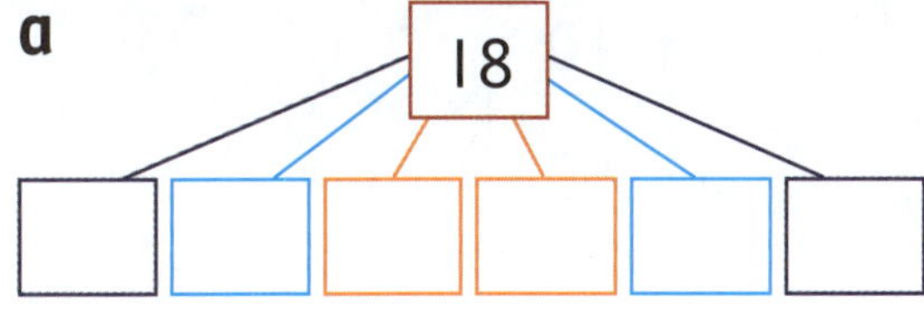

___ × ___ = ___

___ × ___ = ___

___ × ___ = ___

b 7

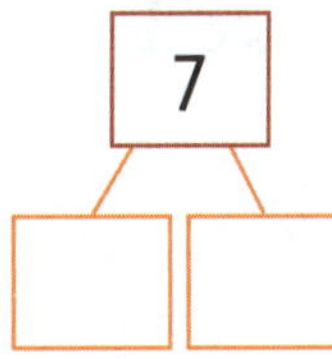

___ × ___ = ___

c 24

___ × ___ = ___

___ × ___ = ___

___ × ___ = ___

___ × ___ = ___

d 16

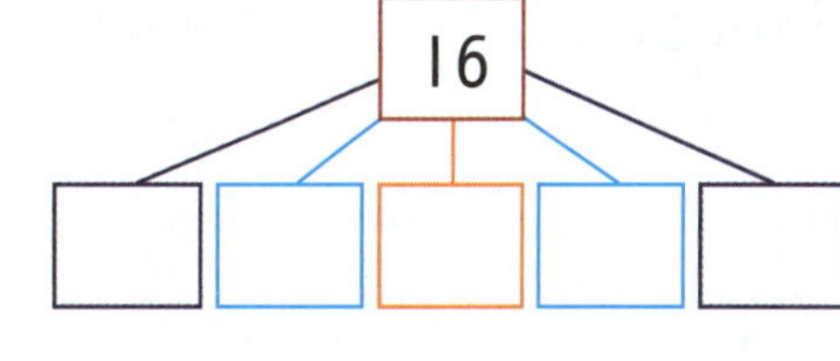

___ × ___ = ___

___ × ___ = ___

___ × ___ = ___

2 Write the product.

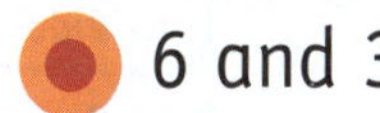

6 and 3 — 18

a 5 and 4 

b 8 and 7 

c 9 and 12 

d 11 and 12

e 9 and 8 

3 Write the next 5 multiples.

5: 15, 20, 25, 30, 35, 40

a 6: 36, ___, ___, ___, ___, ___

b 7: 21, ___, ___, ___, ___, ___

c 9: 27, ___, ___, ___, ___, ___

4 Write the missing numbers.

8 × 7 = 56

a 9 × ___ = 36

b ___ × 6 = 48

c 5 × 12 = ___

d 12 × ___ = 132

e 7 × ___ = 77

f ___ × 4 = 32

g 7 × ___ = 42

h 4 × ___ = 0

MULTIPLYING 2-DIGIT BY 1-DIGIT NUMBERS

Here are three different ways to multiply two-digit numbers by one-digit numbers.

Example 1: 13 × 5

Using known facts

13 × 5

10 × 5 = 50

50 + 5 + 5 + 5
(3 lots of 5)

= 65

Multiplying the tens and then the ones

13 × 5

= 5 tens + 5 threes

= 50 + 15

= 65

Using an area model

13 × 5

	10	3
5	50	15

= 50 + 15

= 65

Example 2: 23 × 4

Using known facts

23 × 4

20 × __ = 80

80 + __ + __ + __
(3 lots of 4)

= ____

Multiplying the tens and then the units

23 × 4

= __ tens × 4
+ __ threes

= 80 + ___

= ____

Using an area model

23 × 4

	20	3
4	80	

= 80 + ___

= ____

Check your answer on the video!

Your turn Solve using the three different methods.

32 × 8

30 × 8 = ________

____

______

32 × 8

3 tens × 8 + ________

____

______

32 × 8

	30	2
__		

____

______

SELF CHECK Tick how you feel

Got it!	Need help...	I don't get it
 ☐	 ☐	  ☐

Check your answers
How many did you get correct? ☐

CATCH UP MATHS YEAR 5 BOOK A © PASCAL PRESS ISBN: 9781925726169

PRACTICE

1 Solve using the different methods of multiplication.

63 × 5 Use known facts

60 × 5 = 300

300 + 5 + 5 + 5

= 315

63 × 5 Multiply tens then ones

6 tens × 5 + 5 threes

300 + 15 = 315

63 × 5 Use an area model

	60	3
5	300	15

300 + 15 = 315

b 67 × 8 Use known facts

____

____

= ______

67 × 8 Multiply tens then ones

____

________ = ______

67 × 8 Use an area model

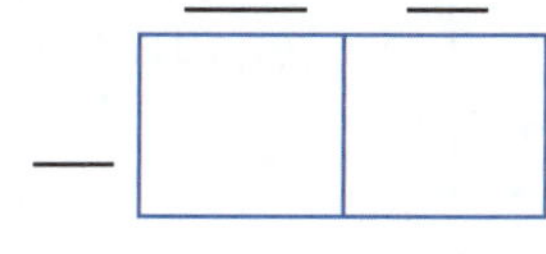

________ = ______

a 34 × 5 Use known facts

____

____

= ______

34 × 5 Multiply tens then ones

____

________ = ______

34 × 5 Use an area model

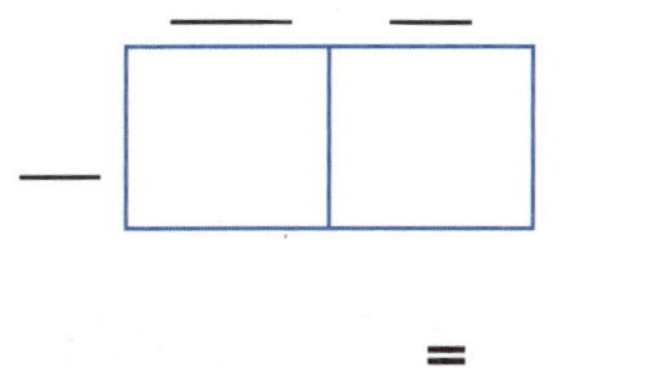

________ = ______

c 41 × 3 Use known facts

____

____

= ______

41 × 3 Multiply tens then ones

________ = ______

41 × 3 Use an area model

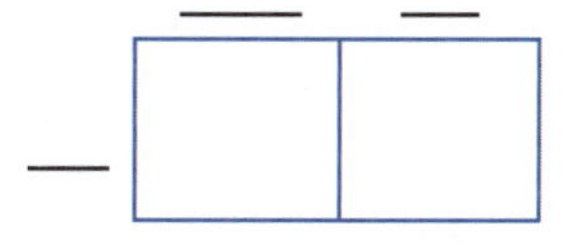

________ = ______

FORMAL ALGORITHMS

We can solve multiplication problems using an algorithm.

Example 1:

First multiply the ones: 2 × 2.

Then multiply the tens: 4 × 2.

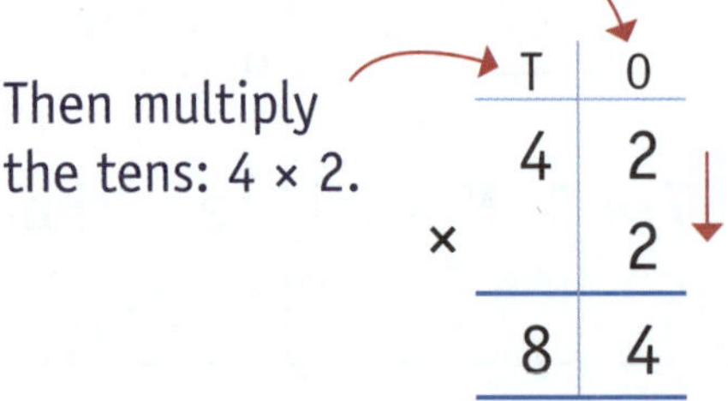

	T	O
	4	2
×		2
	8	4

Example 2:

	H	T	O
		+1 6	3
×			5
	3	1	5

3 × 5 = 15 so carry over to the tens to add 1.

Example 3:

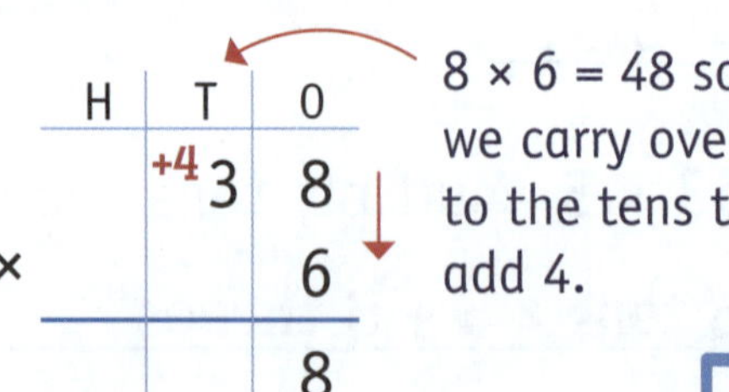

	H	T	O
		+4 3	8
×			6
			8

8 × 6 = 48 so we carry over to the tens to add 4.

Check your answer on the video!

Example 4:

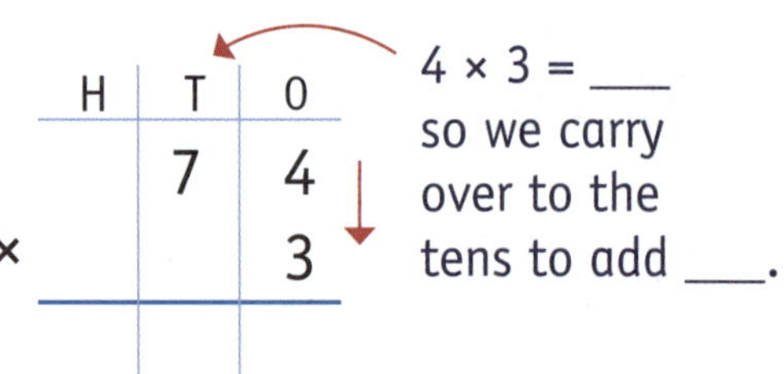

	H	T	O
		7	4
×			3

4 × 3 = ___ so we carry over to the tens to add ___.

Solve these multiplication algorithms.

● (example)

	H	T	O
		+5 4	7
×			8
	3	7	6

a

	H	T	O
		3	5
×			2

b

	H	T	O
		6	1
×			8

c

	H	T	O
		7	2
×			3

d

	H	T	O
		2	3
×			7

e

	H	T	O
		8	9
×			6

SELF CHECK Tick how you feel

Got it!	Need help...	I don't get it
☐	☐	☐

Check your answers

How many did you get correct? ☐

CATCH UP MATHS YEAR 5 BOOK A © PASCAL PRESS ISBN: 9781925726169

PRACTICE

1 Solve these multiplication algorithms.

Example:

	H	T	O
		+2 2	7
×			3
		8	1

a

	H	T	O
		3	2
×			4

b

	H	T	O
		5	8
×			5

c

	H	T	O
		6	4
×			6

d

	H	T	O
		8	1
×			9

e

	H	T	O
		9	2
×			8

f

	H	T	O
		4	0
×			7

g

	H	T	O
		5	5
×			2

2 This is Lina's test. Work out the answers, mark the test and write the score.

a) 23 × 2 = 46 ✓

b) 41 × 3 = 123

c) 72 × 5 = 350

d) +4 87 × 7 = 609

e) +4 55 × 8 = 440

f) 35 × 4 = 120

g) +2 54 × 6 = 324

h) 61 × 8 = 468

☐ out of 8

3 Try solving these algorithms with three-digit numbers.

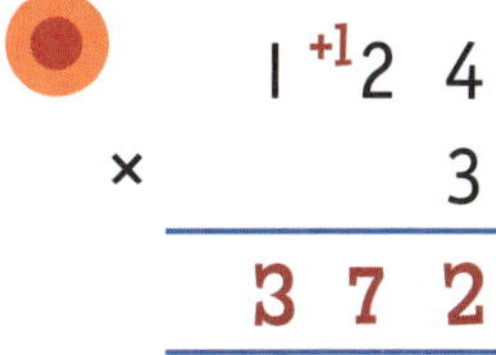

Example: 1 +1 2 4 × 3 = 372

a) 272 × 4 = ____

b) 329 × 5 = ____

c) 403 × 7 = ____

d) 516 × 6 = ____

e) 892 × 8 = ____

MULTIPLYING 3-DIGIT AND 4-DIGIT NUMBERS BY 1-DIGIT NUMBERS

Here are three ways to multiply three-digit or four-digit numbers by one-digit numbers.

Example 1:

Multiplying thousands, hundreds, tens, ones

673 × 4

= (600 × 4) + (70 × 4) + (3 × 4)

= 2400 + 280 + 12

= 2692

Formal Algorithm

673 × 4

$$\begin{array}{r} {}^{2}6\ {}^{1}7\ 3 \\ \times \quad\quad 4 \\ \hline 2\ 6\ 9\ 2 \\ \hline \end{array}$$

Area Model

673 × 4

	600	70	3
4	2400	280	12

2400 + 280 + 12

= 2692

Why is it useful to learn different ways of multiplying?

Example 2:

Multiplying thousands, hundreds, tens, ones

1432 × 5

= (______ × 5) + (_____ × 5) + (30 × 5) + (2 × 5)

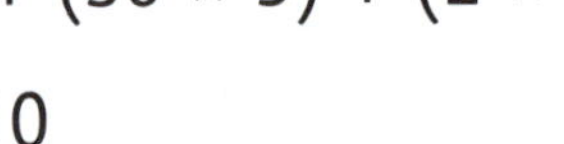

= ______ + ______ + 150 + 10

= ______

Formal Algorithm

1432 × 5

$$\begin{array}{r} {}^{2}1\ {}^{1}4\ {}^{1}3\ 2 \\ \times \quad\quad\quad 5 \\ \hline 0 \\ \hline \end{array}$$

Area Model

1432 × 5

	1000	400	30	2
5				

______ + ______ + ______ + ___

= ______

CATCH UP MATHS YEAR 5 BOOK A © PASCAL PRESS ISBN: 9781925726169

Your turn

Solve.

5382 × 4

= (5000 × 4) + (300 × 4) + (80 × 4) + (2 × 4)

= 20 000 + 1200 + 320 + 8

= 21 528

```
  ¹5 ³3  8  2
×           4
--------------
   2  1  5  2  8
```

5382 × 4

	5000	300	80	2
4	20 000	1200	320	8

20 000 + 1200 + 320 + 8

= 21 528

a 631 × 4

= ______________________

= ______________________

= __________

```
    6  3  1
×         4
-----------

-----------
```

631 × 4

	___	___	___
___			

____

= __________

b 742 × 8

= ______________________

= ______________________

= __________

```
    7  4  2
×         8
-----------

-----------
```

742 × 8

	___	___	___
___			

____

= __________

SELF CHECK Tick how you feel

Got it!	Need help...	I don't get it
 ☐	 ☐	 ☐

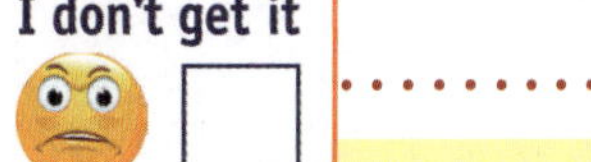

Check your answers
How many did you get correct? ☐

PRACTICE

1 Solve by multiplying thousands, hundreds, tens and ones.

$7134 \times 3 = (7000 \times 3) + (100 \times 3) + (30 \times 3) + (4 \times 3)$
$= 21\,000 + 300 + 90 + 12$
$= 21\,402$

a $326 \times 6 =$ ____________
= ____________
= ____________

b $1526 \times 5 =$ ____________
= ____________
= ____________

c $439 \times 7 =$ ____________
= ____________
= ____________

d $5963 \times 8 =$ ____________
= ____________
= ____________

2 Solve with formal algorithms.

$$\begin{array}{r} {}^{2}3\;{}^{1}5\;{}^{2}3\;4 \\ \times \quad\quad 5 \\ \hline 1\;7\;6\;7\;0 \\ \hline \end{array}$$

b
$$\begin{array}{r} 1\;5\;3\;2 \\ \times \quad\quad 6 \\ \hline \\ \hline \end{array}$$

d
$$\begin{array}{r} 4\;9\;3\;6 \\ \times \quad\quad 8 \\ \hline \\ \hline \end{array}$$

a
$$\begin{array}{r} 6\;1\;5 \\ \times \quad 4 \\ \hline \\ \hline \end{array}$$

c
$$\begin{array}{r} 7\;2\;5 \\ \times \quad 7 \\ \hline \\ \hline \end{array}$$

e
$$\begin{array}{r} 5\;0\;3 \\ \times \quad 7 \\ \hline \\ \hline \end{array}$$

CATCH UP MATHS YEAR 5 BOOK A © PASCAL PRESS ISBN: 9781925726169

3 Solve using the area models.

5123 × 3

	5000	100	20	3
3	15 000	300	60	9

15 000 + 300 + 60 + 9

= 15 369

a 732 × 4

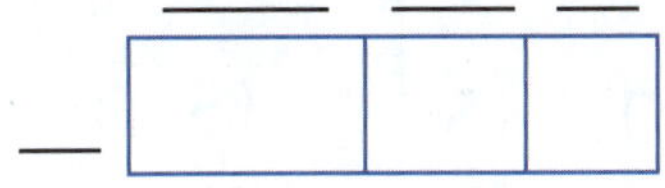

= ______

b 4940 × 6

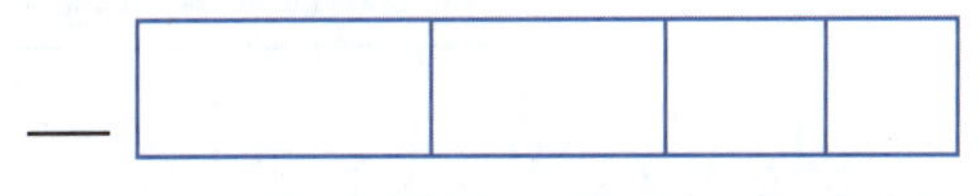

= ______

c 2935 × 5

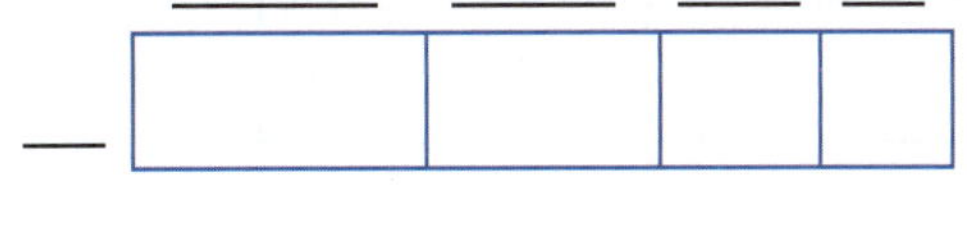

= ______

d 419 × 8

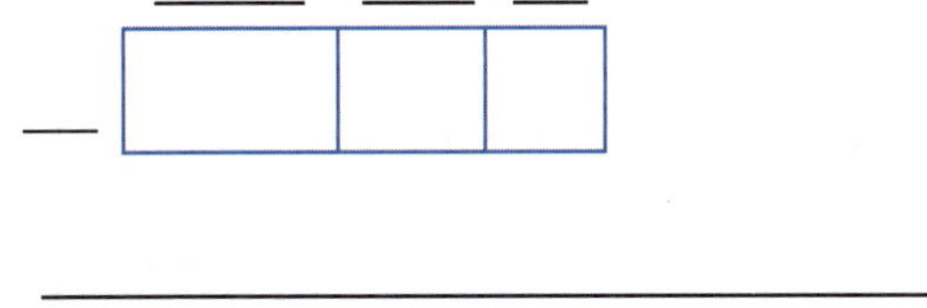

= ______

e 506 × 7

= ______

f 827 × 3

= ______

g 1470 × 9

= ______

h 7495 × 6

= ______

i 5678 × 7

= ______

MULTIPLYING 2-DIGIT AND 3-DIGIT NUMBERS BY 2-DIGIT NUMBERS

Here are two ways of multiplying two-digit and three-digit numbers with two-digit numbers.

Example 1: 43 × 25

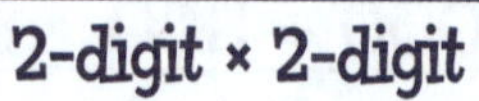

Long Multiplication

	Th	H	T	O
			$^{1}4$	3
×			2	5
		2	1	5
+		8	6	0
	1	0	7	5

First do 3 × 5.
Then do 4 × 5.

Now write zero in the ones.

Next do 3 × 2, then 4 × 2.

Finally, add.

Area Model

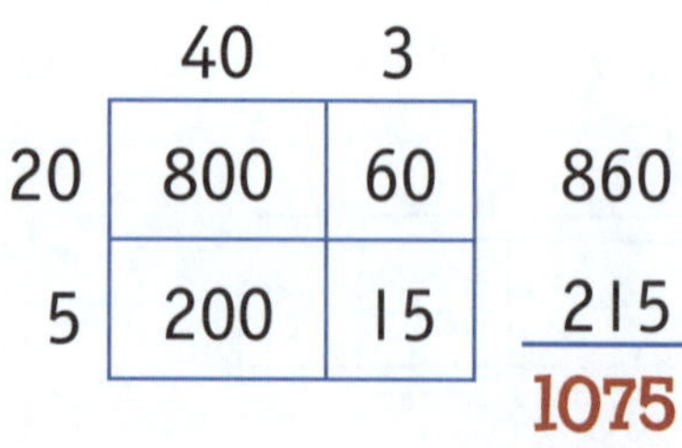

	40	3	
20	800	60	860
5	200	15	215
			1075

860 + 215 = 1075

Example 2: 412 × 36

3-digit × 2-digit

	TT	Th	H	T	O
			4	$^{1}1$	2
×				3	6
		2	$^{1}4$	7	2
+	1	2	3	6	0
	1	4	8	3	2

Steps:
2 × 6
1 × 6
4 × 6
Put 0 in the ones.
2 × 3
1 × 3
4 × 3
Add.

	400	10	2	
30	12 000	300	60	12 360
6	2400	60	12	2 472
				14 832

12 360 + 2472 = 14 832

Example 3: 748 × 42

	TT	Th	H	T	O
			7	$^{1}4$	8
×				4	2
		1	4	9	6
+					0

Steps:
8 × 2
4 × 2
7 × 2
Put 0 in the ones.
8 × 4
4 × 4
7 × 4
Add.

	700	40	8	
40	28 000		320	______
2		80		______

______ + ______ = ______

Check your answer on the video!

CATCH UP MATHS YEAR 5 BOOK A © PASCAL PRESS ISBN: 9781925726169

Your turn

Complete these multiplications.

52 × 10

```
    5 2
×   1 0
-------
    0 0
+ 5 2 0
-------
  5 2 0
```

	50	2
10	500	20
0	0	0

520
0
———
520

a 64 × 13

```
    6 4
×   1 3
-------

+
-------

-------
```

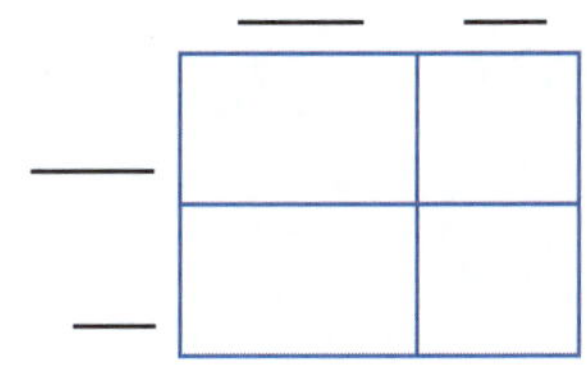

b 176 × 49

```
  1 7 6
×   4 9
-------

+
-------

-------
```

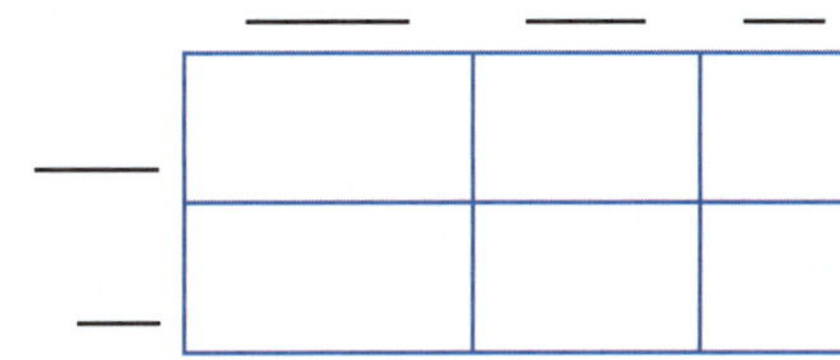

c 843 × 70

```
  8 4 3
×   7 0
-------

+
-------

-------
```

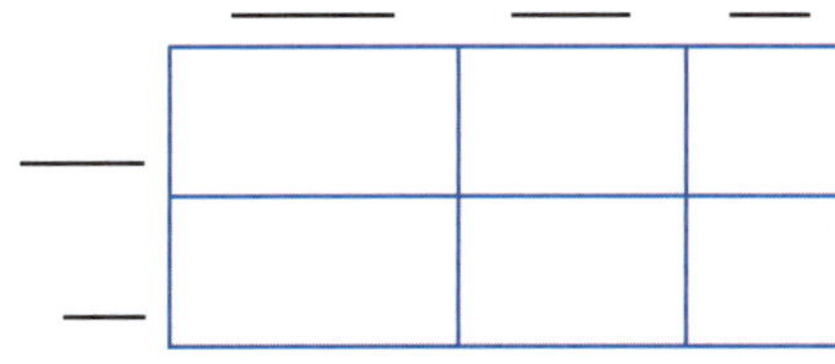

SELF CHECK Tick how you feel

Got it!	Need help...	I don't get it
 ☐	 ☐	☐

Check your answers
How many did you get correct? ☐

PRACTICE

1 Solve using long multiplication.

Example:

	Th	H	T	O
		[1]3	[2]2	7
×			1	4
	1	3	0	8
+	3	2	7	0
	4	5	7	8

a

	Th	H	T	O
		4	2	4
×			1	5
+				

b

	Th	H	T	O
			6	9
×			2	7
+				

2 Solve using long multiplication.

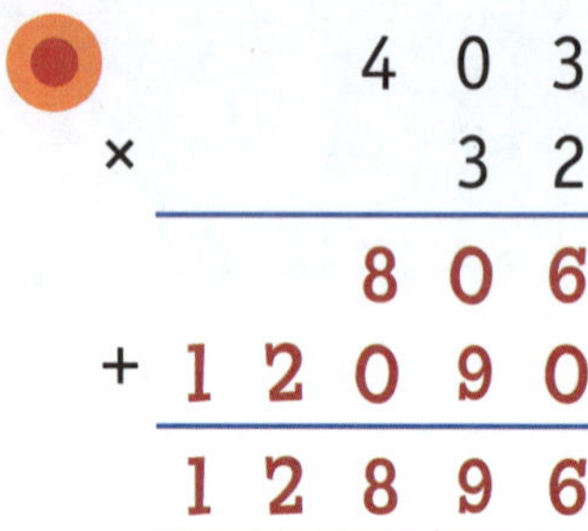

a 231 × 12 = ____

b 72 × 87 = ____

c 510 × 32 = ____

d 88 × 43 = ____

e 694 × 68 = ____

f 594 × 21 = ____

g 679 × 49 = ____

h 876 × 69 = ____

CATCH UP MATHS YEAR 5 BOOK A © PASCAL PRESS ISBN: 9781925726169

3 Solve using area models.

347 × 24

	300	40	7
20	6000	800	140
4	1200	160	28

$$\begin{array}{r} {}^{1}6\,{}^{1}9\,4\,0 \\ +\ 1\,3\,8\,8 \\ \hline 8\,3\,2\,8 \end{array}$$

a 27 × 49

b 67 × 84

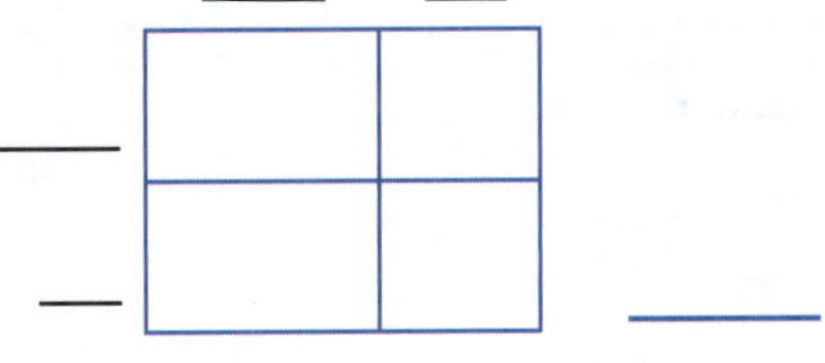

c 684 × 72

d 924 × 35

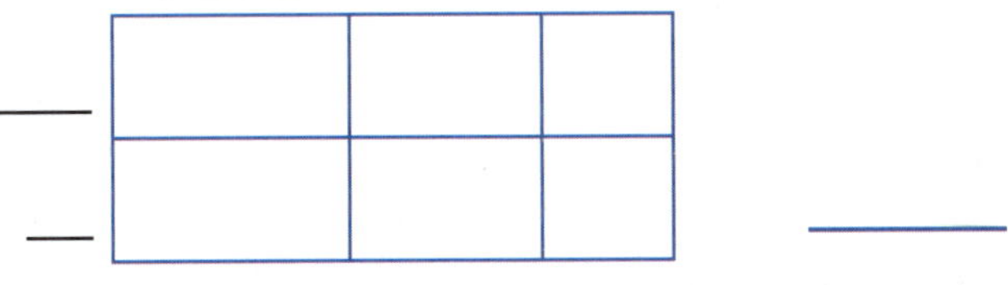

e 93 × 74

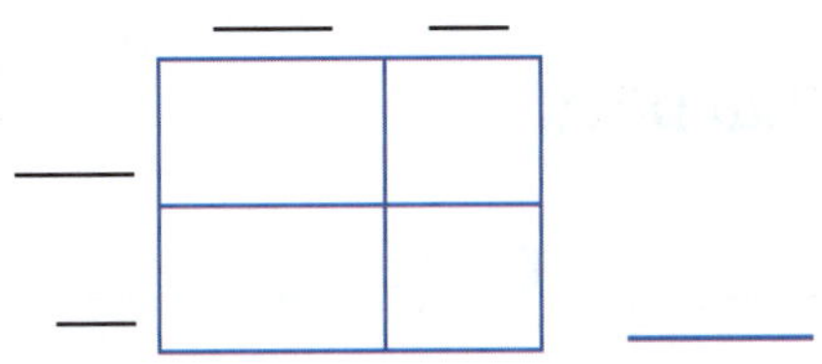

MULTIPLICATION REVIEW

1 Count forwards by:

a 3: 12, ___, ___, ___, ___, ___, ___

b 6: 18, ___, ___, ___, ___, ___, ___

c 10: 50, ___, ___, ___, ___, ___, ___

d 9: 27, ___, ___, ___, ___, ___, ___

2 Count backwards by:

a 4: 32, ___, ___, ___, ___, ___, ___

b 7: 56, ___, ___, ___, ___, ___, ___

c 8: 72, ___, ___, ___, ___, ___, ___

d 12: 132, ___, ___, ___, ___, ___, ___

3 Write the product of each pair of numbers.

a 6 and 9 ____

b 7 and 7 ____

c 8 and 6 ____

d 12 and 4 ____

e 8 and 9 ____

f 5 and 6 ____

4 Write the first 6 multiples of each number.

a 6: ___, ___, ___, ___, ___, ___

b 8: ___, ___, ___, ___, ___, ___

c 9: ___, ___, ___, ___, ___, ___

d 4: ___, ___, ___, ___, ___, ___

e 7: ___, ___, ___, ___, ___, ___

5 Write all the factors of these numbers.

a 36 ______________________

b 48 ______________________

c 15 ______________________

d 12 ______________________

CATCH UP MATHS YEAR 5 BOOK A © PASCAL PRESS ISBN: 9781925726169

6 Fill in the missing numbers.

a 11 × ___ = 110
b 6 × ___ = 24
c ___ × 8 = 72
d ___ × 12 = 48

e 7 × 8 = ___
f 12 × ___ = 60
g 3 × ___ = 27
h 6 × ___ = 36

i 9 × 9 = ___
j 4 × ___ = 20
k ___ × 7 = 63
l 8 × ___ = 32

7 Solve using formal algorithms.

a 73 × 7 = ___
b 49 × 3 = ___
c 82 × 9 = ___

d 79 × 4 = ___
e 37 × 6 = ___
f 51 × 7 = ___

g 123 × 7 = ___
h 563 × 2 = ___
i 649 × 7 = ___

j 828 × 6 = ___
k 257 × 3 = ___
l 403 × 5 = ___

8 Solve using known facts.

14 × 5

10 × 5 = 50

50 + 5 + 5 + 5 + 5

= 70

a 24 × 6

___

___

= ___

b 32 × 8

___

___

= ___

c 43 × 7

___

___

= ___

REVIEW

9 Solve by multiplying the tens and then multiplying the ones.

● 14 × 5

5 tens + 5 fours

50 + 20 = 70

b 81 × 7

____

__________ = ______

a 78 × 9

____

__________ = ______

c 97 × 3

____

__________ = ______

10 Solve using area models.

● 14 × 5

	10	4
5	50	20

50 + 20 = 70

b 49 × 6 = ____

	__	__
__		

__________ = ______

a 28 × 3

	__	__
__		

__________ = ______

c 54 × 7 = ____

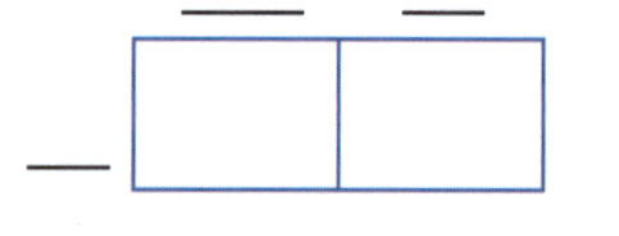

__________ = ______

11 Solve by multiplying thousands, hundreds, tens and ones.

● 2674 × 3 = (2000 × 3) + (600 × 3) + (70 × 3) + (4 × 3)

= 6000 + 1800 + 210 + 12

= 8022

a 362 × 4 = ______________________

= ______________________

= __________

CATCH UP MATHS YEAR 5 BOOK A © PASCAL PRESS ISBN: 9781925726169

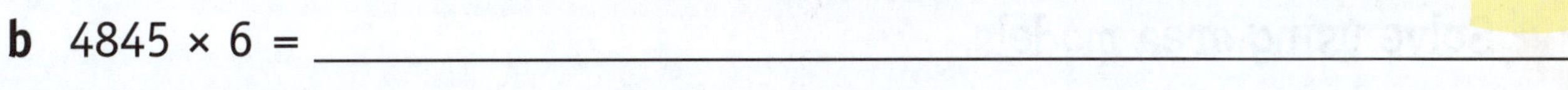

b 4845 × 6 = ______
= ______
= ______

c 573 × 5 = ______
= ______
= ______

12 Solve using area models.

2674 × 3

	2000	600	70	4
3	6000	1800	210	12

6000 + 1800 + 210 + 12
= 8022

b 5903 × 4

	___	___	___	___
___				

______
= ______

a 326 × 8

	___	___	___
___			

______
= ______

c 682 × 6

	___	___	___
___			

______
= ______

13 Solve the following.

a
```
  6 7 3
×     4
-------

-------
```

c
```
  8 9 3
×     5
-------

-------
```

e
```
  5 9 7
×     3
-------

-------
```

b
```
  1 2 7 5
×       6
---------

---------
```

d
```
  4 3 9 1
×       7
---------

---------
```

f
```
  5 8 0 3
×       9
---------

---------
```

REVIEW

14 Solve using area models.

32 × 59

	30	2	
50	1500	100	1600
9	270	18	288
			1888

a 46 × 28

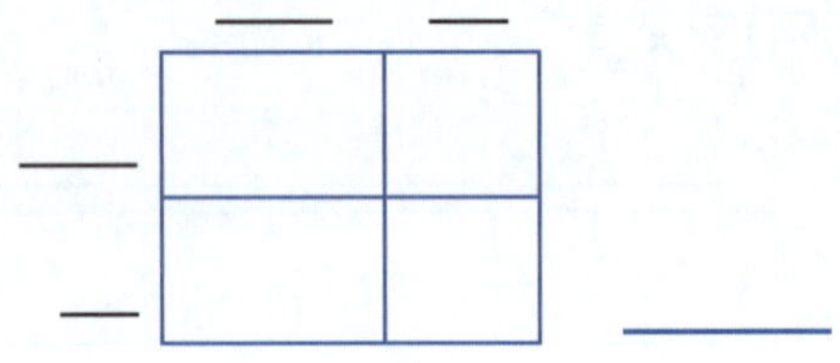

b 82 × 63

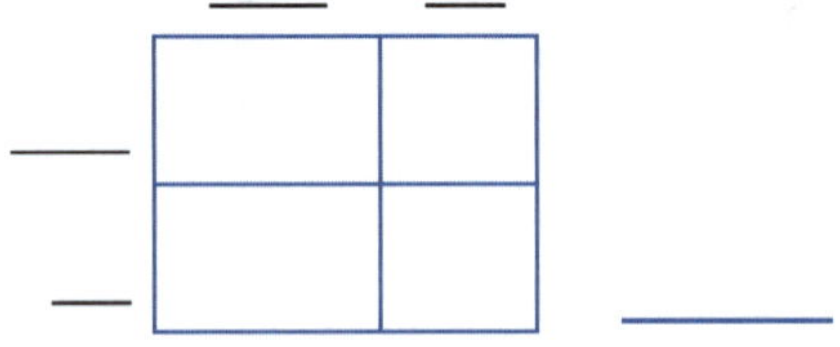

c 70 × 15

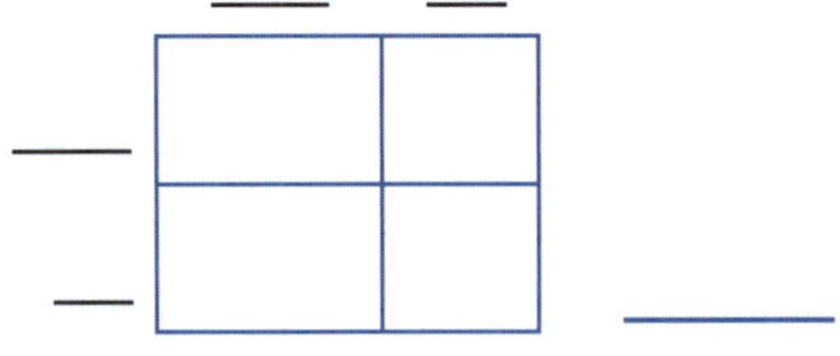

d 96 × 46

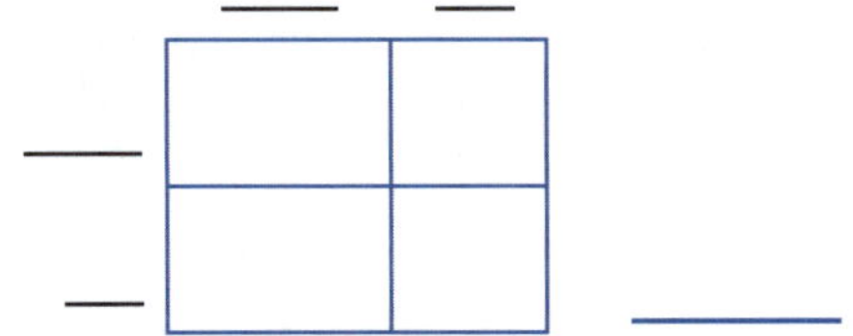

e 413 × 26

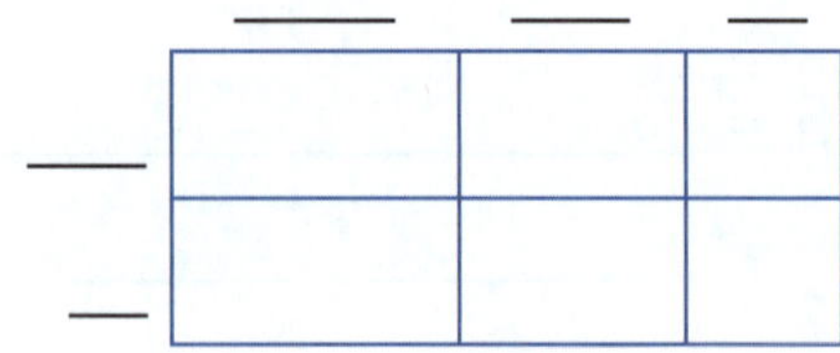

f 599 × 43

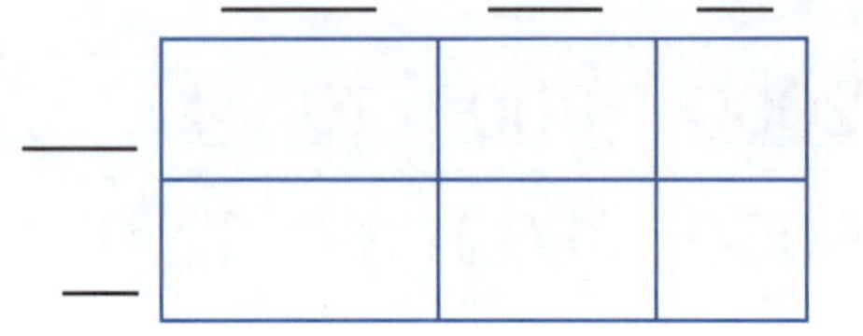

g 703 × 64

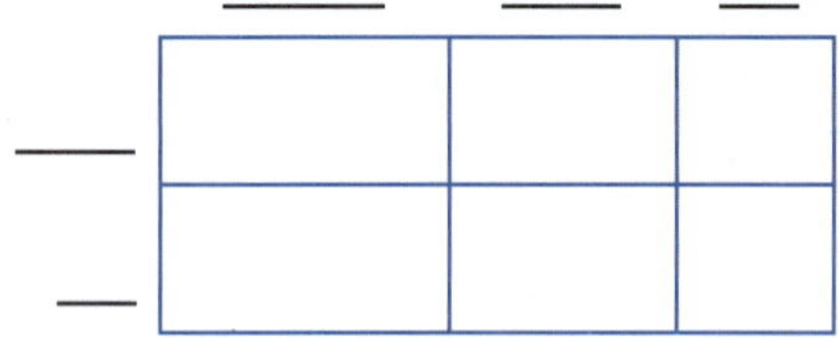

h 930 × 56

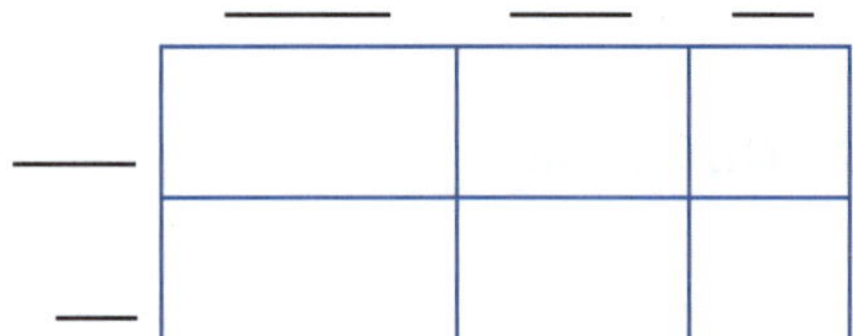

i 876 × 54

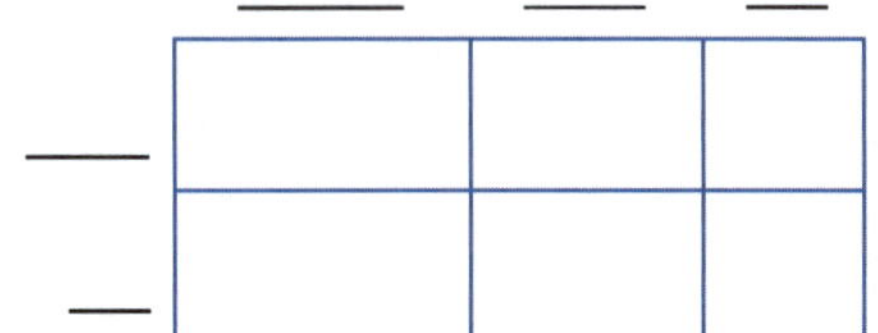

CATCH UP MATHS YEAR 5 BOOK A © PASCAL PRESS ISBN: 9781925726169

15 Solve using long multiplication.

a
$$\begin{array}{r} 231 \\ \times\ 12 \\ \hline + \quad \\ \hline \\ \hline \end{array}$$

b
$$\begin{array}{r} 38 \\ \times\ 95 \\ \hline + \quad \\ \hline \\ \hline \end{array}$$

c
$$\begin{array}{r} 959 \\ \times\ 34 \\ \hline + \quad \\ \hline \\ \hline \end{array}$$

d
$$\begin{array}{r} 64 \\ \times\ 82 \\ \hline + \quad \\ \hline \\ \hline \end{array}$$

e
$$\begin{array}{r} 28 \\ \times\ 36 \\ \hline + \quad \\ \hline \\ \hline \end{array}$$

f
$$\begin{array}{r} 711 \\ \times\ 49 \\ \hline + \quad \\ \hline \\ \hline \end{array}$$

g
$$\begin{array}{r} 20 \\ \times\ 74 \\ \hline + \quad \\ \hline \\ \hline \end{array}$$

h
$$\begin{array}{r} 830 \\ \times\ 56 \\ \hline + \quad \\ \hline \\ \hline \end{array}$$

i
$$\begin{array}{r} 507 \\ \times\ 30 \\ \hline + \quad \\ \hline \\ \hline \end{array}$$

16 Mark as correct (✓) or incorrect (✗).

a
$$\begin{array}{r} {}^{3}2\ 5 \\ \times\ 7 \\ \hline 1\ 7\ 5 \\ \hline \end{array}$$

b
$$\begin{array}{r} 8\ 4 \\ \times\ 3 \\ \hline 2\ 4\ 2 \\ \hline \end{array}$$

c
$$\begin{array}{r} {}^{6}6\ 8 \\ \times\ 8 \\ \hline 5\ 4\ 4 \\ \hline \end{array}$$

d
$$\begin{array}{r} {}^{2}1\ {}^{2}3\ 4 \\ \times\ 2\ 6 \\ \hline {}^{1}\ 8\ 0\ 4 \\ +\ 2\ 6\ 8\ 0 \\ \hline 3\ 4\ 8\ 4 \\ \hline \end{array}$$

e
$$\begin{array}{r} {}^{4}5\ 5 \\ \times\ 2\ 8 \\ \hline 4\ 0\ 0 \\ +\ 1\ 0\ 0 \\ \hline 5\ 0\ 0 \\ \hline \end{array}$$

f
$$\begin{array}{r} {}^{1}7\ {}^{3}2\ 7 \\ \times\ 3\ 5 \\ \hline {}^{1}3\ 6\ 3\ 5 \\ +\ 2\ 1\ 8\ 1\ 0 \\ \hline 2\ 5\ 4\ 4\ 5 \\ \hline \end{array}$$

g
$$\begin{array}{r} {}^{4}7\ 5 \\ \times\ 9\ 4 \\ \hline {}^{1}\ {}^{1}2\ 8\ 0 \\ +\ 6\ 7\ 5\ 0 \\ \hline 7\ 0\ 3\ 0 \\ \hline \end{array}$$

GROUPS AND EQUAL ROWS

Grouping is sharing (or dividing) objects into groups of the same size. Equal rows have the same number in each row.

Groups

Example 1:
Share 15 balls among 3 children.

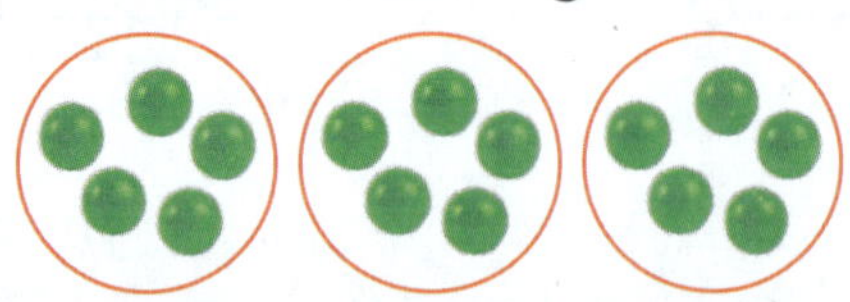

$15 \div 3 = 5$ 3 groups of 5 = 15

Example 2:
Share 24 balls among 6 children.

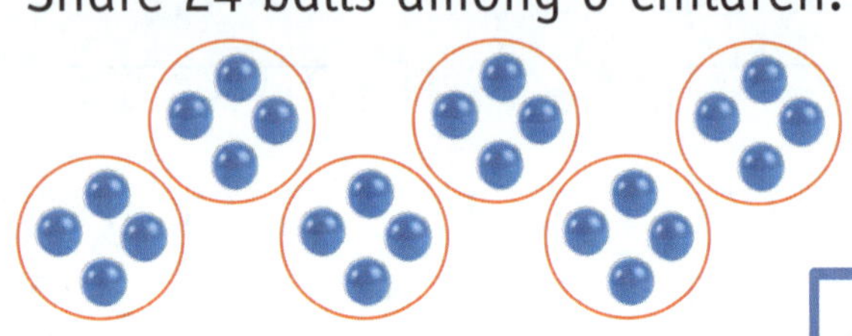

___ $\div 6 = 4$

6 groups of __ = ___

Equal Rows

Example 3:
Share 15 balls among 3 children.

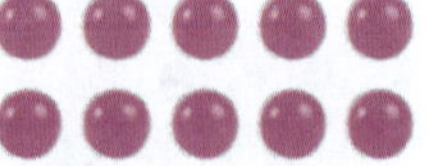

$15 \div 3 = 5$

3 rows of 5 = 15

Example 4:
Share 24 balls among 6 children.

___ $\div 6 =$ __

6 rows of __ = ___

Check your answer on the video!

1 Draw the ■ then complete the sentence.

Share 10 ■ among 5 children.
Each child will get 2 ■.

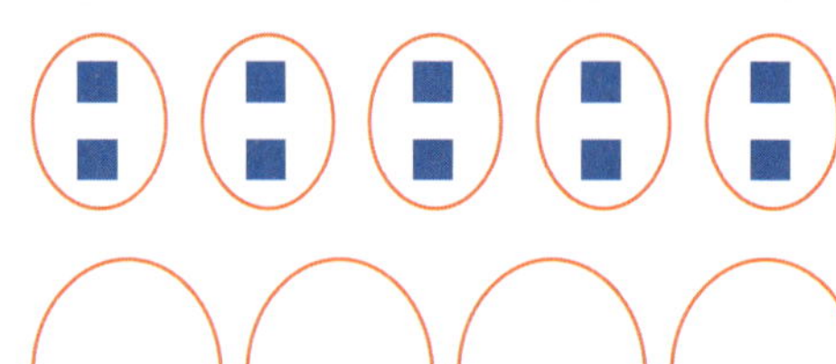

a Divide 20 ■ among 4 children.
Each child will get __ ■.

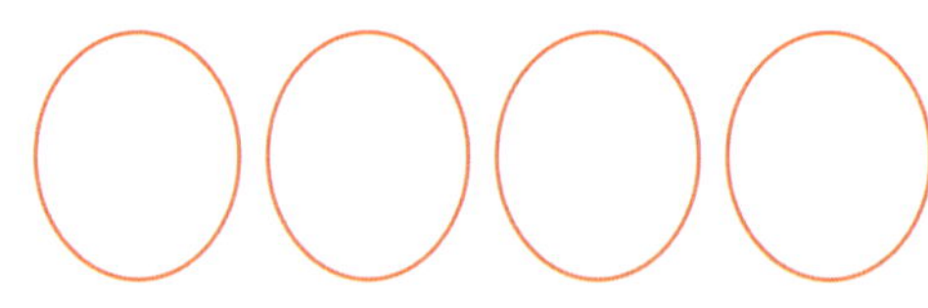

2 Draw the ● in equal rows.

12 ● in 3 rows

a 16 ● in 2 rows

b 9 ● in 3 rows

SELF CHECK Tick how you feel

Got it!	Need help...	I don't get it
 ☐	 ☐	 ☐

Check your answers
How many did you get correct?

CATCH UP MATHS YEAR 5 BOOK A © PASCAL PRESS ISBN: 9781925726169

PRACTICE

Make equal groups, then complete the number sentences.

Example	b	d
Groups of 4  3 groups of 4 12 ÷ 3 = 4	Groups of 4 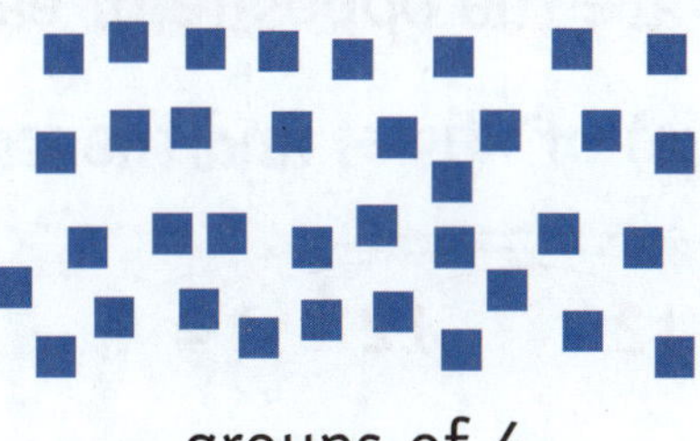 __ groups of 4 36 ÷ __ = 4	Groups of 2 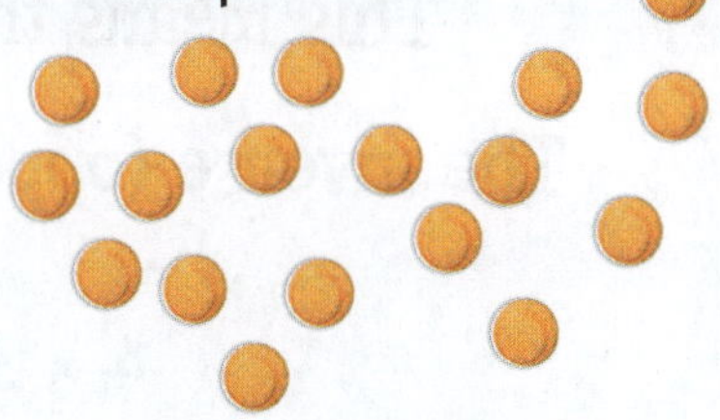 __ groups of 2 18 ÷ __ = 2
a Groups of 5 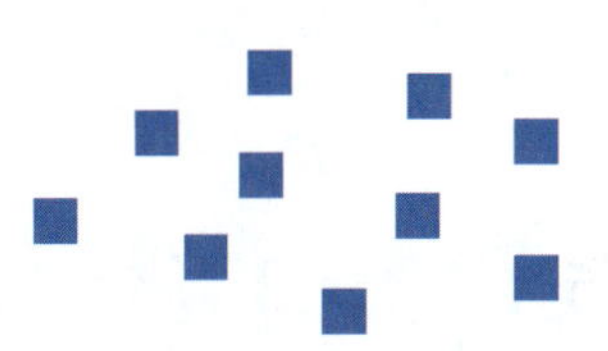 __ groups of 5 10 ÷ __ = 5	**c** Groups of 6 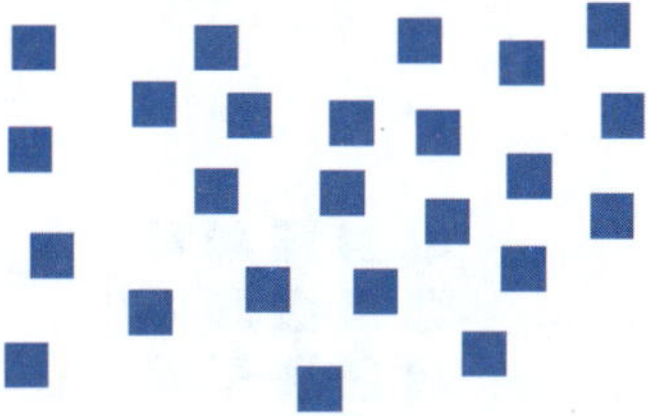 __ groups of 6 24 ÷ __ = 6	**e** Groups of 10 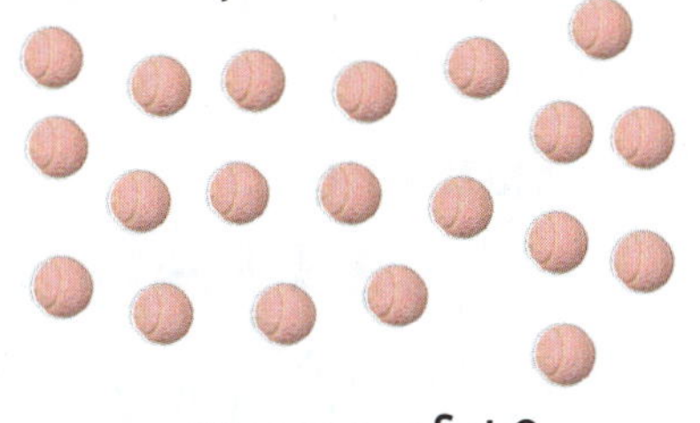 __ groups of 10 20 ÷ __ = 10

2 **Draw ▲ in equal rows, then complete the number sentences.**

Example	b	d
10 ▲ in 2 equal rows 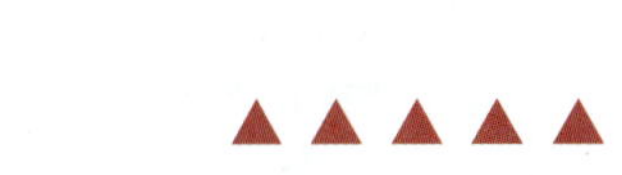 2 rows of 5 = 10 10 ÷ 2 = 5	40 ▲ in 5 equal rows 5 rows of __ = 40 40 ÷ 5 = __	12 ▲ in 6 equal rows 6 rows of __ = 12 12 ÷ 6 = __
a 6 ▲ in 2 equal rows 2 rows of __ = 6 6 ÷ 2 = __	**c** 4 ▲ in 4 equal rows 4 rows of __ = 4 4 ÷ 4 = __	**e** 24 ▲ in 3 equal rows 3 rows of __ = 24 24 ÷ 3 = __

RELATING × TO ÷

Multiplication and division are inverse operations. This means they are the opposite of each other.

The inverse (opposite) of × is ÷, and the inverse of ÷ is ×.

4 × 3 = 12   12 ÷ 3 = 4

Example 1:

4 × 3 = 12 (4: Number of rows; 3: How many in each row; 12: How many in total)

12 ÷ 3 = 4 (12: How many in total; 3: How many in each row; 4: Number of rows)

Example 2:

3 × 5 = ___ (3: Number of rows; 5: How many in each row; ___: How many in total)

15 ÷ ___ = 3 (15: How many in total; ___: How many in each row; 3: Number of rows)

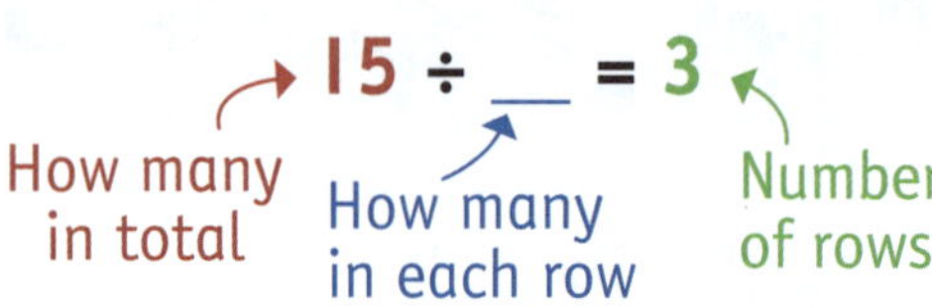

Example 3:

5 × 2 = ___   ___ ÷ ___ = 5

Your turn

Fill in the boxes.

- 6 × [8] = 48 is the inverse of 48 ÷ [8] = 6

a [ ] × 9 = 54 is the inverse of 54 ÷ 9 = [ ]

b 3 × 7 = [ ] is the inverse of [ ] ÷ 7 = 3

c 10 × 10 = [ ] is the inverse of [ ] ÷ 10 = 10

d 11 × [ ] = 121 is the inverse of 121 ÷ [ ] = 11

e 8 × [ ] = 56 is the inverse of 56 ÷ [ ] = 8

SELF CHECK Tick how you feel

Got it!	Need help...	I don't get it
[ ]	[ ]	[ ]

Check your answers

How many did you get correct? [ ]

CATCH UP MATHS YEAR 5 BOOK A © PASCAL PRESS ISBN: 9781925726169

1 Match the inverse operations.

(example)	8 × 3 = 24	64 ÷ 8 = 8
a	9 × 5 = 45	45 ÷ 5 = 9
b	2 × 1 = 2	24 ÷ 3 = 8
c	8 × 8 = 64	42 ÷ 7 = 6
d	6 × 7 = 42	2 ÷ 1 = 2

2 Write facts for each set of numbers.

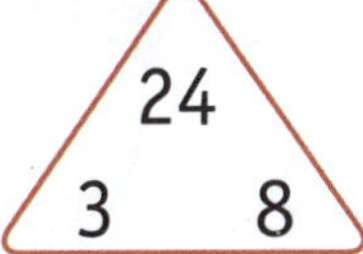

3 × 8 = 24
8 × 3 = 24
24 ÷ 8 = 3
24 ÷ 3 = 8

b 18; 2, 9

___ × ___ = ____
___ × ___ = ____
____ ÷ ___ = ___
____ ÷ ___ = ___

d 72; 9, 8

___ × ___ = ____
___ × ___ = ____
____ ÷ ___ = ___
____ ÷ ___ = ___

a

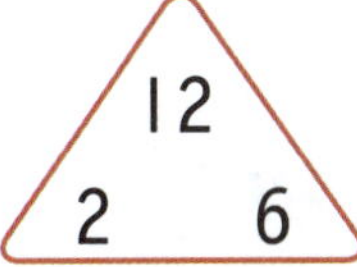

___ × ___ = ____
___ × ___ = ____
____ ÷ ___ = ___
____ ÷ ___ = ___

c 20; 4, 5

___ × ___ = ____
___ × ___ = ____
____ ÷ ___ = ___
____ ÷ ___ = ___

e 120; 12, 10

___ × ___ = ____
___ × ___ = ____
____ ÷ ___ = ___
____ ÷ ___ = ___

3 Write one multiplication fact that relates to each division.

27 ÷ 9
9 × 3

b 45 ÷ 5 __________

d 28 ÷ 7 __________

a 90 ÷ 9 __________

c 40 ÷ 8 __________

e 30 ÷ 6 __________

QUOTIENT, DIVISOR AND DIVIDEND

dividend ÷ divisor = quotient

The **dividend** is the number you are dividing.
The **divisor** is the number you are dividing by.
The **quotient** is the answer you get when you divide.

36 ÷ 12 = 3

dividend ↑ divisor ↑ quotient ↑

It's easier to talk about division and maths when you know what words to use.

Example 1:
Circle the quotient.

35 ÷ 7 = (5)

Example 2:
Circle the dividend.

(32) ÷ 4 = 8

Example 3:
Circle the divisor.

326 ÷ (2) = 163

Example 4:
What is the quotient?

42 ÷ 6 = __

1 Trace over the dividend in blue, the divisor in red and the quotient in green.

● 10 ÷ 1 = 10 **a** 22 ÷ 2 = 11 **b** 40 ÷ 8 = 5

2 Complete the table.

	Division	Dividend	Divisor	Quotient
●	30 ÷ 5 = 6	30	5	6
a	48 ÷ 6 = 8			
b	72 ÷ 8 = 9			
c		10	5	2
d		20	10	2

SELF CHECK Tick how you feel

Got it! ☐ Need help... ☐ I don't get it ☐

Check your answers
How many did you get correct? ☐

CATCH UP MATHS YEAR 5 BOOK A © PASCAL PRESS ISBN: 9781925726169

PRACTICE

1 Circle the quotient.

- 8 ÷ 4 = (2)
- a 77 ÷ 11 = 7
- b 24 ÷ 12 = 2
- c 80 ÷ 8 = 10
- d 36 ÷ 9 = 4
- e 21 ÷ 3 = 7

2 Tick the labels that have a quotient of 4.

4 ÷ 1 ✓	24 ÷ 6	16 ÷ 4	20 ÷ 2	22 ÷ 11
24 ÷ 8	24 ÷ 4	28 ÷ 7	18 ÷ 9	

3 Use red to circle the dividend and blue to circle the divisor. Then write the quotient.

- 8 ÷ 2 = 4
- a 12 ÷ 4 = ___
- b 42 ÷ 7 = ___
- c 27 ÷ 3 = ___
- d 32 ÷ 8 = ___
- e 30 ÷ 6 = ___
- f 28 ÷ 7 = ___
- g 81 ÷ 9 = ___
- h 64 ÷ 8 = ___

4 Complete the tables: **Dividend ÷ Divisor = Quotient**

	Dividend	Divisor	Quotient
●	12	1	12
a	27	9	
b	36	12	
c	56	8	
d	66	11	
e	49	7	
f	110	10	
g	42	6	
h	36	4	
i	25	5	
j	18	9	

	Dividend	Divisor	Quotient
k	7	7	
l	22	2	
m	33	3	
n	40	8	
o	40	4	
p	6	1	
q	9	3	
r	30	6	
s	10	1	
t	63	7	
u	60	5	

FORMAL DIVISION

Formal division is where division problems are written using the $\overline{)}$ symbol instead of ÷.

You can use multiplication to check the answer:

$4 \times 5 = 20$

Example 1:

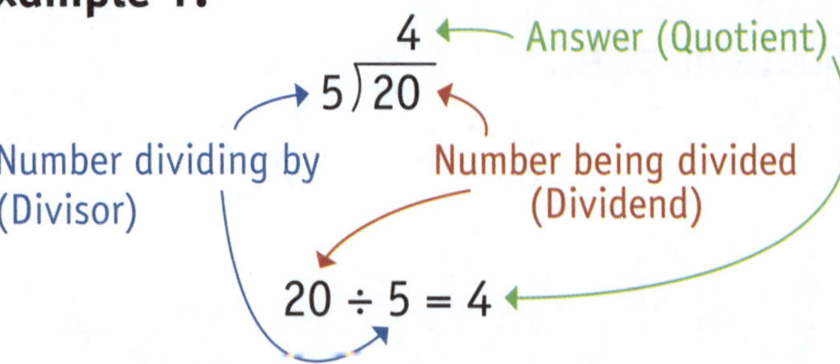

Example 2:

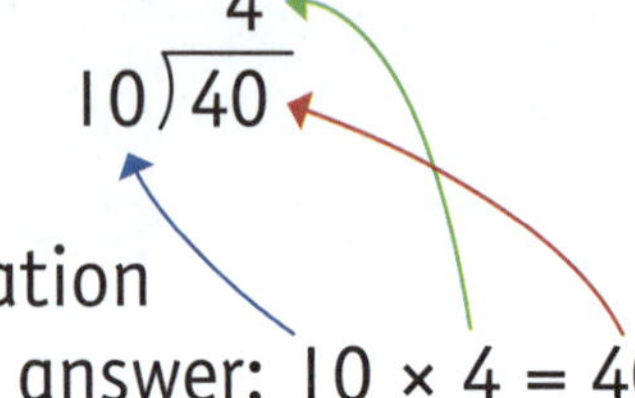

Use multiplication to check your answer: $10 \times 4 = 40$

Example 3:

$8\overline{)24}$

Check using multiplication: $8 \times$ __ $= 24$

Example 4:

$9\overline{)63}$

Check using multiplication: $9 \times$ __ $=$ ___

Example 5:

$\overline{)56}$ with 8 above

Check using multiplication: __ $\times 8 = 56$

Solve.

● $2\overline{)8}$ (answer 4)	**c** $5\overline{)30}$	**f** $9\overline{)36}$
a $7\overline{)14}$	**d** $5\overline{)40}$	**g** $7\overline{)42}$
b $5\overline{)15}$	**e** $10\overline{)30}$	**h** $9\overline{)9}$

SELF CHECK Tick how you feel

Got it!	Need help...	I don't get it
☐	☐	☐

Check your answers

How many did you get correct?

CATCH UP MATHS YEAR 5 BOOK A © PASCAL PRESS ISBN: 9781925726169

1 Solve then check your answer with multiplication.

- 3)9 with 3 above — Check: 3 × 3 = 9
- **a** 4)16 Check: ___ × ___ = ___
- **b** 2)18 Check: ___ × ___ = ___
- **c** 5)20 Check: ___ × ___ = ___
- **d** 6)24 Check: ___ × ___ = ___
- **e** 4)28 Check: ___ × ___ = ___

2 Fill in the missing numbers.

- 6)30 with 5 above
- **a** ___)32 with 4 above
- **b** 11)___ with 4 above
- **c** 7)56
- **d** ___)48 with 6 above
- **e** 5)50
- **f** 12)___ with 4 above
- **g** 9)72
- **h** ___)66 with 11 above

3 Use the numbers in the multiplication to complete the division algorithms.

4 × 9 = 36

36 ÷ 9 = 4   9)36 with 4 above

a 1 × 8 = 8

___ ÷ ___ = ___   ___)___

b 4 × 5 = 20

___ ÷ ___ = ___   ___)___

c 9 × 7 = 63

___ ÷ ___ = ___   ___)___

d 12 × 10 = 120

___ ÷ ___ = ___   ___)___

e 5 × 2 = 10

___ ÷ ___ = ___   ___)___

f 3 × 11 = 33

___ ÷ ___ = ___   ___)___

g 7 × 10 = 70

___ ÷ ___ = ___   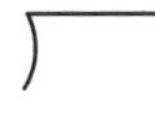

h 9 × 8 = 72

___ ÷ ___ = ___   ___)___

i 6 × 7 = 42

___ ÷ ___ = ___   ___)___

  ISBN: 9781925726169

DIFFERENT WAYS TO WRITE DIVISION

Here are three different ways of writing division.

Example 1:
Write 25 divided by 4 in three different ways.

$25 \div 4$ $\quad 4\overline{)25}$ $\quad \frac{25}{4}$

These all mean the same thing: 25 shared among 4.

Example 2:
Write 52 divided by 3 in three different ways.

$52 \div 3$ $\quad 3\overline{)52}$ $\quad \frac{52}{3}$

Example 3:
Write 27 divided by 9 in three different ways.

$27 \div 9$ $\quad \_\_\overline{)27}$ $\quad \frac{27}{\_\_}$

Check your answer on the video!

Example 4:
Write 60 divided by 10 in three different ways.

$60 \div 10$ $\quad 10\overline{)\_\_}$ $\quad \frac{\_\_}{10}$

Match the divisions.

	$23 \div 7$	$4\overline{)52}$
a	$43 \div 2$	$7\overline{)23}$
b	$74 \div 3$	$\frac{74}{3}$
c	$52 \div 4$	$\frac{43}{2}$

Check your answers
How many did you get correct?

CATCH UP MATHS YEAR 5 BOOK A © PASCAL PRESS ISBN: 9781925726169

PRACTICE

1 Write the divisions using the $\overline{)\ \ }$ symbol.

- 36 ÷ 5 $5\overline{)36}$
- **a** 42 ÷ 6 ______
- **b** $\frac{93}{10}$ ______
- **c** 81 ÷ 9 ______
- **d** $\frac{63}{7}$ ______
- **e** $\frac{75}{25}$ ______
- **f** 230 ÷ 4 ______
- **g** $\frac{121}{11}$ ______
- **h** 182 ÷ 5 ______

2 Record the divisions as fractions.

- 26 ÷ 2 $\frac{26}{2}$
- **a** 53 ÷ 4 ______
- **b** $3\overline{)72}$ ______
- **c** 37 ÷ 4 ______
- **d** $3\overline{)82}$ ______
- **e** 51 ÷ 2 ______
- **f** $3\overline{)95}$ ______
- **g** $7\overline{)64}$ ______
- **h** 42 ÷ 6 ______

3 Write the divisions using the ÷ symbol.

- $\frac{53}{5}$ 53 ÷ 5
- **a** $3\overline{)27}$ ______
- **b** $\frac{47}{4}$ ______
- **c** $5\overline{)64}$ ______
- **d** $\frac{58}{4}$ ______
- **e** $7\overline{)83}$ ______
- **f** $\frac{62}{3}$ ______
- **g** $4\overline{)73}$ ______
- **h** $8\overline{)97}$ ______

4 Fill in the tables.

	Fraction	$\overline{)\ \ }$	÷
	$\frac{57}{3}$	$3\overline{)57}$	57 ÷ 3
a		$4\overline{)61}$	
b			25 ÷ 3
c	$\frac{72}{6}$		
d	$\frac{49}{7}$		

	Fraction	$\overline{)\ \ }$	÷
e		$9\overline{)81}$	
f			74 ÷ 5
g			69 ÷ 9
h			22 ÷ 3
i		$4\overline{)16}$	

DIVISION WITH REMAINDERS

When a number cannot be divided exactly, the leftover is called the remainder.

SCAN to watch video

Example 1:

17 ÷ 3 = 5 remainder 2

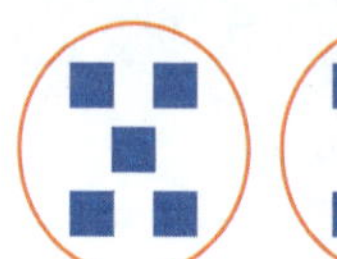

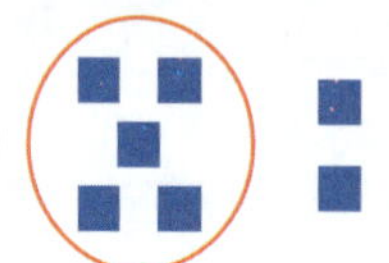

You can use multiplication to check your answer, then add the remainder at the end:

5 × 3 = 15

Add the remainder, 2:

15 + 2 = 17

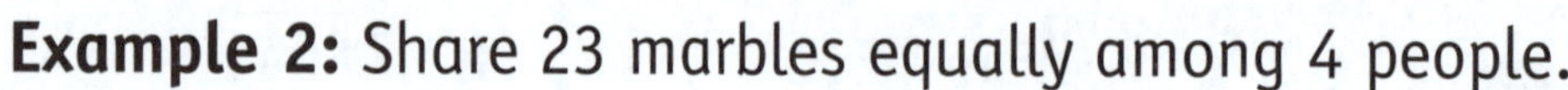

Example 2: Share 23 marbles equally among 4 people.

23 ÷ 4 = 5 remainder 3

Check: 5 × 4 = 20

20 + 3 = 23

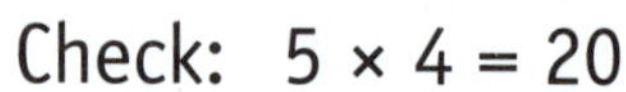

Add the remainder

Example 3: Share 22 stars equally among 6 people.

22 ÷ 6 = __ remainder __

Check: __ × 6 = ___

___ + 4 = ___

Check your answer on the video!

Your turn

Solve these divisions with remainders.

- 32 ÷ 10 = 3 remainder 2

a 24 ÷ 7 = ___ remainder ___

b 44 ÷ 5 = ___ remainder ___

c 55 ÷ 9 = ___ remainder ___

SELF CHECK Tick how you feel		
Got it! ☐	Need help... ☐	I don't get it ☐

Check your answers

How many did you get correct? ☐

CATCH UP MATHS YEAR 5 BOOK A © PASCAL PRESS ISBN: 9781925726169

1 Solve the following.

- 29 ÷ 3 = 9 remainder 2

a 63 ÷ 8 = ___ remainder ___

b 45 ÷ 7 = ___ remainder ___

c 75 ÷ 9 = ___ remainder ___

d 62 ÷ 8 = ___ remainder ___

e 17 ÷ 2 = ___ remainder ___

f $9\overline{)83}$ r

g $8\overline{)58}$ r

h $10\overline{)94}$ r

i $8\overline{)47}$ r

2 Use the first number sentence to complete the division.

- (6 × 3) + 2 = 20 → 20 ÷ 3 = 6 remainder 2

a (4 × 11) + 1 = 45 → 45 ÷ 11 = ___ remainder ___

b (7 × 9) + 2 = 65 → 65 ÷ 9 = ___ remainder ___

c (5 × 4) + 3 = 23 → 23 ÷ 4 = ___ remainder ___

d (9 × 7) + 5 = 68 → 68 ÷ 7 = ___ remainder ___

3 Write the number sentence, then solve.

- 17 jellybeans shared among 2 people 17 ÷ 2 = 8 remainder 1

a 25 balls shared among 4 people ___ ÷ ___ = ___ remainder ___

b 70 cakes shared among 6 people ___ ÷ ___ = ___ remainder ___

c 84 balls shared among 9 people ___ ÷ ___ = ___ remainder ___

d 148 pencils shared among 12 people ___ ÷ ___ = ___ remainder ___

e 106 shirts shared among 10 people ___ ÷ ___ = ___ remainder ___

f 62 fish shared among 5 children ___ ÷ ___ = ___ remainder ___

g 29 cans shared among 6 people ___ ÷ ___ = ___ remainder ___

h 78 apples shared among 8 people ___ ÷ ___ = ___ remainder ___

DIVISION OF 2-DIGIT NUMBERS

Two-digit numbers can be divided by single-digit numbers.

SCAN to watch video

Example 1:

Share 72 lollies among 3 people.

$3\overline{)72}$

Give 20 lollies to each person.

2 tens → 2

$3\overline{)7^{1}2}$

Trade 1 ten for 10 ones

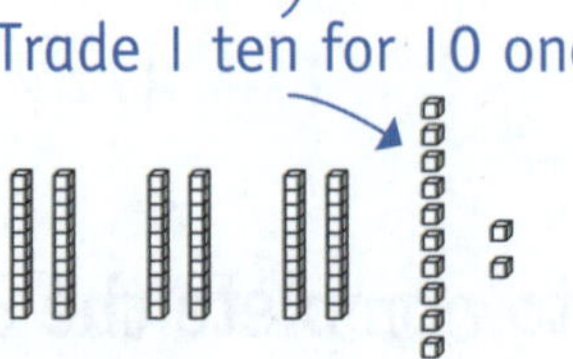

12 lollies remain. So 12 lollies among 3 people is 4 each.

$\begin{array}{r} 24 \\ 3\overline{)7^{1}2} \end{array}$

Example 2:

Share 64 balls among 4 people.

$4\overline{)64}$

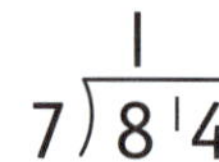

Give __ balls to each person.

1 ten → 1

$4\overline{)6^{2}4}$

Trade 2 tens for ___ ones

24 balls remain. So 24 balls among 4 people is __ each.

$\begin{array}{r} 16 \\ 4\overline{)6^{2}4} \end{array}$

Example 3:

$\begin{array}{r} 1 \\ 7\overline{)8^{1}4} \end{array}$

Example 4:

$4\overline{)52}$

Example 5:

$6\overline{)96}$

Check your answer on the video!

Your turn

Solve the following.

● $\begin{array}{r} 12 \\ 2\overline{)24} \end{array}$

b $4\overline{)84}$

d $4\overline{)88}$

a $3\overline{)36}$

c $6\overline{)66}$

e $3\overline{)33}$

SELF CHECK Tick how you feel

Got it!	Need help...	I don't get it
☐	☐	☐

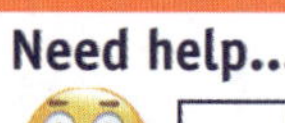

Check your answers

How many did you get correct?

CATCH UP MATHS YEAR 5 BOOK A © PASCAL PRESS ISBN: 9781925726169

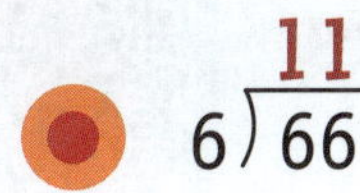

PRACTICE

1 Solve.

●	$\overset{11}{6\overline{)66}}$	b	$3\overline{)39}$	d	$5\overline{)55}$	f	$4\overline{)48}$
a	$7\overline{)77}$	c	$4\overline{)44}$	e	$2\overline{)22}$	g	$2\overline{)26}$

2 Solve these division questions. You will need to trade.

●	$\overset{27}{2\overline{)5^{1}4}}$	d	$8\overline{)96}$	h	$5\overline{)75}$	l	$5\overline{)90}$
a	$3\overline{)51}$	e	$5\overline{)85}$	i	$3\overline{)84}$	m	$6\overline{)96}$
b	$6\overline{)84}$	f	$7\overline{)91}$	j	$6\overline{)84}$	n	$3\overline{)81}$
c	$4\overline{)68}$	g	$3\overline{)87}$	k	$7\overline{)91}$	o	$4\overline{)72}$

3 Solve, then multiply to check your answer.

●	$\overset{16}{5\overline{)8^{3}0}}$ ✓	$^{+3}16 \times 5 = 80$	d	$6\overline{)90}$	$\times\ 6$
a	$4\overline{)76}$	$\times\ 4$	e	$3\overline{)96}$	$\times\ 3$
b	$3\overline{)78}$	$\times\ 3$	f	$5\overline{)65}$	$\times\ 5$
c	$2\overline{)74}$	$\times\ 2$	g	$9\overline{)99}$	$\times\ 9$

DIVISION OF 3-DIGIT NUMBERS

Three-digit numbers can be divided by single-digit numbers.

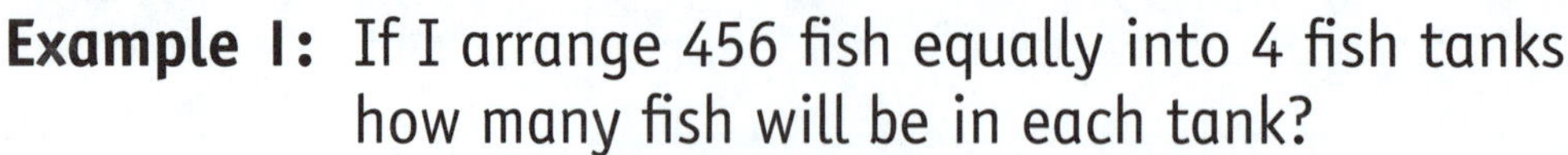

Example 1: If I arrange 456 fish equally into 4 fish tanks, how many fish will be in each tank?

4)4 5 6

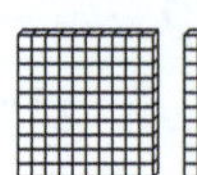 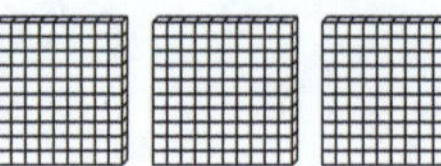 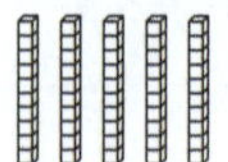

Put 100 fish in each tank. That leaves 56 fish.	Put 10 fish in each tank. That leaves 16 fish.	16 fish into 4 tanks is 4 in each tank.
1 4)4 5 ¹6	1 1 4)4 5 ¹6	1 1 4 4)4 5 ¹6

There will be 114 fish in each tank.

Example 2:

1 2 1
4)4 8 4

Example 4:

4 9
5)2 4 ⁴5

Example 6:

2)8 6 4

Example 8:

7)3 6 4

Example 3:

1 4 0
2)2 8 0

Example 5:

5 1
6)3 0 6

Example 7:

3)3 6 9

Check your answer on the video!

Solve these divisions.

 1 2 4
2)2 4 8

a 3)6 3 6

b 2)4 8 4

c 5)5 5 5

d 4)8 4 4

e 2)6 8 2

f 3)3 6 0

g 4)4 8 0

Check your answers
How many did you get correct?

CATCH UP MATHS YEAR 5 BOOK A © PASCAL PRESS ISBN: 9781925726169

PRACTICE

1 Solve the following.

●	2)104 = **52**	**d**	3)963	**h**	7)189
a	7)945	**e**	5)710	**i**	8)448
b	3)372	**f**	9)882	**j**	6)588
c	4)636	**g**	6)504	**k**	9)585

2 Solve these division problems and multiply to check your answer.

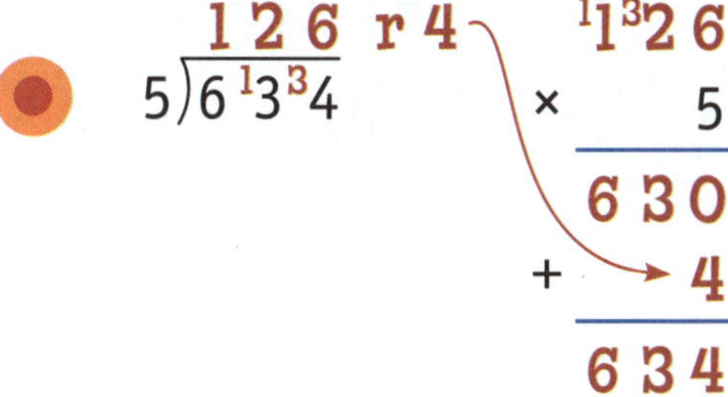

● 5)6 [1]3 [3]4 = **126 r 4**; check: 126 × 5 = **630**; + **4** = **634**

b 4)752 × 4 = ____ + ____ = ____

a 8)937 × 8 = ____ + ____ = ____

c 5)753 × 5 = ____ + ____ = ____

3 Use the rule to complete the table.

● ■ = ▲ ÷ 40

▲	400	240	160	120	280
■	**10**	**6**	**4**	**3**	**7**

b ▲ = ★ ÷ 9

★	810	180	720	630	540
■					

a ★ = ● ÷ 3

●	270	300	360	120	900
★					

c ★ = ■ ÷ 7

■	490	770	210	420	350
★					

RECORDING REMAINDERS AS FRACTIONS AND DECIMALS

The remainder from a division can also be written as a fraction or a decimal.

Remainder as a fraction

$4\overline{)25} = 6 \text{ r } 1$

Here, the remainder is 1.

You can write it as $\frac{1}{4}$.

$4\overline{)25} = 6\frac{1}{4}$

Remainder as a decimal

$4\overline{)25} = 6 \text{ r } 1$

$\frac{1}{4}$ as a decimal is 0.25.

$4\overline{)25} = 6.25$

Put the remainder over the number you divide by to make the fraction.

Example 1:

$$\begin{array}{r} 7\ 2 \text{ r } 3 \\ 4\overline{)2\ 9^{1}1} \end{array}$$

$= 72\frac{3}{4}$ or 72.75

Some fraction and decimal equivalents

$\frac{1}{8} = 0.125$	$\frac{2}{5} = 0.4$
$\frac{1}{5} = 0.2$	$\frac{1}{2} = 0.5$
$\frac{1}{4} = 0.25$	$\frac{2}{3} = 0.67$
$\frac{1}{3} = 0.33$	$\frac{3}{4} = 0.75$

Example 2:

$$\begin{array}{r} 4\ 9 \text{ r } 2 \\ 3\overline{)1\ 4^{2}9} \end{array}$$

$= 49\frac{2}{3}$ or 49.67

Example 3:

$$\begin{array}{r} 5\ 4 \text{ r } 2 \\ 4\overline{)2\ 1^{1}8} \end{array}$$

$= 54\frac{—}{—} = 54\frac{—}{—}$ or ____

Solve and record the remainders as fractions and decimals.

● $\begin{array}{r} 2\ 7 \text{ r } 1 \\ 8\overline{)2\ 1^{5}7} \end{array}$

= $27\frac{1}{8}$ or 27.125

a $2\overline{)1\ 4\ 9}$

= ________ or ________

b $3\overline{)1\ 3\ 6}$

= ________ or ________

c $5\overline{)6\ 4\ 7}$

= ________ or ________

SELF CHECK Tick how you feel

Got it!	Need help...	I don't get it
 ☐	 ☐	☐

Check your answers

How many did you get correct? ☐

CATCH UP MATHS YEAR 5 BOOK A © PASCAL PRESS ISBN: 9781925726169

1 Solve and record the remainder as a fraction.

- 94 r 3
  8)7 5³5
  = 94 remainder 3 = $94\frac{3}{8}$

a 3)1 4 8
= ____ remainder __ = ☐

b 4)4 4 7
= ____ remainder __ = ☐

c 2)3 4 7
= ____ remainder __ = ☐

d 4)2 5 4
= ____ remainder __ = ☐

e 3)5 1 5
= ____ remainder __ = ☐

2 Solve and record the remainder as a decimal.

- 68 r 6
  8)5 5⁷0
  = 68 remainder 6 = 68.75

a 4)2 1 9
= ___ remainder __ = _____

b 5)3 1 2
= ___ remainder __ = _____

c 8)3 4 9
= ___ remainder __ = _____

3 Fill in the table.

	Dividend	Divisor	Quotient	Remainder	Quotient and Remainder as a Fraction	Quotient and Remainder as a Decimal
●	25	4	6	1	$6\frac{1}{4}$	6.25
a	15	2				
b	32	5				
c	57	8				
d	38	3				

DIVISION REVIEW

1 Make equal groups and complete the number sentences.

a Groups of 3	b Groups of 4	d Groups of 8

__ groups of 3	__ groups of 4	__ groups of 8
15 ÷ __ = 3	___ ÷ __ = 4	___ ÷ __ = 8

2 Draw ▲ in equal rows, then complete the number sentences.

a 21 ▲ in 3 equal rows	b 20 ▲ in 4 equal rows	c 12 ▲ in 2 equal rows
3 rows of __ = 21	4 rows of __ = 20	2 rows of __ = 12
21 ÷ 3 = __	20 ÷ 4 = __	12 ÷ 2 = __

3 Write an inverse.

a 8 × 6 = 48

____

b 9 × 8 = 72

____

c 5 × 2 = 10

____

d 4 × 7 = 28

____

e 120 ÷ 10 = 12

____

f 36 ÷ 12 = 3

____

g 64 ÷ 8 = 8

____

h 48 ÷ 12 = 4

____

i 63 ÷ 7 = 9

____

j 144 ÷ 12 = 12

____

k 7 × 1 = 7

____

l 42 ÷ 6 = 7

____

CATCH UP MATHS YEAR 5 BOOK A © PASCAL PRESS ISBN: 9781925726169

4 Write facts for each set of numbers.

a

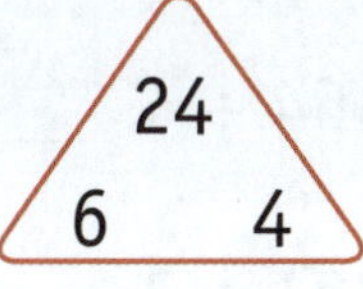

___ × ___ = ___

___ × ___ = ___

___ ÷ ___ = ___

___ ÷ ___ = ___

b

21
7 3

___ × ___ = ___

___ × ___ = ___

___ ÷ ___ = ___

___ ÷ ___ = ___

c

56
7 8

___ × ___ = ___

___ × ___ = ___

___ ÷ ___ = ___

___ ÷ ___ = ___

5 Write one multiplication fact that relates to each division.

a 42 ÷ 6 ____________

b 99 ÷ 11 ____________

c 16 ÷ 2 ____________

d 48 ÷ 8 ____________

e 40 ÷ 5 ____________

f 49 ÷ 7 ____________

g 32 ÷ 4 ____________

h 36 ÷ 6 ____________

i 121 ÷ 11 ____________

6 What is the quotient?

a 32 ÷ 8 _____

b 72 ÷ 9 _____

c 30 ÷ 3 _____

d 96 ÷ 12 _____

e 132 ÷ 11 _____

f 96 ÷ 8 _____

g 24 ÷ 6 _____

h 40 ÷ 4 _____

i 56 ÷ 8 _____

7 Tick the labels that have a quotient of 12.

12 ÷ 6	18 ÷ 3	36 ÷ 3	12 ÷ 1	14 ÷ 2
24 ÷ 3	12 ÷ 4	12 ÷ 12	24 ÷ 2	24 ÷ 4

REVIEW

8 Write the divisions using the $\overline{)}$ symbol.

a 42 ÷ 6 ______

b 27 ÷ 9 ______

c 81 ÷ 9 ______

d 127 ÷ 2 ______

e 464 ÷ 4 ______

f 721 ÷ 3 ______

9 Record the divisions as fractions.

a 25 ÷ 5 ____

b 36 ÷ 4 ____

c $6\overline{)24}$ ____

d $8\overline{)72}$ ____

e 9 ÷ 9 ____

f 63 ÷ 7 ____

g 40 ÷ 4 ____

h $8\overline{)32}$ ____

i 10 ÷ 2 ____

10 Write the divisions using the ÷ symbol.

a $\frac{24}{3}$ ______

b $6\overline{)42}$ ______

c $7\overline{)49}$ ______

d $\frac{121}{11}$ ______

e $\frac{80}{8}$ ______

f $9\overline{)90}$ ______

11 Solve these divisions.

a $9\overline{)36}$

b $5\overline{)30}$

c $7\overline{)42}$

d $8\overline{)56}$

e $8\overline{)96}$

f $8\overline{)40}$

g $5\overline{)45}$

h $6\overline{)60}$

12 Solve these divisions with remainders.

a 36 ÷ 10 = ___ remainder ___

b 62 ÷ 6 = ___ remainder ___

c 24 ÷ 8 = ___ remainder ___

d 19 ÷ 4 = ___ remainder ___

e 11 ÷ 9 = ___ remainder ___

f 16 ÷ 3 = ___ remainder ___

g 23 ÷ 4 = ___ remainder ___

h 44 ÷ 6 = ___ remainder ___

i 16 ÷ 4 = ___ remainder ___

j 73 ÷ 8 = ___ remainder ___

CATCH UP MATHS YEAR 5 BOOK A © PASCAL PRESS ISBN: 9781925726169

13 Fill in the blank spaces.

a $(6 \times 2) + 3 = 15$ → $15 \div 2 =$ ___ remainder ___

b $(8 \times 3) + 2 = 26$ → $26 \div 3 =$ ___ remainder ___

c $(9 \times 9) + 6 = 87$ → $87 \div 9 =$ ___ remainder ___

d $(4 \times 7) + 5 = 33$ → $33 \div 7 =$ ___ remainder ___

14 Write the number sentence, then solve.

a 18 jellybeans shared among 4 people ___ ÷ ___ = ___ remainder __

b 134 pencils shared among 12 people ___ ÷ ___ = ___ remainder __

c 109 lollies shared among 10 people ___ ÷ ___ = ___ remainder __

d 79 apples shared among 8 people ___ ÷ ___ = ___ remainder __

e 28 cans shared among 6 people ___ ÷ ___ = ___ remainder __

15 Solve these divisions.

a $2\overline{)62}$ d $3\overline{)63}$ g $7\overline{)91}$ j $3\overline{)87}$

b $4\overline{)48}$ e $2\overline{)86}$ h $5\overline{)85}$ k $6\overline{)96}$

c $2\overline{)48}$ f $8\overline{)96}$ i $5\overline{)75}$ l $6\overline{)84}$

16 Solve, then check your answer with multiplication.

a $5\overline{)85}$ × 5 ___

c $6\overline{)84}$ × 6 ___

b $4\overline{)80}$ × 4 ___

d $4\overline{)72}$ × 4 ___

REVIEW

Solve, then check your answer with multiplication. Make sure you add on the remainder.

a $5\overline{)27}$ r × 5 +

d $7\overline{)68}$ r × 7 +

b $6\overline{)94}$ r × 6 +

e $3\overline{)42}$ r × 3 +

c $9\overline{)84}$ r × 9 +

f $6\overline{)49}$ r × 6 +

18 **Solve.**

a $2\overline{)842}$

c $4\overline{)488}$

e $2\overline{)468}$

b $3\overline{)639}$

d $5\overline{)555}$

f $3\overline{)996}$

Solve these divisions with remainders.

a $3\overline{)964}$ r

c $7\overline{)836}$ r

e $2\overline{)687}$ r

b $8\overline{)934}$ r

d $4\overline{)725}$ r

f $5\overline{)649}$ r

CATCH UP MATHS YEAR 5 BOOK A © PASCAL PRESS ISBN: 9781925726169

20 Solve, then check your answer.

a $5\overline{)637}$ r

× 5

+

b $4\overline{)473}$ r

× 4

+

c $8\overline{)849}$ r

× 8

+

d $7\overline{)893}$ r

× 7

+

21 Use the rule to complete the table.

a ■ = ▲ ÷ 4

▲	16	24	32	56	88
■					

b ★ = ● ÷ 6

●	36	6	42	12	66
★					

c ▲ = ■ ÷ 9

■	9	27	54	72	45
▲					

d ⬢ = ★ ÷ 7

★	7	28	56	70	63
⬢					

22 Solve and record the remainder as a fraction.

a $4\overline{)623}$

= ______ remainder __ = ☐

b $5\overline{)532}$

= ______ remainder __ = ☐

c $7\overline{)882}$

= ______ remainder __ = ☐

d $6\overline{)937}$

= ______ remainder __ = ☐

REVIEW

e $9\overline{)8\ 6\ 3}$
= ___ remainder __ = ☐

i $2\overline{)6\ 9\ 4}$
= ___ remainder __ = ☐

f $4\overline{)7\ 7\ 5}$
= ___ remainder __ = ☐

j $7\overline{)8\ 5\ 2}$
= ___ remainder __ = ☐

g $5\overline{)4\ 1\ 6}$
= ___ remainder __ = ☐

k $8\overline{)5\ 1\ 4}$
= ___ remainder __ = ☐

h $3\overline{)3\ 2\ 8}$
= ___ remainder __ = ☐

l $6\overline{)2\ 9\ 1}$
= ___ remainder __ = ☐

23 **Solve and record the remainder as a decimal.**

a $2\overline{)3\ 9\ 5}$
= ___ remainder __ = _______

e $4\overline{)8\ 9\ 9}$
= ___ remainder __ = _______

b $5\overline{)7\ 8\ 4}$
= ___ remainder __ = _______

f $3\overline{)3\ 7\ 3}$
= ___ remainder __ = _______

c $4\overline{)6\ 7\ 5}$
= ___ remainder __ = _______

g $8\overline{)4\ 1\ 8}$
= ___ remainder __ = _______

d $2\overline{)5\ 6\ 3}$
= ___ remainder __ = _______

h $5\overline{)3\ 2\ 4}$
= ___ remainder __ = _______

CATCH UP MATHS YEAR 5 BOOK A © PASCAL PRESS ISBN: 9781925726169

i $5\overline{)981}$

= ___ remainder __ = _______

k $3\overline{)454}$

= ___ remainder __ = _______

j $4\overline{)634}$

= ___ remainder __ = _______

l $2\overline{)121}$

= ___ remainder __ = _______

24 **Use multiplication to check the answers and then mark each one as correct (✓) or incorrect (✗).**

a $\begin{array}{r} 154 \text{ r } 1 \\ 3\overline{)4^{1}6^{1}3} \end{array}$ × 3, + ___, ___

e $\begin{array}{r} 72 \text{ r } 4 \\ 6\overline{)42^{1}6} \end{array}$ × 6, + ___, ___

b $\begin{array}{r} 64 \text{ r } 2 \\ 4\overline{)25^{1}8} \end{array}$ × 4, + ___, ___

f $\begin{array}{r} 92 \text{ r } 7 \\ 9\overline{)83^{2}5} \end{array}$ × 9, + ___, ___

c $\begin{array}{r} 90 \text{ r } 3 \\ 7\overline{)635} \end{array}$ × 7, + ___, ___

g $\begin{array}{r} 96 \text{ r } 5 \\ 8\overline{)74^{2}2} \end{array}$ × 8, + ___, ___

d $\begin{array}{r} 112 \text{ r } 3 \\ 5\overline{)554} \end{array}$ × 5, + ___, ___

h $\begin{array}{r} 260 \text{ r } 1 \\ 2\overline{)5^{1}21} \end{array}$ × 2, + ___, ___

PARTS OF A FRACTION

A fraction has three parts: the numerator, the vinculum (the line) and the denominator.

vinculum (the line) → $\frac{1}{4}$

numerator — The top number in a fraction is the number of parts in this fraction.

denominator — The bottom number in a fraction is the total number of parts.

The larger the denominator, the smaller one part is. $\frac{1}{3}$ is bigger than $\frac{1}{50}$.

Example 1:
Write the fraction. $\frac{1}{2}$
denominator = 2
numerator = 1

Example 2:
Write the fraction. $\frac{\square}{\square}$
numerator = 7
denominator = 12

Example 3:
Label the parts of the fraction.

$\frac{3}{5}$ ← __________ ← __________ ↑ __________

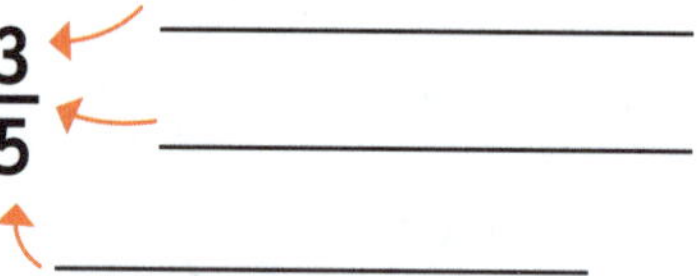

Check your answer on the video!

1 Colour the numerator red, the vinculum green and the denominator blue.

● $\frac{1}{4}$ **a** $\frac{3}{5}$ **b** $\frac{1}{2}$ **c** $\frac{7}{8}$ **d** $\frac{5}{8}$

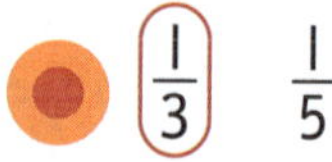

2 Circle the larger fraction.

● ($\frac{1}{3}$) $\frac{1}{5}$ **b** $\frac{1}{10}$ $\frac{11}{100}$ **d** $\frac{1}{5}$ $\frac{1}{10}$

a $\frac{1}{4}$ $\frac{1}{3}$ **c** $\frac{1}{4}$ $\frac{1}{2}$ **e** $\frac{1}{50}$ $\frac{10}{100}$

SELF CHECK Tick how you feel

Got it!	Need help...	I don't get it
☐	☐	☐

Check your answers
How many did you get correct? ☐

CATCH UP MATHS YEAR 5 BOOK A © PASCAL PRESS ISBN: 9781925726169

PRACTICE

1 Write the numerator in each fraction.

● $\frac{3}{4}$ 3	b $\frac{3}{100}$ ___	d $\frac{7}{10}$ ___	f $\frac{1}{5}$ ___
a $\frac{2}{5}$ ___	c $\frac{4}{5}$ ___	e $\frac{1}{4}$ ___	g $\frac{2}{8}$ ___

2 Circle the denominator in each fraction.

● $\frac{3}{(5)}$	b $\frac{1}{100}$	d $\frac{2}{10}$	f $\frac{4}{5}$	h $\frac{3}{4}$
a $\frac{2}{3}$	c $\frac{4}{8}$	e $\frac{6}{8}$	g $\frac{3}{8}$	i $\frac{1}{10}$

3 Write the fractions in words.

● $\frac{3}{5}$ three-fifths	d $\frac{18}{100}$ ___
a $\frac{2}{4}$ ___	e $\frac{4}{5}$ ___
b $\frac{1}{8}$ ___	f $\frac{7}{8}$ ___
c $\frac{1}{10}$ ___	g $\frac{2}{3}$ ___

4 Cross out the fraction with the different denominator.

● $\frac{1}{4}, \frac{3}{4}, \frac{3}{8}, \frac{2}{4}$	b $\frac{4}{8}, \frac{2}{5}, \frac{1}{8}, \frac{5}{8}$	d $\frac{1}{3}, \frac{3}{5}, \frac{3}{3}, \frac{2}{3}$
a $\frac{1}{5}, \frac{3}{5}, \frac{4}{5}, \frac{3}{4}$	c $\frac{1}{10}, \frac{10}{100}, \frac{3}{10}, \frac{4}{10}$	e $\frac{1}{2}, \frac{2}{8}, \frac{4}{8}, \frac{6}{8}$

5 Write the name of each fraction you crossed out in question 4.

● three-eighths	c ___
a ___	d ___
b ___	e ___

HALVES & HALVES OF COLLECTIONS

When an object or group of objects is shared into equal parts, each part is a fraction.

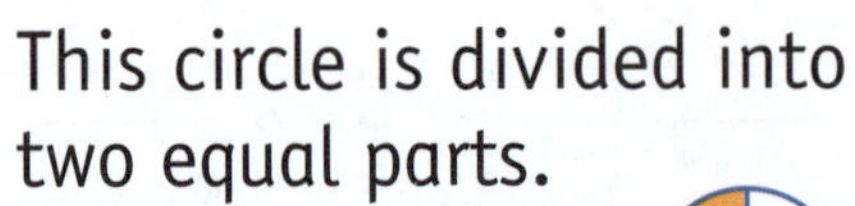

This circle is divided into two equal parts.

Half ($\frac{1}{2}$) of this circle is coloured.

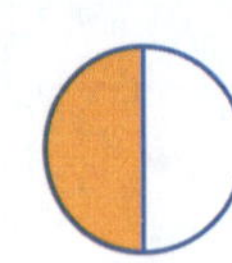

Half of the group of objects is circled. 5 out of the 10 circles are circled. Half of 10 is 5.

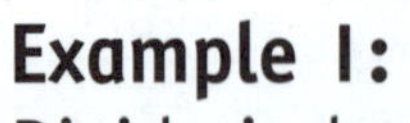

Example 1:
Divide in half.

Example 3:
Divide into two equal parts.

Example 2:
Divide into halves.

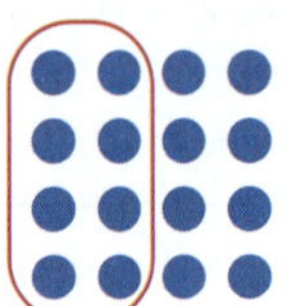

Example 4:
Circle half.

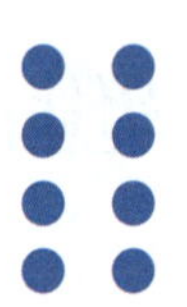

Your turn

1 Circle the objects that are divided into two equal parts.

a 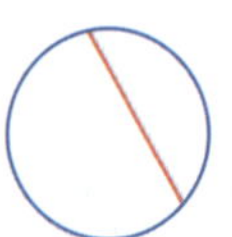

b 

c 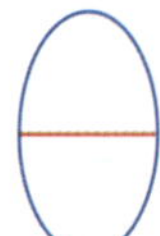

2 Circle half of each group and then write the numbers.

Half of 6 = 3

a Half of ___ = ___

b Half of ___ = ___

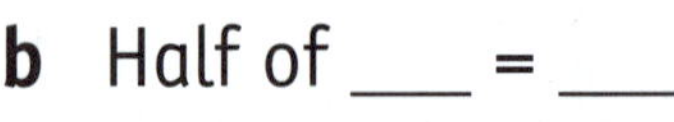

c Half of ___ = ___

Got it! 

Need help... 

I don't get it 

Check your answers
How many did you get correct?

CATCH UP MATHS YEAR 5 BOOK A © PASCAL PRESS ISBN: 9781925726169

PRACTICE

1 Cut these shapes into halves. Then colour one half.

  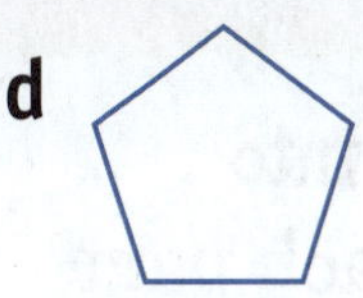 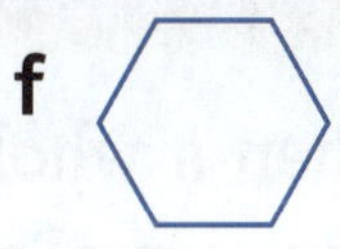

 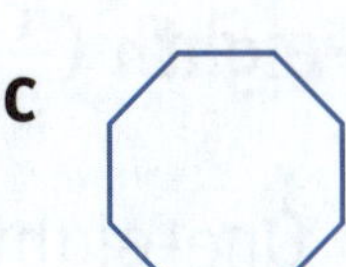 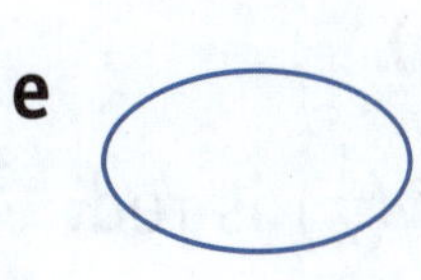 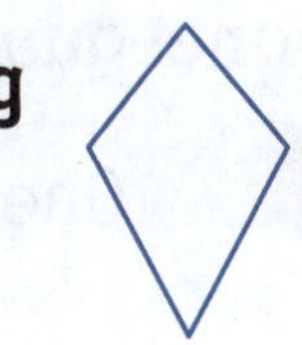

2 Cross out the shapes that are NOT cut into halves.

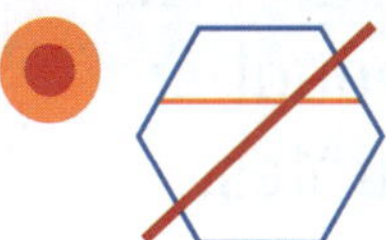  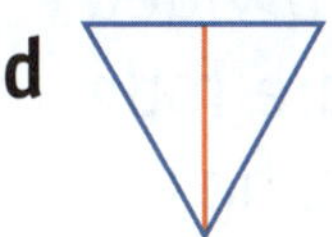 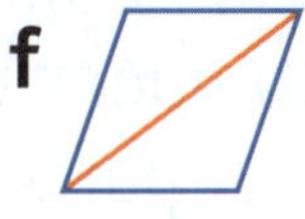

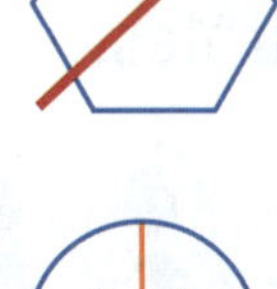 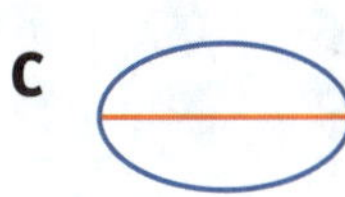  

3 Circle half of each collection and fill in the missing numbers.

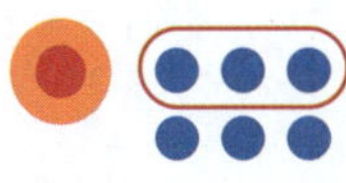

$\frac{1}{2}$ of 6 = 3

a 

$\frac{1}{2}$ of __ = __

b 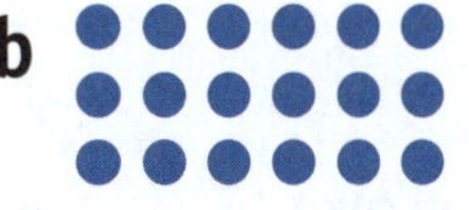

$\frac{1}{2}$ of __ = __

c 

$\frac{1}{2}$ of __ = __

4 These groups have been divided into halves. Fill in the missing numbers.

 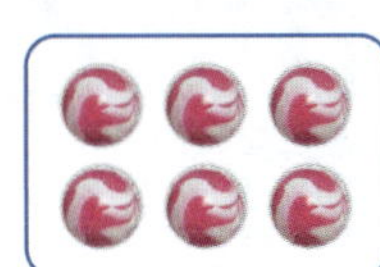

Each group has 6 of the 12 marbles. Half of 12 is 6.

b  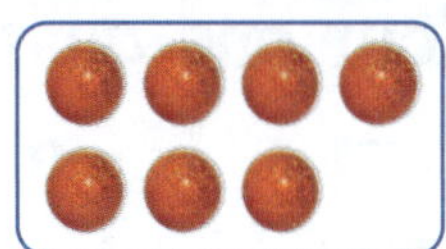 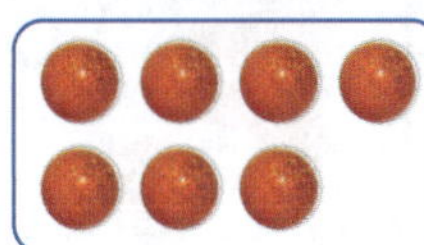

Each group has ___ of the ___ marbles. Half of ___ is ___.

a 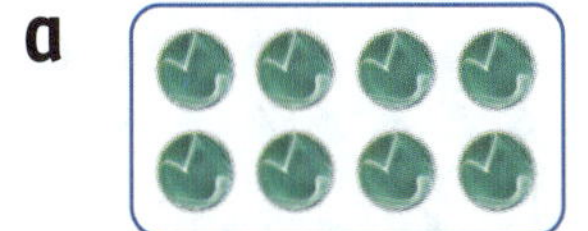 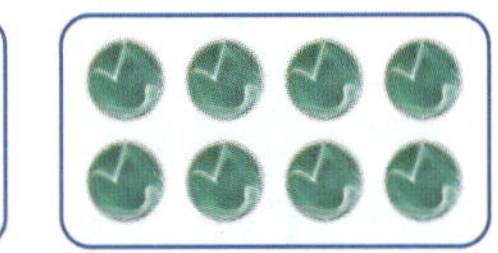

Each group has ___ of the ___ marbles. Half of ___ is ___.

c  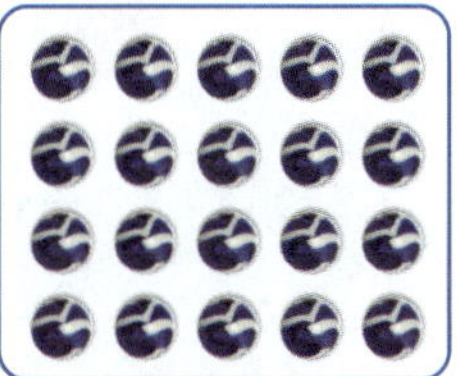

Each group has ___ of the ___ marbles. Half of ___ is ___.

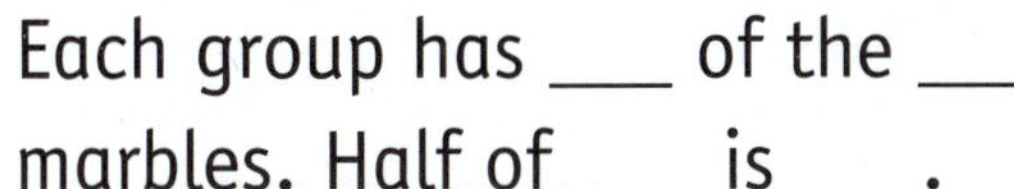

QUARTERS AND EIGHTHS – FRACTIONS AND COLLECTIONS

When a whole is cut into four equal parts, each part is one-quarter ($\frac{1}{4}$).

One-quarter ($\frac{1}{4}$) is red.

Example 1:
Here, 12 marbles are divided into quarters. There are four equal parts. Each part has 3 marbles.

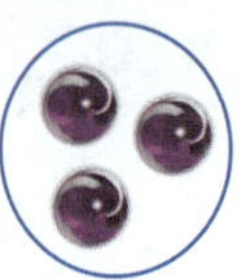

One-quarter of 12 is 3.
$\frac{1}{4}$ of 12 = 3

Example 2:
Divide into quarters.

$\frac{1}{4}$ of __ = __

When a whole is cut into eight equal parts, each part is one-eighth ($\frac{1}{8}$).

SCAN to watch video

One-eighth ($\frac{1}{8}$) is orange.

Example 3:
Here, 16 marbles are divided into eighths. There are eight equal parts. Each part has 2 marbles.

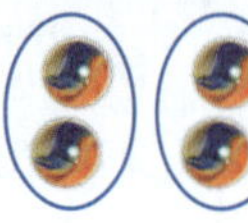

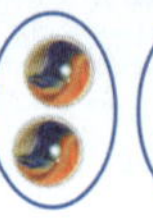

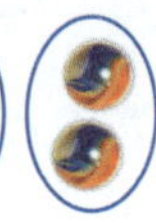

One-eighth of 16 is 2.
$\frac{1}{8}$ of 16 = 2

Example 4:
Divide into eighths.

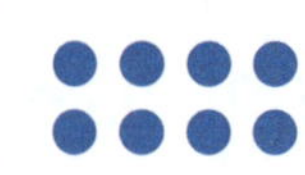

$\frac{1}{8}$ of __ = __

Check your answer on the video!

1 Colour one-quarter.

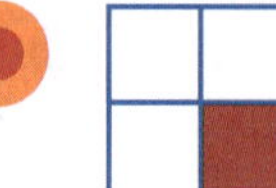

a 

b 

c 

2 What is one-eighth? Circle, then answer.

$\frac{1}{8}$ of 8 = 1

a 

$\frac{1}{8}$ of ___ = ___

b 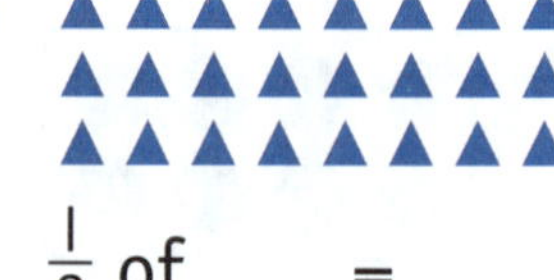

$\frac{1}{8}$ of ___ = ___

SELF CHECK Tick how you feel

Got it!	Need help...	I don't get it
 ☐	 ☐	☐

Check your answers
How many did you get correct?

CATCH UP MATHS YEAR 5 BOOK A © PASCAL PRESS ISBN: 9781925726169

PRACTICE

Colour $\frac{1}{4}$ of each shape.

 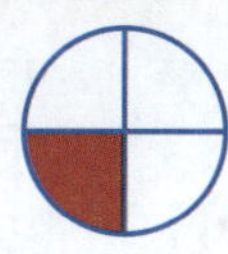

a 

b 

c 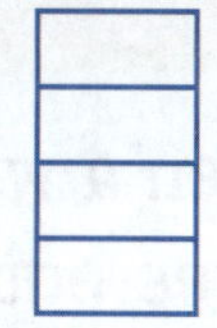

2 Colour $\frac{1}{8}$ of each shape.

a 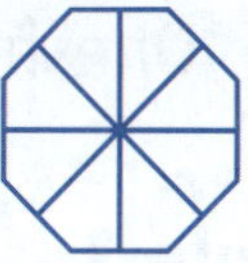

b 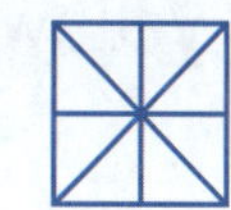

c 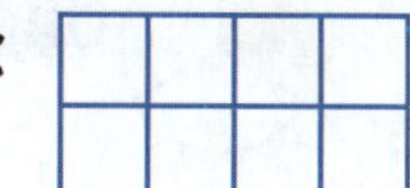

3 Complete the table.

	Fraction name	Fraction	Picture
●	three-eighths	$\frac{3}{8}$	
a		$\frac{2}{4}$	
b		$\frac{7}{8}$	
c			
d	five-eighths		
e		$\frac{2}{8}$	

4 What is one-quarter of each number? (Divide by 4.)

● 12 3   b 24 ___   d 4 ___   f 48 ___

a 16 ___   c 32 ___   e 40 ___   g 36 ___

5 What is one-eighth of each number? (Divide by 8.)

● 16 2   b 8 ___   d 48 ___   f 40 ___

a 24 ___   c 32 ___   e 56 ___   g 64 ___

THIRDS AND FIFTHS – FRACTIONS AND COLLECTIONS

When a whole is cut into three equal parts, each part is one-third ($\frac{1}{3}$).

One-third ($\frac{1}{3}$) is yellow.

Example 1:
Here, 12 marbles are divided into thirds. There are three equal parts. Each part has 4 marbles.

 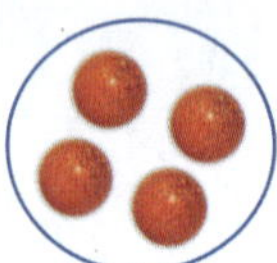 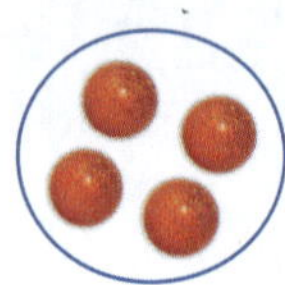

One-third of 12 is 4.

$\frac{1}{3}$ of 12 = 4

Example 2:
Divide into thirds.

$\frac{1}{3}$ of __ = __

When a whole is cut into five equal parts, each part is one-fifth ($\frac{1}{5}$).

SCAN to watch video

One-fifth ($\frac{1}{5}$) is green.

Example 3:
Here, 15 marbles are divided into fifths. There are 5 equal parts. Each part has 3 marbles.

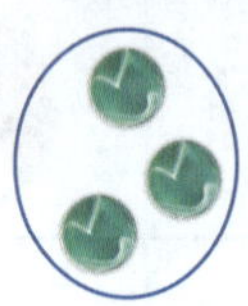 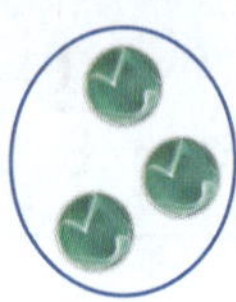  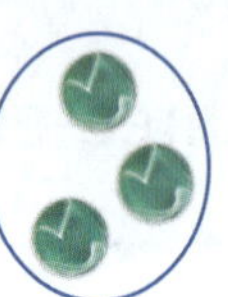 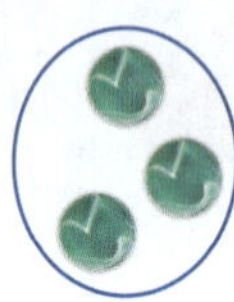

One-fifth of 15 is 3.

$\frac{1}{5}$ of 15 = 3

Check your answer on the video!

Example 4:
Divide into fifths.

$\frac{1}{5}$ of __ = __

1 Colour one-fifth.

a 

b 

c 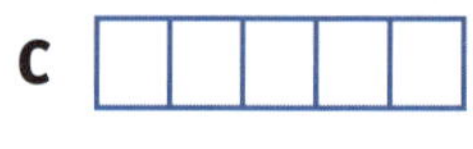

2 What is one-third? Circle, then answer.

 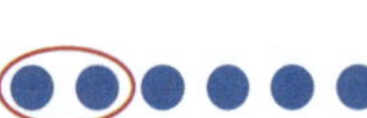

$\frac{1}{3}$ of 6 = 2

a 

$\frac{1}{3}$ of __ = __

b 

$\frac{1}{3}$ of __ = __

SELF CHECK	Tick how you feel	
Got it!	Need help...	I don't get it

Check your answers

How many did you get correct?

CATCH UP MATHS YEAR 5 BOOK A © PASCAL PRESS ISBN: 9781925726169

PRACTICE

1 Colour two-thirds ($\frac{2}{3}$) of each shape.

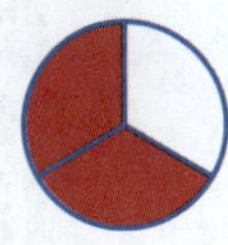

a b c

2 Colour three-fifths ($\frac{3}{5}$) of each shape.

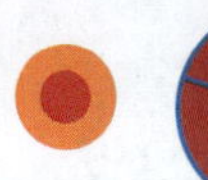
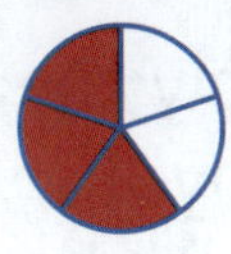

a 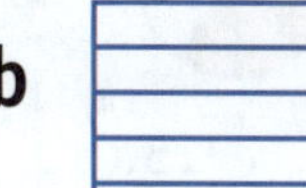 b c 

3 Complete the table.

	Fraction name	Fraction	Picture
•	two-thirds	$\frac{2}{3}$	
a		$\frac{3}{5}$	
b		$\frac{1}{3}$	
c	two-fifths		
d			
e			

4 What is one-third of each number? (Divide by 3.)

• 12 4   b 33 ___   d 27 ___   f 6 ___

a 18 ___   c 36 ___   e 3 ___   g 12 ___

5 What is one-fifth of each number? (Divide by 5.)

• 15 3   b 5 ___   d 60 ___   f 10 ___

a 25 ___   c 40 ___   e 50 ___   g 35 ___

  ISBN: 9781925726169

EQUIVALENT FRACTIONS

Equivalent means equal or the same.
Equivalent fractions are equal or the same value.

Example 1:

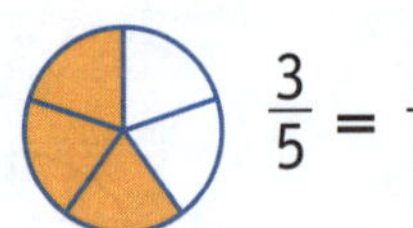

$\frac{3}{5} = \frac{6}{10}$

Both the 3 and the 5 in $\frac{3}{5}$ are multiplied by 2 to get $\frac{6}{10}$.

Example 2:

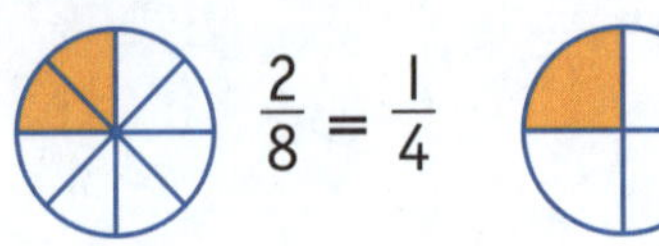

$\frac{2}{8} = \frac{1}{4}$

Both the 2 and the 8 in $\frac{2}{8}$ are divided by 2 to get $\frac{1}{4}$.

Example 3:

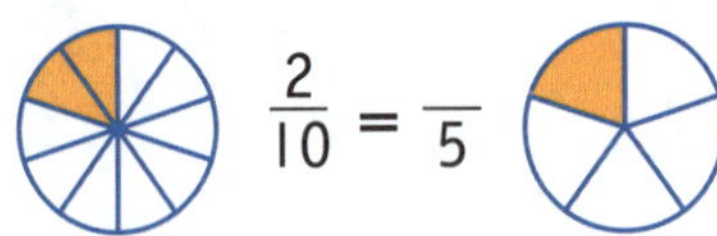

$\frac{2}{10} = \frac{\_\_}{5}$

Both the 2 and the 10 in $\frac{2}{10}$ are divided by __ to get $\frac{\_\_}{5}$.

Example 4:

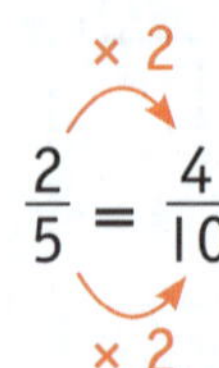

$\frac{2}{5} = \frac{4}{10}$ (× 2 numerator, × 2 denominator)

If we multiply the numerator by 2, we must multiply the denominator by 2.

Example 5:

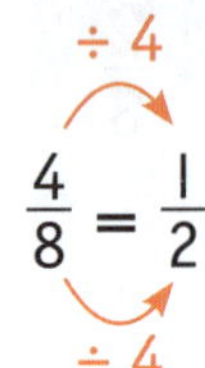

$\frac{4}{8} = \frac{1}{2}$ (÷ 4 numerator, ÷ 4 denominator)

If we divide the numerator by 4, we must divide the denominator by 4.

Example 6:

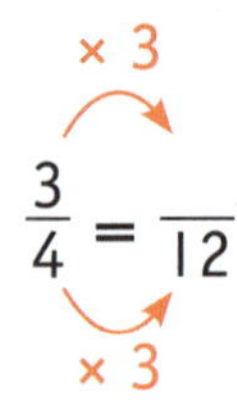

$\frac{3}{4} = \frac{\_\_}{12}$ (× 3 numerator, × 3 denominator)

If we multiply the denominator by __, we must multiply the numerator by __.

Your turn

Colour to show the equivalent fraction and then write the fraction.

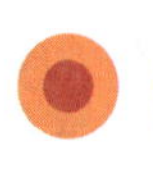 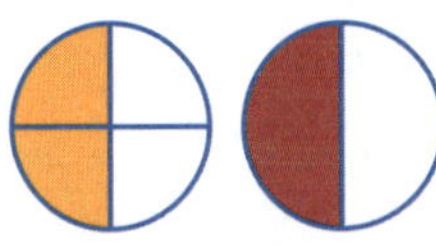

$\frac{2}{4} = \frac{1}{2}$

a  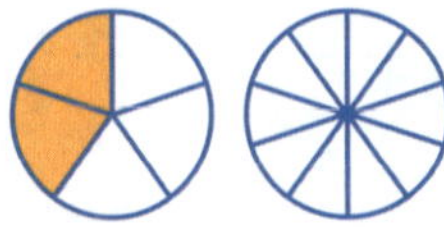

$\frac{2}{5} = \_\_$

b  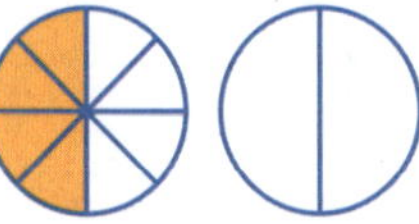

$\frac{4}{8} = \_\_$

SELF CHECK Tick how you feel

Got it!	Need help...	I don't get it
 ☐	 ☐	☐

Check your answers
How many did you get correct? ☐

CATCH UP MATHS YEAR 5 BOOK A © PASCAL PRESS ISBN: 9781925726169

PRACTICE

1 Colour to show the equivalent fraction, then complete the fraction.

● $\frac{3}{5}$ is equivalent to $\frac{6}{10}$

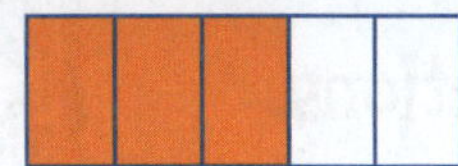 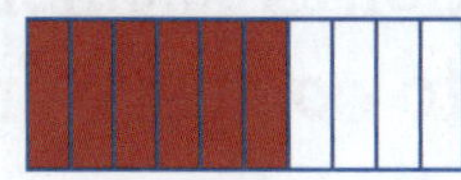

d $\frac{2}{8}$ is equivalent to $\frac{}{4}$

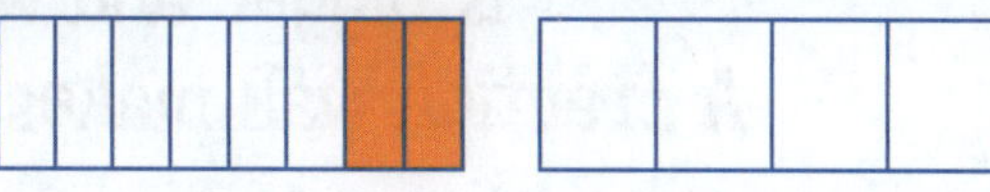

a $\frac{1}{2}$ is equivalent to $\frac{}{4}$

 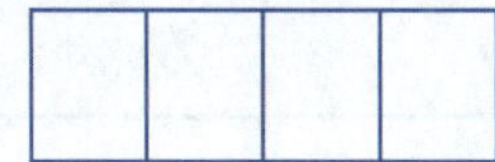

e $\frac{6}{8}$ is equivalent to $\frac{}{4}$

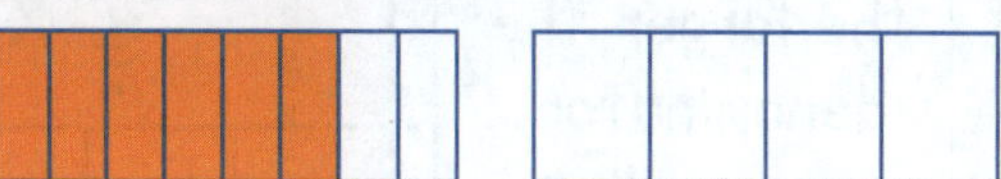

b $\frac{3}{4}$ is equivalent to $\frac{}{8}$

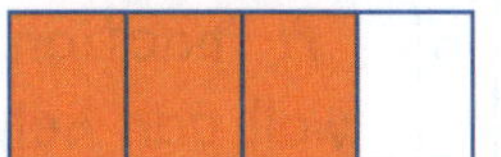 

f $\frac{2}{10}$ is equivalent to $\frac{}{5}$

c $\frac{5}{10}$ is equivalent to $\frac{}{2}$

g $\frac{10}{10}$ is equivalent to $\frac{}{5}$

2 Complete the equivalent fractions.

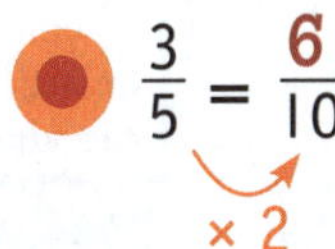

● $\frac{3}{5} = \frac{6}{10}$ (× 2)

a $\frac{3}{4} = \frac{}{8}$ (× 2)

b $\frac{2}{5} = \frac{6}{}$ (× 3)

c $\frac{4}{8} = \frac{}{2}$ (÷ 4)

d $\frac{1}{4} = \frac{}{8}$

e $\frac{1}{5} = \frac{}{10}$

f $\frac{2}{3} = \frac{}{6}$

g $\frac{2}{6} = \frac{}{3}$

h $\frac{3}{4} = \frac{15}{}$

i $\frac{3}{8} = \frac{9}{}$

j $\frac{5}{6} = \frac{}{30}$

k $\frac{2}{8} = \frac{4}{}$

COMPARING FRACTIONS

Comparing fractions means deciding which fraction is bigger and which fraction is smaller.
A fraction wall makes it easier to compare fractions.

When the numerator is 1, the larger the denominator, the smaller the fraction.

1 whole									
$\frac{1}{2}$					$\frac{1}{2}$				
$\frac{1}{4}$	$\frac{1}{4}$	$\frac{1}{4}$	$\frac{1}{4}$						
$\frac{1}{8}$	$\frac{1}{8}$	$\frac{1}{8}$	$\frac{1}{8}$	$\frac{1}{8}$	$\frac{1}{8}$	$\frac{1}{8}$	$\frac{1}{8}$		
$\frac{1}{3}$	$\frac{1}{3}$	$\frac{1}{3}$							
$\frac{1}{5}$	$\frac{1}{5}$	$\frac{1}{5}$	$\frac{1}{5}$	$\frac{1}{5}$					
$\frac{1}{10}$	$\frac{1}{10}$	$\frac{1}{10}$	$\frac{1}{10}$	$\frac{1}{10}$	$\frac{1}{10}$	$\frac{1}{10}$	$\frac{1}{10}$	$\frac{1}{10}$	$\frac{1}{10}$

A fraction wall can help you work out which fractions are equal (equivalent).

Example 1: Two-eighths ($\frac{2}{8}$) is equal to one-quarter ($\frac{1}{4}$).

Example 2: One-fifth ($\frac{1}{5}$) is equal to two-tenths ($\frac{2}{10}$).

Example 3: Three-fifths ($\frac{3}{5}$) is equal to _____-tenths ($\frac{\;}{10}$).

Example 4: Four-eighths ($\frac{4}{8}$) is equal to _____-quarters ($\frac{\;}{4}$).

1 Write a smaller fraction.

● $\frac{1}{2}$ $\underline{\frac{1}{4}}$ **a** $\frac{1}{3}$ ___ **b** $\frac{1}{5}$ ___ **c** $\frac{1}{4}$ ___

2 Write an equivalent (equal) fraction.

● $\frac{4}{10} = \frac{2}{5}$ **a** $\frac{2}{4} = \frac{\;}{2}$ **b** $\frac{1}{4} = \frac{\;}{8}$ **c** $\frac{4}{6} = \frac{\;}{3}$

SELF CHECK Tick how you feel

Got it!	Need help...	I don't get it
  ☐	 ☐	 ☐

Check your answers
How many did you get correct? ☐

CATCH UP MATHS YEAR 5 BOOK A © PASCAL PRESS ISBN: 9781925726169

PRACTICE

1 Complete each fraction. Then number the boxes to order the fractions from smallest (1) to largest (6).

a

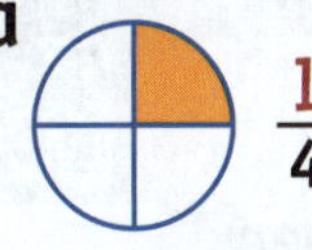
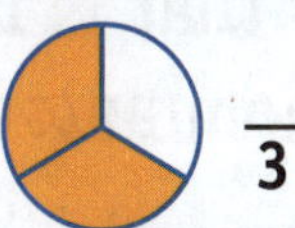

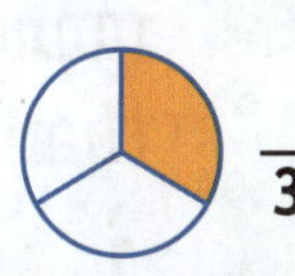

$\frac{1}{4}$ [1] $\frac{\square}{3}$ [ ] $\frac{\square}{5}$ [ ] $\frac{\square}{8}$ [ ] $\frac{3}{4}$ [6] $\frac{\square}{3}$ [ ]

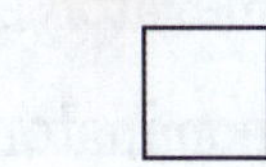

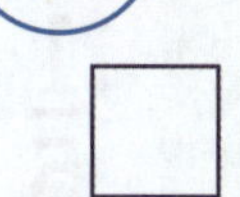

b

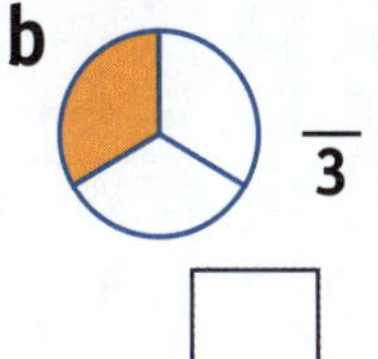
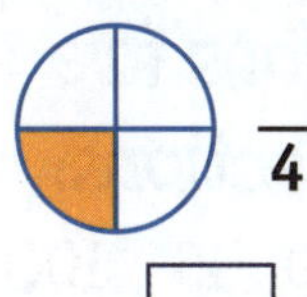
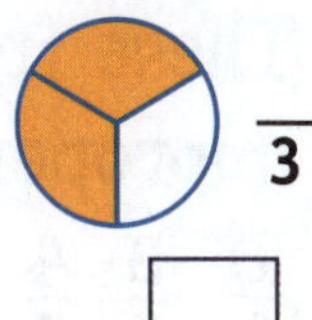
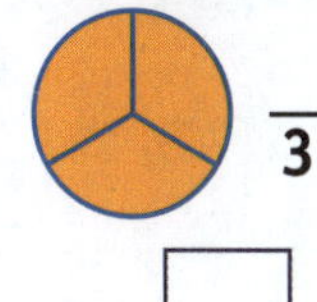
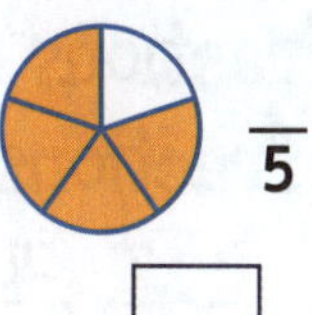
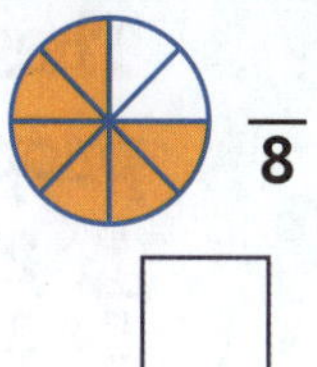

$\frac{\square}{3}$ [ ] $\frac{\square}{4}$ [ ] $\frac{\square}{3}$ [ ] $\frac{\square}{3}$ [ ] $\frac{\square}{5}$ [ ] $\frac{\square}{8}$ [ ]

2 Draw a smaller fraction and then write the fraction in the box.

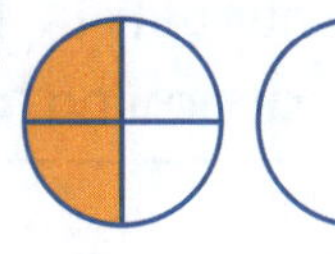

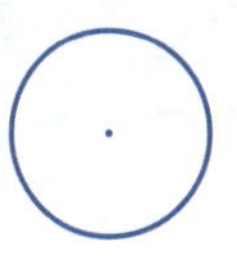

Example: $\frac{1}{3}$ [$\frac{1}{8}$] **a** $\frac{2}{4}$ [ ] **b** $\frac{3}{8}$ [ ] **c** $\frac{4}{5}$ [ ]

3 Draw a larger fraction and then write the fraction in the box.

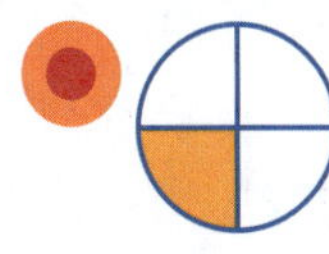

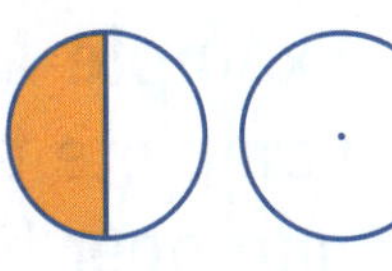
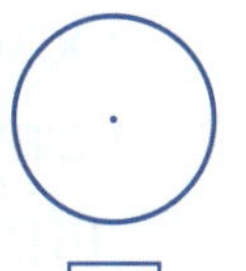
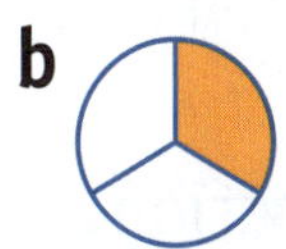
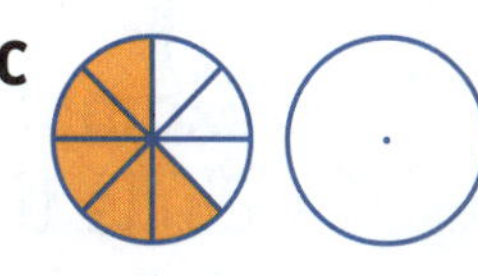

Example: $\frac{1}{4}$ [$\frac{1}{2}$] **a** $\frac{1}{2}$ [ ] **b** $\frac{1}{3}$ [ ] **c** $\frac{5}{8}$ [ ]

4 Write the correct symbol: > (greater than), < (less than) or = (equal to).

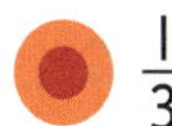

Example: $\frac{1}{3}$ [>] $\frac{1}{4}$	**c** $\frac{7}{8}$ [ ] $\frac{3}{4}$	**f** $\frac{1}{3}$ [ ] $\frac{3}{8}$	**i** $\frac{7}{10}$ [ ] $\frac{1}{8}$
a $\frac{2}{5}$ [ ] $\frac{3}{10}$	**d** $\frac{1}{2}$ [ ] $\frac{6}{8}$	**g** $\frac{10}{10}$ [ ] $\frac{5}{5}$	**j** $\frac{8}{10}$ [ ] $\frac{7}{8}$
b $\frac{4}{10}$ [ ] $\frac{2}{5}$	**e** $\frac{6}{10}$ [ ] $\frac{3}{5}$	**h** $\frac{2}{5}$ [ ] $\frac{5}{10}$	**k** $\frac{2}{3}$ [ ] $\frac{4}{5}$

PROPER & IMPROPER FRACTIONS

A proper fraction has a numerator that is smaller than the denominator.

$\frac{1}{5}$ ← The numerator is smaller. ← The denominator is larger.

These fractions are proper fractions:

$\frac{1}{5}$, $\frac{2}{3}$, $\frac{5}{10}$, $\frac{3}{5}$, $\frac{7}{8}$, $\frac{20}{100}$

An improper fraction has a numerator that is larger than the denominator.

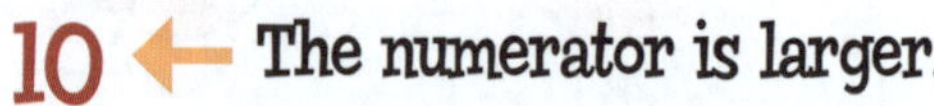

$\frac{10}{8}$ ← The numerator is larger. ← The denominator is smaller.

These fractions are improper fractions:

$\frac{10}{8}$, $\frac{3}{2}$, $\frac{4}{3}$, $\frac{5}{4}$, $\frac{9}{7}$, $\frac{15}{10}$, $\frac{100}{50}$

SCAN to watch video

The top number is the numerator.

The bottom number is the denominator.

Example 1:
Circle the improper fraction.

$\frac{1}{2}$ $\frac{3}{4}$ $\frac{4}{3}$ (circled) $\frac{7}{12}$

Example 3:
Complete to show a proper fraction.

$\frac{2}{\ }$

Example 2:
Circle the proper fraction.

$\frac{2}{1}$ $\frac{8}{4}$ $\frac{9}{3}$ $\frac{11}{40}$ (circled)

Example 4:
Complete to show an improper fraction.

$\frac{\ }{2}$

Improper fractions have a value greater than one.

Use green to circle the proper fractions and red to circle the improper fractions.

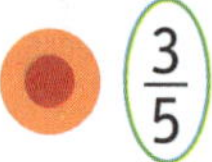

$\frac{3}{5}$ (circled green) $\frac{4}{3}$ (circled red) $\frac{2}{8}$ $\frac{8}{2}$ $\frac{4}{9}$ $\frac{7}{3}$ $\frac{5}{10}$ $\frac{10}{4}$ $\frac{1}{4}$ $\frac{9}{6}$

SELF CHECK Tick how you feel

Got it!	Need help...	I don't get it

Check your answers
How many did you get correct?

CATCH UP MATHS YEAR 5 BOOK A © PASCAL PRESS ISBN: 9781925726169

1 Circle the proper fraction.

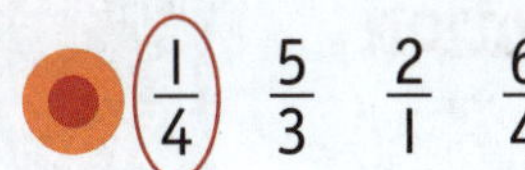

(example) $\frac{1}{4}$ $\frac{5}{3}$ $\frac{2}{1}$ $\frac{6}{4}$

a $\frac{5}{4}$ $\frac{3}{2}$ $\frac{1}{2}$ $\frac{4}{2}$

b $\frac{2}{3}$ $\frac{4}{2}$ $\frac{10}{5}$ $\frac{8}{2}$

c $\frac{7}{3}$ $\frac{2}{8}$ $\frac{16}{3}$ $\frac{14}{2}$

d $\frac{21}{3}$ $\frac{16}{4}$ $\frac{12}{24}$ $\frac{13}{4}$

e $\frac{53}{60}$ $\frac{73}{30}$ $\frac{28}{10}$ $\frac{47}{20}$

2 Cross out the improper fraction.

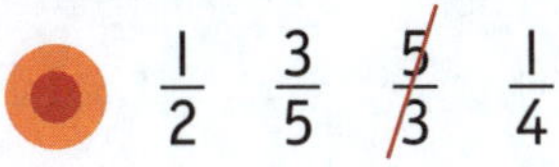

(example) $\frac{1}{2}$ $\frac{3}{5}$ $\frac{5}{3}$ $\frac{1}{4}$

a $\frac{2}{4}$ $\frac{3}{6}$ $\frac{8}{5}$ $\frac{5}{8}$

b $\frac{1}{3}$ $\frac{3}{1}$ $\frac{2}{4}$ $\frac{10}{20}$

c $\frac{16}{20}$ $\frac{5}{3}$ $\frac{20}{30}$ $\frac{3}{5}$

d $\frac{16}{20}$ $\frac{14}{18}$ $\frac{81}{41}$ $\frac{24}{30}$

e $\frac{27}{50}$ $\frac{53}{24}$ $\frac{24}{35}$ $\frac{29}{51}$

3 Write three proper fractions that match the description.

a 5 is the denominator

$\frac{2}{5}$

c 10 is the denominator

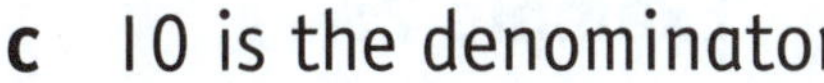

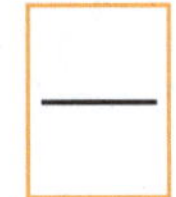 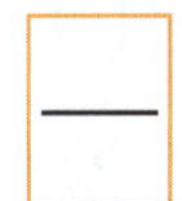 

b 4 is the numerator

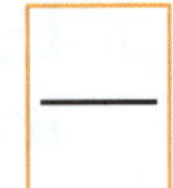  

d 3 is the denominator

 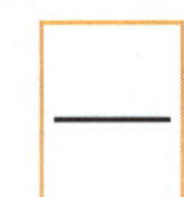 

4 Write three improper fractions that match the description.

a 7 is the numerator

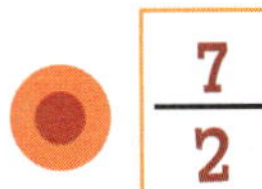

$\frac{7}{2}$

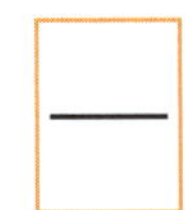 

d 5 is the denominator

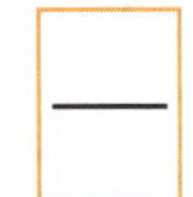  

b 3 is the denominator

e 6 is the numerator

 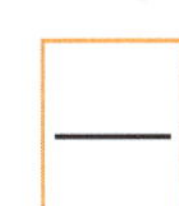 

c 8 is the numerator

f 12 is the numerator

MIXED NUMBERS

SCAN to watch video

A mixed number has a whole number and a proper fraction written together.

Whole number → $3\frac{1}{5}$ ← Proper fraction

↑ Mixed number

Example 1:

$3\frac{1}{5} = 3 + \frac{1}{5}$

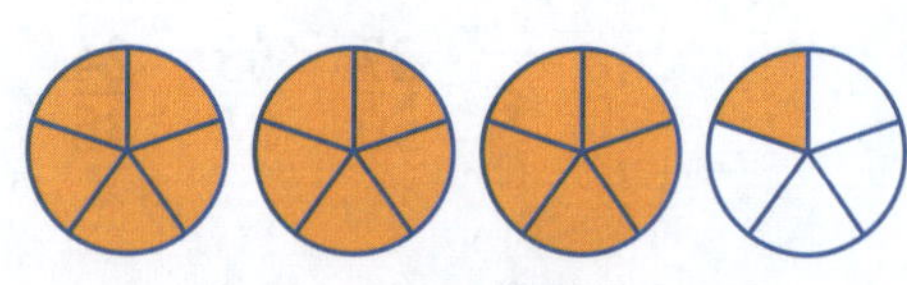

Example 3:

$\_\_\frac{}{} = \_\_ + \frac{}{}$

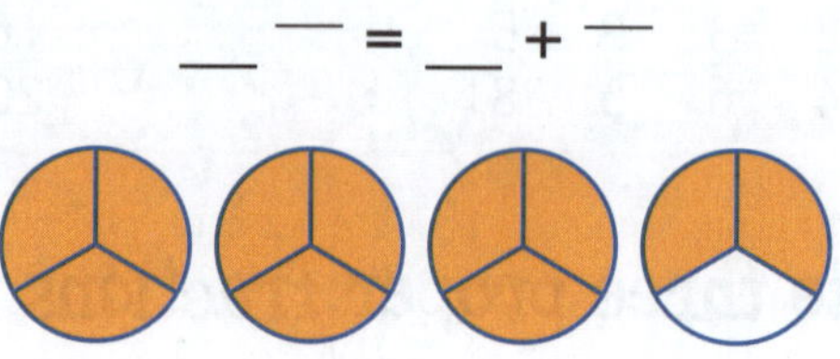

Example 2:

$1\frac{4}{6} = 1 + \frac{4}{6}$

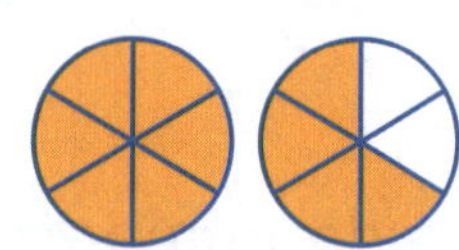

Example 4:

$2\frac{}{} = \_\_ + \frac{}{}$

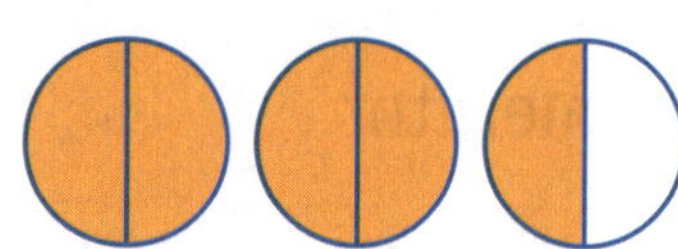

Check your answer on the video!

Your turn

Colour to show the mixed number.

● $2\frac{1}{4}$

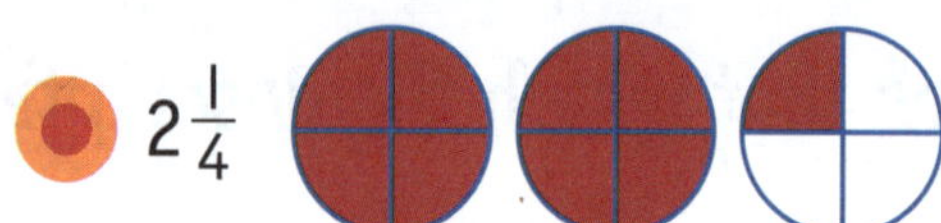

a $3\frac{2}{3}$

b $6\frac{1}{2}$

c $1\frac{5}{10}$

SELF CHECK Tick how you feel		
Got it! ☐	Need help... ☐	I don't get it ☐

Check your answers

How many did you get correct? ☐

CATCH UP MATHS YEAR 5 BOOK A © PASCAL PRESS ISBN: 9781925726169

PRACTICE

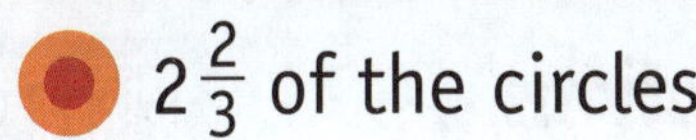

1 Colour shapes to show the mixed numbers.

● $2\frac{2}{3}$ of the circles

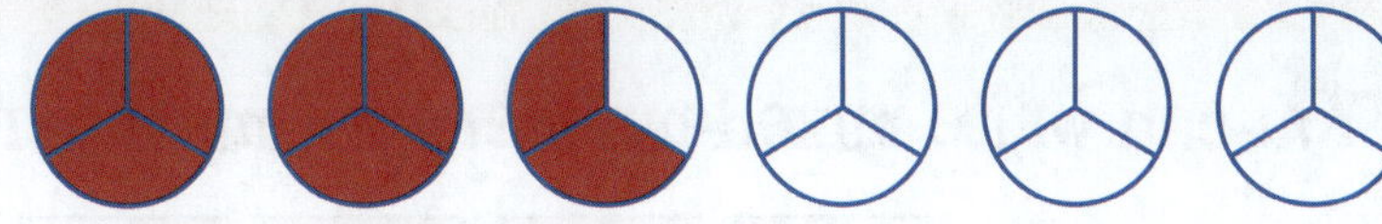

a $4\frac{3}{4}$ of the squares

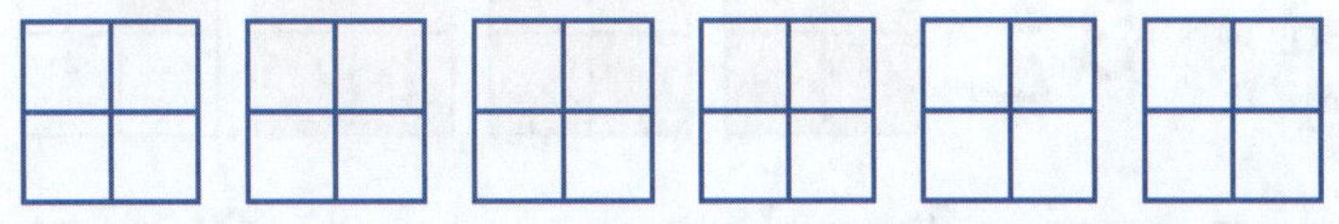

b $5\frac{1}{2}$ of the triangles

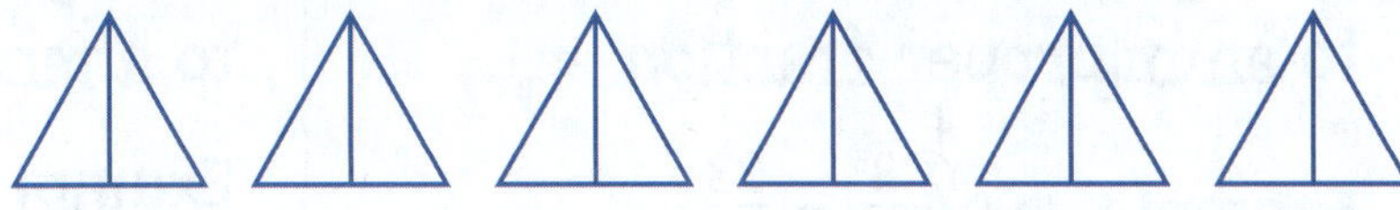

c $3\frac{4}{5}$ of the rectangles

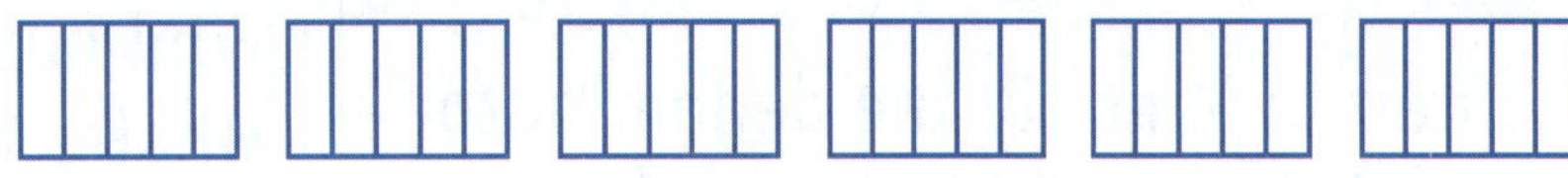

2 Write the mixed numbers to show how many shapes are coloured orange.

● $2\frac{4}{6}$ of the hexagons are orange.

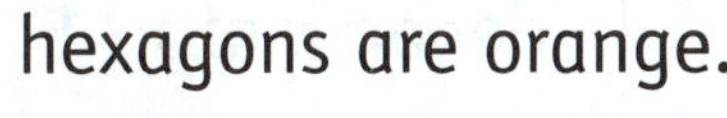

    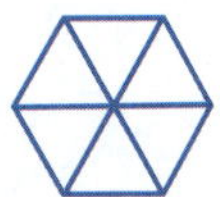 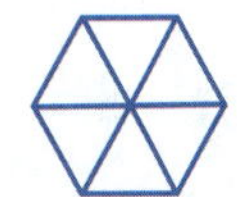

a ☐ of the pentagons are orange.

b ☐ of the octagons are orange.

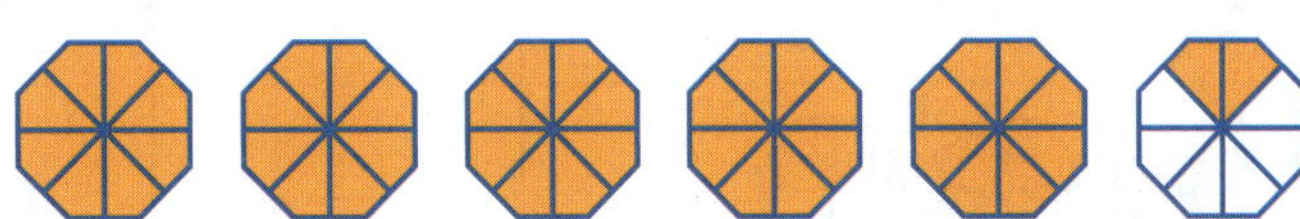

c ☐ of the triangles are orange.

 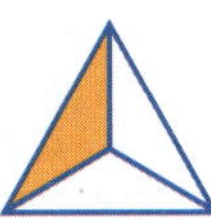 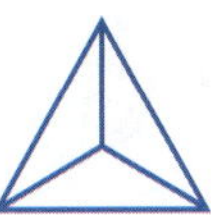 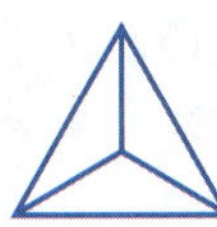 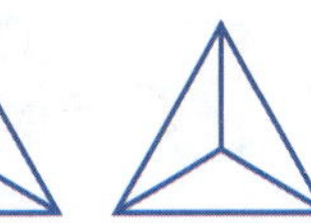 

  ISBN: 9781925726169

IMPROPER FRACTIONS AND MIXED NUMBERS

You can write mixed numbers as improper fractions.

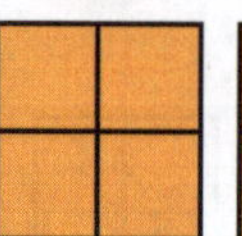

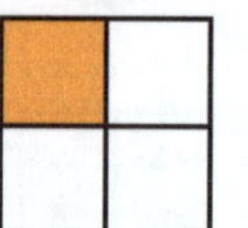

Change a mixed number to an improper fraction

Example 1: $4\frac{3}{5} = \frac{23}{5}$ (+, ×)

Step 1: Multiply the denominator (5) by the whole number (4): $5 \times 4 = 20$

Step 2: Add the answer to the numerator (3): $20 + 3 = 23$. Keep the same denominator, 5.

Example 2: $3\frac{2}{3} = \frac{}{3}$

$3 \times 3 + 2 =$ ___

Change an improper fraction to a mixed number

Example 3: $\div\left(\frac{23}{5}\right) = 4\frac{3}{5}$

Find out how many wholes there are in $\frac{23}{5}$. How many times can 5 go into 23?

$23 \div 5 = 4$ remainder 3

Write the 3 remainder as the numerator. Keep the same denominator, 5.

Example 4: $\div\left(\frac{29}{3}\right) =$ ___ $\frac{}{3}$

$29 \div 3 =$ ___ remainder ___

Check your answer on the video!

Your turn

Complete the table.

	Diagram	Improper Fraction	Mixed Number
●	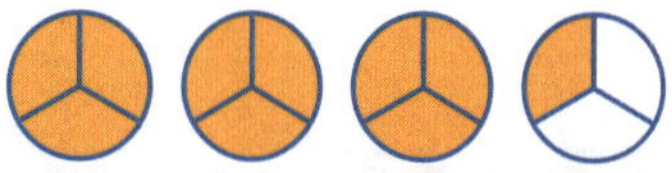 (4 circles in thirds, 10 shaded)	$\frac{10}{3}$	$3\frac{1}{3}$
a	 (5 circles in quarters)	$\frac{}{4}$	___ $\frac{}{}$
b	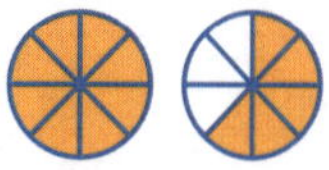 (2 circles in eighths)	$\frac{}{8}$	___ $\frac{}{}$
c	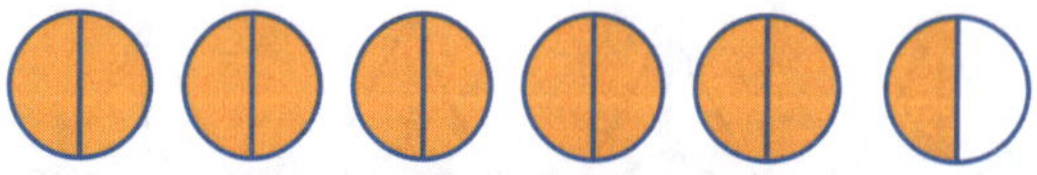 (6 circles in halves)	$\frac{}{2}$	___ $\frac{}{}$

SELF CHECK Tick how you feel

Got it! ☐ Need help... ☐ I don't get it ☐

Check your answers

How many did you get correct? ☐

CATCH UP MATHS YEAR 5 BOOK A © PASCAL PRESS ISBN: 9781925726169

PRACTICE

1 Change these mixed numbers to improper fractions

Example: $2\frac{3}{4} = \frac{11}{4}$

a $3\frac{1}{2} =$ ___	**d** $1\frac{3}{4} =$ ___	**g** $4\frac{1}{2} =$ ___	**j** $2\frac{7}{8} =$ ___
b $2\frac{4}{6} =$ ___	**e** $5\frac{2}{3} =$ ___	**h** $2\frac{2}{5} =$ ___	**k** $1\frac{6}{8} =$ ___
c $3\frac{2}{3} =$ ___	**f** $6\frac{1}{2} =$ ___	**i** $8\frac{1}{4} =$ ___	

2 Change these improper fractions to mixed numbers.

Example: $\div\frac{5}{4} = 1\frac{1}{4}$

a $\frac{10}{3} =$ ___	**d** $\frac{9}{2} =$ ___	**g** $\frac{21}{4} =$ ___	**j** $\frac{25}{3} =$ ___
b $\frac{7}{5} =$ ___	**e** $\frac{6}{4} =$ ___	**h** $\frac{32}{3} =$ ___	**k** $\frac{43}{7} =$ ___
c $\frac{5}{3} =$ ___	**f** $\frac{12}{10} =$ ___	**i** $\frac{14}{3} =$ ___	

3 Circle the mixed number that matches the improper fraction.

	Fraction	Options
Example	$\frac{12}{5}$	$2\frac{3}{5}$ ($2\frac{2}{5}$ circled) $1\frac{1}{5}$
a	$\frac{25}{7}$	$3\frac{4}{7}$ $3\frac{7}{4}$ $1\frac{25}{7}$
b	$\frac{50}{9}$	$5\frac{5}{9}$ $5\frac{4}{5}$ $5\frac{1}{9}$
c	$\frac{35}{10}$	$3\frac{1}{10}$ $3\frac{5}{10}$ $3\frac{4}{10}$
d	$\frac{67}{8}$	$8\frac{3}{8}$ $8\frac{7}{8}$ $9\frac{3}{8}$
e	$\frac{45}{6}$	$7\frac{3}{6}$ $5\frac{3}{6}$ $6\frac{5}{6}$

4 Circle the improper fraction that matches the mixed number.

	Mixed number	Options
Example	$2\frac{3}{4}$	$\frac{15}{4}$ ($\frac{11}{4}$ circled) $\frac{12}{4}$
a	$3\frac{3}{5}$	$\frac{18}{5}$ $\frac{17}{5}$ $\frac{19}{5}$
b	$4\frac{7}{8}$	$\frac{36}{8}$ $\frac{37}{8}$ $\frac{39}{8}$
c	$6\frac{2}{3}$	$\frac{21}{3}$ $\frac{20}{3}$ $\frac{22}{3}$
d	$1\frac{3}{9}$	$\frac{12}{9}$ $\frac{15}{9}$ $\frac{13}{9}$
e	$5\frac{4}{7}$	$\frac{39}{7}$ $\frac{40}{7}$ $\frac{35}{7}$
f	$7\frac{1}{5}$	$\frac{30}{5}$ $\frac{35}{5}$ $\frac{36}{5}$
g	$2\frac{5}{8}$	$\frac{25}{8}$ $\frac{21}{8}$ $\frac{19}{8}$

ADD AND SUBTRACT FRACTIONS WITH THE SAME DENOMINATOR

Look at the denominators. If the denominators are the same, you can add or subtract the numerators to find the answer.

SCAN to watch video

Adding fractions

Example 1: $\frac{2}{4} + \frac{1}{4}$

First look at the denominators. The denominators are both 4.

You can add the numerators: $2 + 1 = 3$

$$\frac{2}{4} + \frac{1}{4} = \frac{3}{4}$$

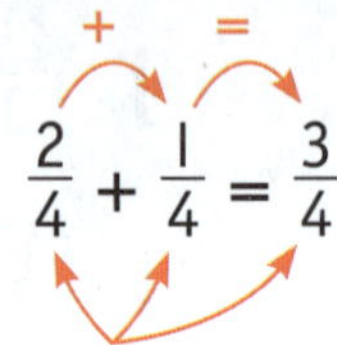

The denominators are the same.

Example 2: $\frac{3}{5} + \frac{1}{5} = \frac{}{5}$

Example 3: $\frac{2}{8} + \frac{3}{8} = \frac{}{8}$

Subtracting fractions

Example 4: $\frac{5}{12} - \frac{2}{12}$

First look at the denominators. The denominators are both 12.

You can subtract the numerators: $5 - 2 = 3$

$$\frac{5}{12} - \frac{2}{12} = \frac{3}{12}$$

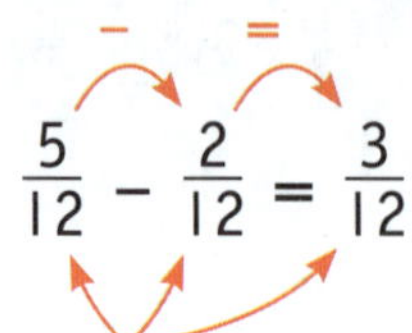

The denominators are the same.

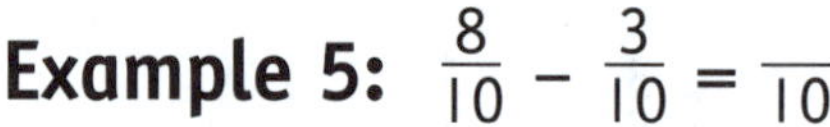

Example 5: $\frac{8}{10} - \frac{3}{10} = \frac{}{10}$

Example 6: $\frac{4}{5} - \frac{2}{5} = \frac{}{5}$

Check your answer on the video!

Solve these additions and subtractions.

$\frac{7}{10} - \frac{3}{10} = \frac{4}{10}$

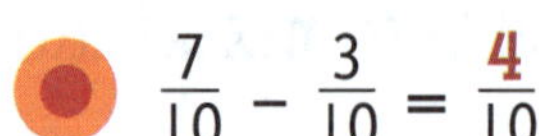

a $\frac{5}{8} - \frac{2}{8} = \frac{}{8}$

b $\frac{5}{6} - \frac{1}{6} = \frac{}{6}$

c $\frac{3}{5} + \frac{1}{5} = \frac{}{5}$

d $\frac{3}{8} + \frac{4}{8} = \frac{}{8}$

e $\frac{2}{4} + \frac{2}{4} = \frac{}{4}$

SELF CHECK Tick how you feel

Got it!	Need help...	I don't get it
 ☐	 ☐	 ☐

Check your answers

How many did you get correct? ☐

CATCH UP MATHS YEAR 5 BOOK A © PASCAL PRESS ISBN: 9781925726169

PRACTICE

1 Solve these additions.

● $\frac{1}{4} + \frac{1}{4} = \frac{2}{4}$

a $\frac{1}{3} + \frac{1}{3} =$ —

b $\frac{4}{8} + \frac{1}{8} =$ —

c $\frac{3}{8} + \frac{2}{8} =$ —

d $\frac{3}{10} + \frac{2}{10} =$ —

e $\frac{1}{2} + \frac{1}{2} =$ —

f $\frac{7}{10} + \frac{1}{10} =$ —

g $\frac{2}{5} + \frac{2}{5} =$ —

h $\frac{3}{6} + \frac{1}{6} =$ —

i $\frac{2}{10} + \frac{7}{10} =$ —

j $\frac{5}{8} + \frac{2}{8} =$ —

k $\frac{6}{10} + \frac{2}{10} =$ —

2 Solve these subtraction problems.

● $\frac{5}{12} - \frac{1}{12} = \frac{4}{12}$

a $\frac{7}{10} - \frac{3}{10} =$ —

b $\frac{7}{8} - \frac{4}{8} =$ —

c $\frac{5}{8} - \frac{1}{8} =$ —

d $\frac{7}{10} - \frac{2}{10} =$ —

e $\frac{2}{3} - \frac{1}{3} =$ —

f $\frac{2}{4} - \frac{1}{4} =$ —

g $\frac{3}{8} - \frac{1}{8} =$ —

h $\frac{4}{5} - \frac{2}{5} =$ —

i $\frac{9}{10} - \frac{3}{10} =$ —

j $\frac{6}{8} - \frac{2}{8} =$ —

k $\frac{11}{12} - \frac{7}{12} =$ —

3 Match the answers.

● $\frac{5}{8} - \frac{2}{8}$	$= \frac{3}{4}$
a $\frac{6}{10} + \frac{1}{10}$	$= \frac{7}{8}$
b $\frac{5}{8} + \frac{2}{8}$	$= \frac{3}{8}$
c $\frac{3}{4} - \frac{1}{4}$	$= \frac{1}{10}$
d $\frac{9}{10} - \frac{8}{10}$	$= \frac{2}{4}$
e $\frac{1}{4} + \frac{2}{4}$	$= \frac{7}{10}$

FRACTIONS REVIEW

1 Colour the numerator red, the vinculum green and the denominator blue.

a $\frac{2}{3}$ b $\frac{7}{8}$ c $\frac{4}{5}$ d $\frac{1}{2}$ e $\frac{10}{12}$ f $\frac{35}{50}$

2 Write the fractions in words.

a $\frac{3}{5}$ ______________________ d $\frac{7}{10}$ ______________________

b $\frac{2}{8}$ ______________________ e $\frac{3}{4}$ ______________________

c $\frac{1}{3}$ ______________________ f $\frac{1}{2}$ ______________________

3 Cut each shape into halves.

a  b c  d e

4 Circle $\frac{1}{2}$ of each collection and fill in the missing numbers.

a Half of ___ = ___

b Half of ___ = ___

c  Half of ___ = ___

5 Fill in the missing numbers.

a 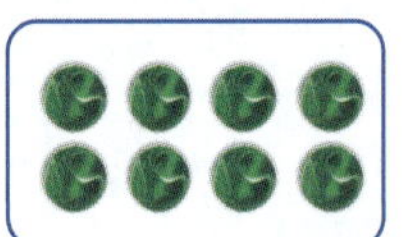

Each group has ___ of the ___ marbles. Half of ___ is ___.

b 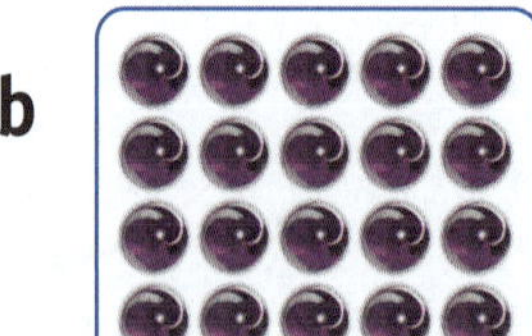 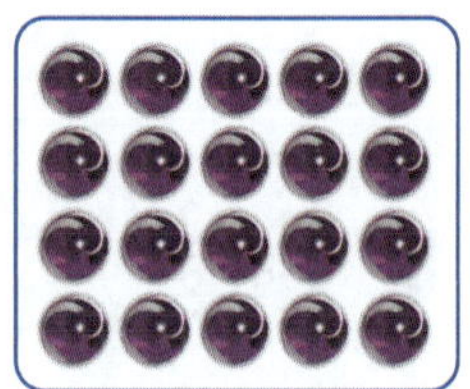

Each group has ___ of the ___ marbles. Half of ___ is ___.

6 Add these fractions.

a $\frac{1}{4} + \frac{2}{4} = $ —

b $\frac{3}{5} + \frac{1}{5} = $ —

c $\frac{3}{8} + \frac{2}{8} = $ —

d $\frac{2}{6} + \frac{1}{6} = $ —

CATCH UP MATHS YEAR 5 BOOK A © PASCAL PRESS ISBN: 9781925726169

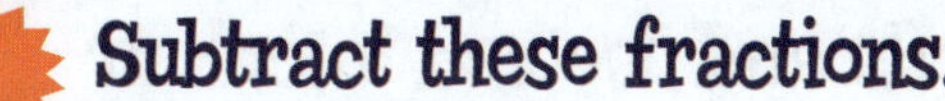

7 Subtract these fractions.

a $\frac{7}{8} - \frac{2}{8} = \_$ b $\frac{3}{5} - \frac{2}{5} = \_$ c $\frac{6}{8} - \frac{1}{8} = \_$ d $\frac{9}{10} - \frac{4}{10} = \_$

8 Cut into quarters and then colour three-quarters.

a 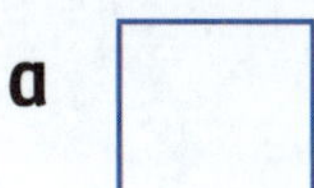 b c d

9 What is one-quarter of each group? Circle, then answer.

a $\frac{1}{4}$ of ___ = ___

b $\frac{1}{4}$ of ___ = ___

c 

$\frac{1}{4}$ of ___ = ___

10 Cut into eighths and then colour two-eighths.

a b c d

11 What is one-eighth of each group? Circle, then answer.

a $\frac{1}{8}$ of ___ = ___

b $\frac{1}{8}$ of ___ = ___

c 

$\frac{1}{8}$ of ___ = ___

12 What is one-quarter of each number?

a 24 ___	c 4 ___	e 44 ___	g 28 ___
b 36 ___	d 16 ___	f 12 ___	h 40 ___

13 What is one-eighth of each number?

a 16 ___	c 56 ___	e 72 ___	g 80 ___
b 32 ___	d 64 ___	f 96 ___	h 48 ___

REVIEW

14 Use blue to colour the shapes that show thirds and use red to colour the shapes that show fifths.

a b c 

d e 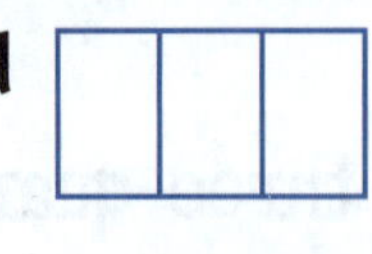

f g 

h 

15 What is one-third of each group? Circle, then answer.

a $\frac{1}{3}$ of ___ = ___

b 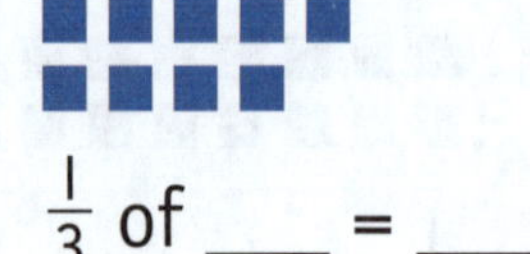

$\frac{1}{3}$ of ___ = ___

c 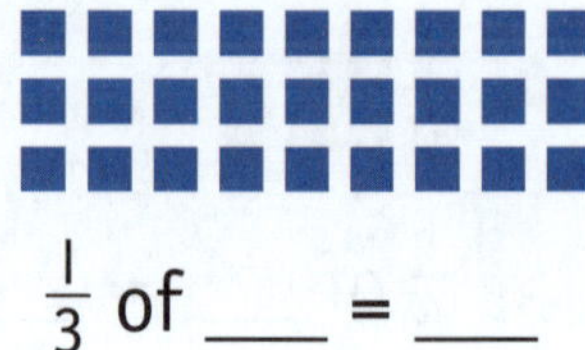

$\frac{1}{3}$ of ___ = ___

16 Write the answer.

a $\frac{1}{3}$ of 24 = ___

b $\frac{1}{3}$ of 30 = ___

c $\frac{1}{3}$ of 15 = ___

17 What is one-fifth of each group? Circle, then answer.

a 

$\frac{1}{5}$ of ___ = ___

b 

$\frac{1}{5}$ of ___ = ___

c 

$\frac{1}{5}$ of ___ = ___

18 Write the answer.

a $\frac{1}{5}$ of 15 = ___

b $\frac{1}{5}$ of 35 = ___

c $\frac{1}{5}$ of 40 = ___

19 Write the equivalent fractions.

a $\frac{3}{4} = \frac{}{8}$

b $\frac{1}{2} = \frac{}{10}$

c $\frac{1}{4} = \frac{}{16}$

d $\frac{4}{10} = \frac{}{5}$

e $\frac{2}{8} = \frac{}{4}$

f $\frac{3}{5} = \frac{}{10}$

g $\frac{2}{10} = \frac{}{5}$

h $\frac{4}{8} = \frac{}{2}$

i $\frac{2}{5} = \frac{}{10}$

j $\frac{2}{4} = \frac{}{2}$

k $\frac{4}{4} = \frac{}{8}$

l $\frac{1}{5} = \frac{}{10}$

m $\frac{4}{5} = \frac{}{10}$

n $\frac{1}{2} = \frac{}{8}$

o $\frac{1}{2} = \frac{}{10}$

CATCH UP MATHS YEAR 5 BOOK A © PASCAL PRESS ISBN: 9781925726169

20 Circle the improper fraction.

a $\frac{2}{3}$ $\frac{3}{2}$ $\frac{1}{3}$

b $\frac{1}{4}$ $\frac{1}{2}$ $\frac{4}{2}$

c $\frac{1}{3}$ $\frac{2}{3}$ $\frac{4}{3}$

d $\frac{1}{5}$ $\frac{5}{1}$ $\frac{5}{10}$

e $\frac{1}{2}$ $\frac{2}{3}$ $\frac{3}{2}$

f $\frac{3}{8}$ $\frac{5}{4}$ $\frac{8}{10}$

g $\frac{22}{26}$ $\frac{45}{30}$ $\frac{32}{50}$

h $\frac{15}{30}$ $\frac{51}{30}$ $\frac{20}{25}$

21 Colour the circles to show the mixed numbers.

a $2\frac{1}{2}$

b $3\frac{1}{3}$

c $4\frac{3}{4}$

d $1\frac{5}{8}$

22 Write the mixed number.

a 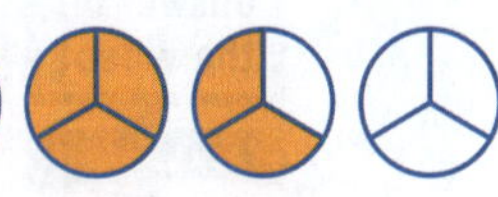 _____ of the circles are coloured.

b _____ of the circles are coloured.

c _____ of the circles are coloured.

d 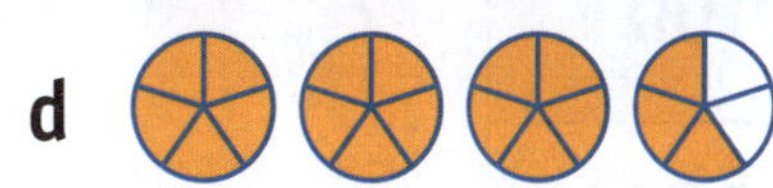 _____ of the circles are coloured.

23 Change these mixed numbers into improper fractions.

a $3\frac{1}{3} = —$

b $2\frac{3}{6} = —$

c $1\frac{5}{8} = —$

d $4\frac{1}{3} = —$

e $10\frac{1}{2} = —$

f $7\frac{3}{8} = —$

g $2\frac{4}{7} = —$

h $2\frac{1}{4} = —$

24 Change these improper fractions to mixed numbers.

a $\frac{10}{4}$ = ___ —

b $\frac{15}{7}$ = ___ —

c $\frac{21}{5}$ = ___ —

d $\frac{32}{10}$ = ___ —

e $\frac{64}{9}$ = ___ —

f $\frac{32}{3}$ = ___ —

g $\frac{49}{5}$ = ___ —

h $\frac{17}{2}$ = ___ —

WRITING DECIMALS

You can write fractions in decimal form.

The 0 in the ones position means there are no whole numbers.

Example 1: 28 out of 100 = $\frac{28}{100}$ = 0.28

two-tenths; eight-hundredths

The 1 in the ones position means there is 1 whole number.

Example 2: 143 out of 100 = $\frac{143}{100}$ = 1.43

four-tenths; three-hundredths

Example 3: 62 out of 100 = $\frac{}{100}$ = ___.______

Example 4: 259 out of 100 = $\frac{}{100}$ = ___.______

Check your answer on the video!

Complete the table.

	Fraction	Decimal fraction	Out of 100
●	$\frac{78}{100}$	0.78	78 out of 100
a		0.32	___ out of ___
b	$\frac{57}{100}$		___ out of ___
c		0.93	___ out of ___
d			22 out of 100
e	$\frac{138}{100}$		___ out of ___
f			365 out of 100

SELF CHECK Tick how you feel

Got it!	Need help...	I don't get it
☐	☐	☐

Check your answers

How many did you get correct? ☐

CATCH UP MATHS YEAR 5 BOOK A © PASCAL PRESS ISBN: 9781925726169

1 **Use blue to circle the whole numbers and red to circle the decimal fractions.**

● 24.63   b 20.34   d 64.39   f 1.34   h 142.53

a 124.35   c 82.10   e 913.26   g 52.45   i 106.50

2 **Write the numbers in the place value table below.**

● 9 ones, 33 hundredths

a 1 ten, 3 ones, 78 hundredths

b 4 tens, 6 ones, 24 hundredths

c 5 tens, 1 one, 3 hundredths

d 3 tens, 8 hundredths

	Tens	Ones	.	Tenths	Hundredths
●		9	.	3	3
a			.		
b			.		
c			.		
d			.		

3 **Write these decimals in place value words.**

● 7.65 7 ones + 6 tenths + 5 hundredths

a 8.23 ______________________

b 4.05 ______________________

c 0.76 ______________________

4 **Write the decimal fractions in ascending order.**

● 2.35, 4.73, 1.29, 5.63, 1.07 1.07, 1.29, 2.35, 4.73, 5.63

a 2.17, 7.12, 2.71, 1.27, 7.21 ______________________

b 3.03, 3.30, 0.33, 0.30, 3.31 ______________________

c 8.15, 1.85, 1.8, 8.51, 5.18 ______________________

d 6.37, 7.63, 6.73, 3.76, 3.67 ______________________

RELATING TENTHS TO HUNDREDTHS

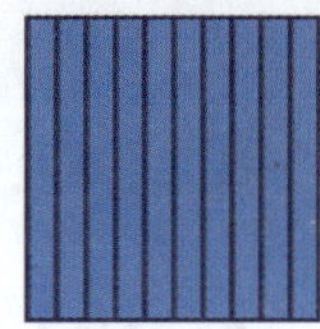

This square is cut into tenths. It has 10 equal parts.

Each tenth is $0.10 = \frac{1}{10}$.

This square is cut into hundredths.

It has 100 equal parts.

Each hundredth is $\frac{1}{100} = 0.01$.

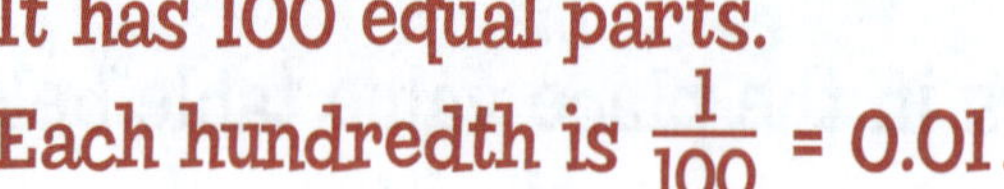

SCAN to watch video

Example 1:

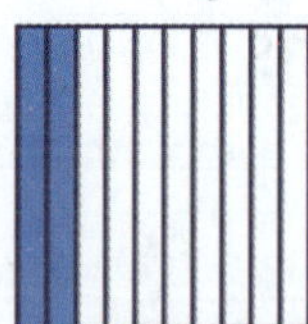

This square shows 2 tenths.

$\frac{2}{10} = 0.20$

Example 2:

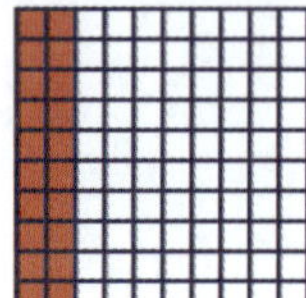

This square shows 20 hundredths.

$\frac{20}{100} = 0.20$

Example 3:

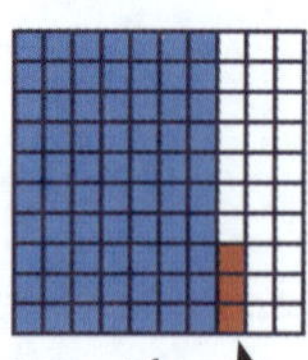

This square shows 73 hundredths.

$\frac{73}{100} = 0.73$

7 tenths

3 hundredths

Example 4:

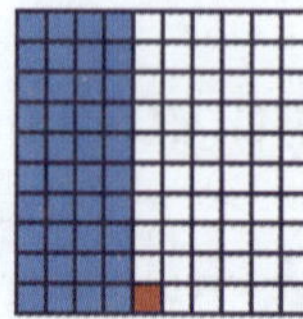

This square shows 41 hundredths.

$\frac{41}{100} = 0.41$

Example 5:

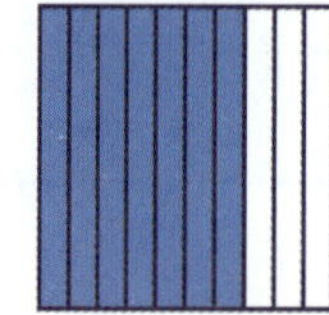

This square shows ___ tenths.

$\frac{}{10} = 0.$___

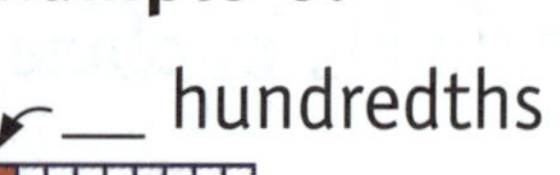

Check your answer on the video!

Example 6:

___ hundredths

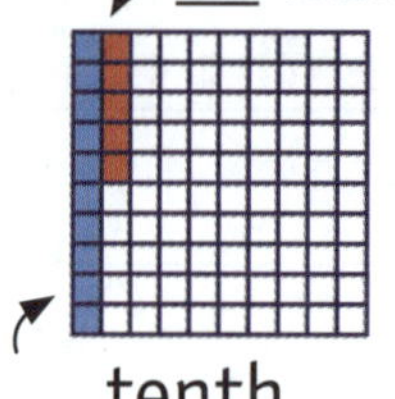

___ tenth

This square shows ___ hundredths.

$\frac{}{100} = 0.$___

Your turn

What fraction is coloured?

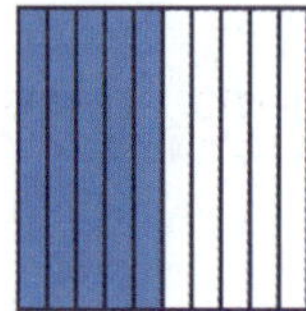

$\frac{5}{10} = 0.50$

a

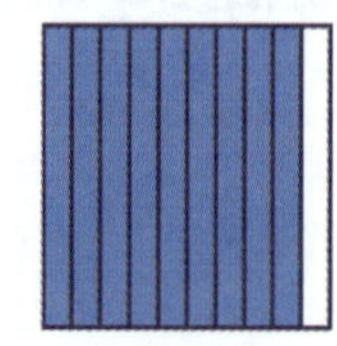

$\frac{}{10} =$ ______

b

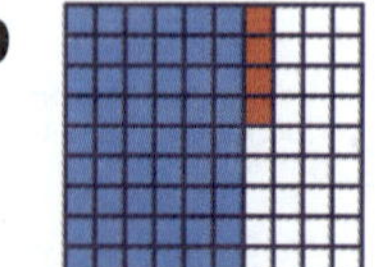

$\frac{}{100} =$ ______

c

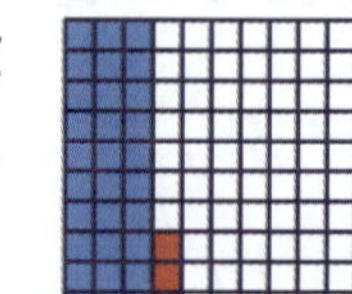

$\frac{}{100} =$ ______

SELF CHECK Tick how you feel

Got it!	Need help...	I don't get it

Check your answers

How many did you get correct?

CATCH UP MATHS YEAR 5 BOOK A © PASCAL PRESS ISBN: 9781925726169

PRACTICE

1 Show the decimal fractions.
Use blue for tenths and red for hundreds.

● 0.63 | a 0.40 | b 0.98 | c 0.03

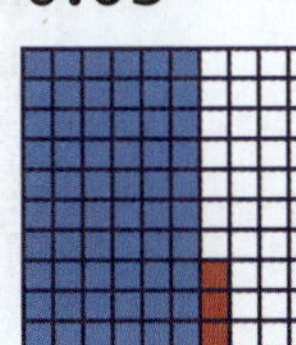
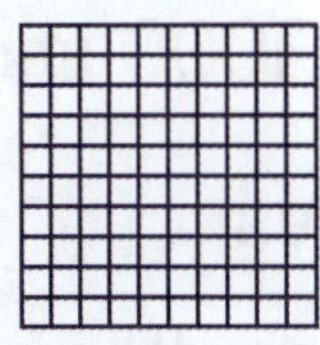
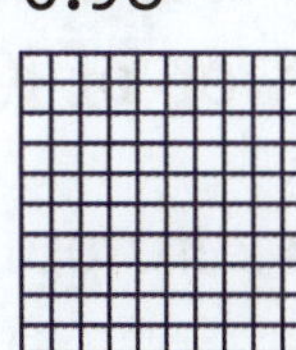
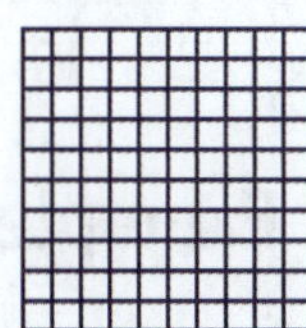

2 Write the decimal fraction.

● 0.24 | a ______ | b ______ | c ______

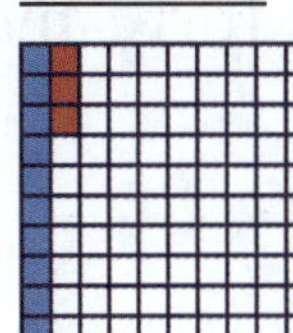
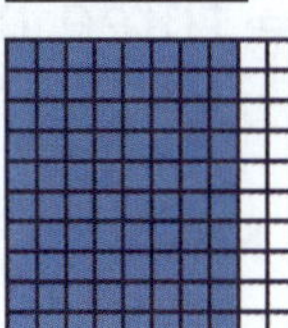
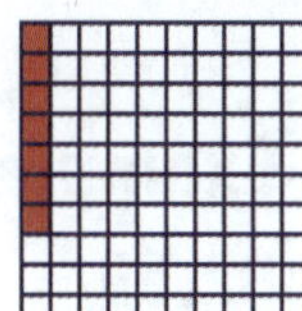

3 Complete the table.

	Words	Diagram	Fraction	Decimal
●	seventy-five hundredths		$\frac{75}{100}$	0.75
a			$\frac{60}{100}$	
b				0.42
c	ninety-three hundredths			
d				0.01
e	eighteen hundredths			

DECIMAL FRACTIONS AND FRACTIONS IN WORDS

A decimal fraction is a fraction with a denominator that is 10, or a power of 10 like 100, 1000 and 10 000.

Example 1: 0.27 = zero point two seven

$= \frac{27}{100}$ = twenty-seven hundredths

Example 2: 3.45 = three point four five

$= \frac{345}{100}$ = three hundred and forty-five hundredths

Example 3:

82.69 = ____________-_____ point _____ _____

$= \frac{\quad}{100}$ = _______ thousand, _____ hundred and _______-_____ hundredths

Example 4:

0.99 = _____ point ______ ______

$= \frac{\quad}{100}$ = ____________-______ hundredths

Your turn

Complete the table.

	Fraction	Fraction in words	Decimal fraction	Decimal in words
●	$\frac{92}{100}$	ninety-two hundredths	0.92	zero point nine two
a	$\frac{84}{100}$			
b			0.36	
c		one hundred and thirty-nine hundredths		

SELF CHECK Tick how you feel

Got it! ☐ Need help... ☐ I don't get it ☐

Check your answers

How many did you get correct? ☐

CATCH UP MATHS YEAR 5 BOOK A © PASCAL PRESS ISBN: 9781925726169

PRACTICE

1 Write these decimal fractions in words.

- 3.05 three point zero five
- a 2.94 ____________________
- b 0.82 ____________________
- c 8.93 ____________________
- d 1.37 ____________________
- e 2.40 ____________________
- f 10.73 ____________________
- g 20.02 ____________________
- h 0.16 ____________________
- i 29.06 ____________________
- j 43.44 ____________________

2 Write as fractions.

- twelve out of 100 $\frac{12}{100}$
- a six out of 100 ____
- b twenty-three out of 100 ____
- c forty-two out of 100 ____
- d eighty-nine out of 100 ____
- e one hundred and two out of 100 ____
- f seven hundred and twelve out of 100 ____
- g nine hundred and sixty-three out of 100 ____
- h eight hundred and ninety out of 100 ____
- i seven hundred and one out of 100 ____
- j five hundred and forty-seven out of 100 ____
- k eight hundred out of 100 ____
- l three hundred and twenty out of 100 ____

THOUSANDTHS

The third decimal place is thousandths.
There are 1000 thousandths in 1 whole.

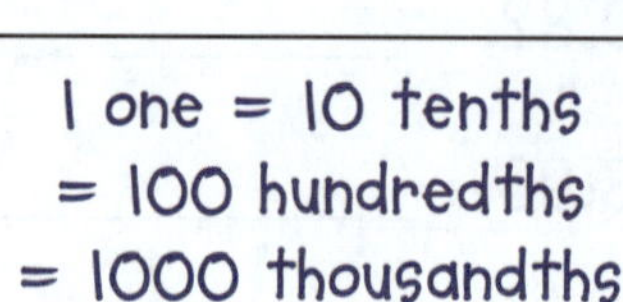

Example 1:

2.451 has:
- 2 ones
- 4 tenths
- 5 hundredths
- 1 thousandth.

Example 2:

6.873 has:
- 6 ones
- 8 tenths
- 7 hundredths
- 3 thousandths.

Example 3:

8.296 has:
- __ ones
- __ tenths
- __ hundredths
- __ thousandths.

Example 4:

47.609 has:
- __ tens
- __ ones
- __ tenths
- __ hundredths
- __ thousandths.

Your turn

Write the missing numbers.

- ● 6.924   6 ones, 9 tenths, 2 hundredths, 4 thousandths
- **a** 7.382   __ ones, __ tenths, __ hundredths, __ thousandths
- **b** 5.410   __ ones, __ tenths, __ hundredth, __ thousandths
- **c** 9.734   __ ones, __ tenths, __ hundredths, __ thousandths
- **d** 8.532   __ ones, __ tenths, __ hundredths, __ thousandths
- **e** 6.593   __ ones, __ tenths, __ hundredths, __ thousandths
- **f** 7.113   __ ones, __ tenth, __ hundredth, __ thousandths

Check your answers
How many did you get correct?

CATCH UP MATHS YEAR 5 BOOK A © PASCAL PRESS ISBN: 9781925726169

1 Write the digit in each place in 6.842.

- tenths place 8
- **a** ones place ___
- **b** thousandths place ___
- **c** hundredths place ___

2 Write the digit in each place in 3.079.

- tenths place 0
- **a** thousandths place ___
- **b** ones place ___
- **c** hundredths place ___

3 Write the numbers.

- 7 in the thousandths place,
  2 in the ones place,
  4 in the tenths place and
  6 in the hundredths place
  2.467
- **a** 8 in the ones place,
  9 in the hundredths place,
  6 in the tenths place and
  1 in the thousandths place
  ____
- **b** 5 in the thousandths place,
  8 in the hundredths place,
  4 in the ones place and
  3 in the tenths place
  ____
- **c** 7 in the tenths place,
  6 in the ones place,
  2 in the thousandths place,
  0 in the hundredths place
  ____

4 Write in decimal form.

- 3674 thousandths = 3.674
- **a** 1672 thousandths = ________
- **b** 5381 thousandths = ________
- **c** 9035 thousandths = ________
- **d** 8354 thousandths = ________
- **e** 7438 thousandths = ________
- **f** 1085 thousandths = ________
- **g** 1006 thousandths = ________
- **h** 6952 thousandths = ________
- **i** 2010 thousandths = ________
- **j** 2340 thousandths = ________
- **k** 5000 thousandths = ________
- **l** 6100 thousandths = ________
- **m** 7467 thousandths = ________

PLACE VALUE

Place value is the value of each digit in a number.
It means how much the digit is worth.

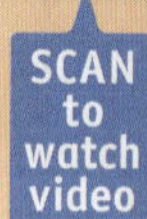

Example 1:
Complete the place value chart for 64.832.

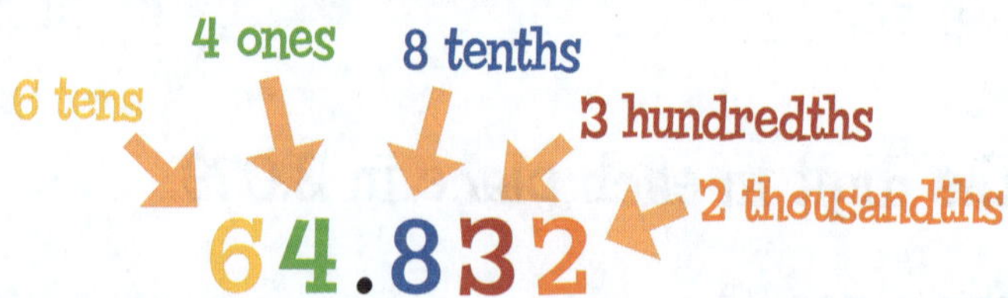

Tens	Ones	.	Tenths	Hundredths	Thousandths	Ten Thousandths
6	4	.	8	3	2	

Example 2: Write 32.5194 in the place value chart.

Tens	Ones	.	Tenths	Hundredths	Thousandths	Ten Thousandths
		.				

Check your answer on the video!

Example 3: Write 14.3109 in the place value chart.

Tens	Ones	.	Tenths	Hundredths	Thousandths	Ten Thousandths
		.				

Your turn

1 Use orange to trace over the thousandths.

24.927 a 62.493 b 25.620 c 7.358

2 Use red to trace over the hundredths.

3.845 a 1.246 b 29.368 c 84.721

3 Use blue to trace over the tenths.

26.429 a 32.739 b 1.435 c 0.583

4 Use green to trace over the ones.

3.246 a 73.282 b 1.122 c 40.475

5 Use yellow to trace over the tens.

32.473 a 10.650 b 74.009 c 21.436

SELF CHECK Tick how you feel

Got it!  Need help...  I don't get it 

Check your answers
How many did you get correct?

CATCH UP MATHS YEAR 5 BOOK A © PASCAL PRESS ISBN: 9781925726169

PRACTICE

What is the place value of the 5?

- 25.2368 ones
- a 63.7454 ______
- b 54.2790 ______
- c 1.3527 ______
- d 4.5796 ______
- e 43.5391 ______
- f 12.2053 ______
- g 2.6521 ______
- h 14.5263 ______
- i 53.2104 ______
- j 13.0506 ______
- k 73.8045 ______

Trace the tens in yellow, ones in green, tenths in blue, hundredths in red, thousandths in orange and ten thousandths in purple.

- 35.3782
- a 21.350
- b 71.4583
- c 0.359
- d 54.9252
- e 99.9999
- f 64.381
- g 82.563
- h 0.5034
- i 17.32
- j 53.1
- k 80.0019

Complete the table.

	Decimal	Tens	Ones	.	Tenths	Hundredths	Thousandths	Ten Thousandths
	20.8351	2	0	.	8	3	5	1
a	43.246							
b	74.9783							
c	87.903							
d	95.0902							
e	16.10							
f		2	4	.	6	4	9	
g		3	0	.	2	5	8	7
h		6	1	.	0	8	2	
i		5	3	.	9	4	2	
j		6	0	.	4	3	6	5

PARTITIONING DECIMALS

A decimal is made up of parts. These parts can be whole numbers and fractions of whole numbers.

$$2.39 = 2\frac{39}{100}$$

This mixed number is made up of 2 wholes + $\frac{3}{10}$ + $\frac{9}{100}$

It can be written as 2 + $\frac{39}{100}$.

Example 1:

$6.876 = 6\frac{876}{1000}$

$= 6 \text{ wholes} + \frac{8}{10} + \frac{7}{100} + \frac{6}{1000}$

$= 6 + \frac{876}{1000}$

Example 2:

$2.493 = 2\frac{493}{1000}$

$= \_\_ \text{ wholes} + \frac{\quad}{10} + \frac{\quad}{100} + \frac{\quad}{1000}$

$= 2 + \frac{\quad}{1000}$

Example 3:

$5.457 = 5\frac{\quad}{1000}$

$= \_\_ \text{ wholes} + \frac{\quad}{10} + \frac{\quad}{100} + \frac{\quad}{1000}$

$= \_\_ + \frac{\quad}{1000}$

There are many different ways to partition and write numbers.

Complete the fractions.

- 1.32 $= 1\frac{32}{100} = 1 + \frac{3}{10} + \frac{2}{100}$

a 2.27 $= 2\frac{27}{100} = 2 + \frac{\quad}{10} + \frac{\quad}{100}$

b 8.327 $= 8\frac{327}{1000} = 8 + \frac{\quad}{10} + \frac{\quad}{100} + \frac{\quad}{1000}$

c 7.43 $= 7\frac{\quad}{100} = 7 + \frac{\quad}{10} + \frac{\quad}{100}$

d 4.103 $= 4\frac{\quad}{1000} = \_\_ + \frac{\quad}{10} + \frac{\quad}{100} + \frac{\quad}{1000}$

Check your answers
How many did you get correct?

CATCH UP MATHS YEAR 5 BOOK A © PASCAL PRESS ISBN: 9781925726169

PRACTICE

1 Complete the table.

	Decimal	Mixed Number	Wholes	Tenths	Hundredths	Thousandths
●	5.213	$5\frac{213}{1000}$	5	$\frac{2}{10}$	$\frac{1}{100}$	$\frac{3}{1000}$
a			4	$\frac{1}{10}$	$\frac{7}{100}$	$\frac{5}{1000}$
b	6.157					
c			1	$\frac{4}{10}$	$\frac{9}{100}$	$\frac{3}{1000}$
d		$8\frac{459}{1000}$				
e			3	$\frac{2}{10}$	$\frac{3}{100}$	$\frac{7}{1000}$
f	7.368					
g			9	$\frac{9}{10}$	$\frac{4}{100}$	$\frac{5}{1000}$

2 Complete the table.

	Decimal	Mixed Number	Wholes	Thousandths
●	1.453	$1\frac{453}{1000}$	1	$\frac{453}{1000}$
a			5	$\frac{324}{1000}$
b	6.173			
c			4	$\frac{159}{1000}$
d		$3\frac{438}{1000}$		
e			6	$\frac{805}{1000}$
f	7.59			
g			9	$\frac{720}{1000}$

3 Look at the number expander, then expand the decimal fraction.

8	2 • 1	6	4	thousandths

= 82 + 164 thousandths = $82 + \frac{164}{1000}$

a

9	0 • 3	8	6	thousandths

= ______ = ______

b

3	8 • 4	tenths	9	hundredths	1	thousandths

= ______ = ______

c

4	6 • 9	3	hundredths	7	thousandths

= ______ = ______

d

3	5	4 • 0	2	5	thousandths

= ______ = ______

e

4	6	0 • 3	tenths	2	hundredths	1	thousandths

= ______ = ______

f

4	6 • 9	3	hundredths	7	thousandths

= ______ = ______

g

9	0 • 3	8	6	thousandths

= ______ = ______

  ISBN: 9781925726169

DECIMALS REVIEW

1 Use blue to circle the whole numbers and red to circle the decimal fractions.

a 24.20   b 63.58   c 0.30   d 58.37   e 70.59

2 Write the numbers in the place value table below.

a 3 tens, 5 ones, 6 tenths, 5 hundredths, 1 thousandth

b 8 ones, 6 hundredths, 5 tenths, 2 thousandths

c 9 thousandths, 6 tens, 3 hundredths, 9 ones, 8 tenths

d 5 tenths, 7 tens, 6 ones, 8 hundredths, 6 thousandths

e 1 tenth, 6 thousandths, 5 hundredths, 9 ones, 3 tens

f 1 ten, 4 tenths, 9 hundredths, 7 thousandths

	Tens	Ones	.	Tenths	Hundredths	Thousandths
a						
b						
c						
d						
e						
f						

3 Write the decimal fractions in descending order.

a 3.42, 2.43, 4.32, 2.34, 3.24 ______________________

b 1.473, 1.734, 3.174, 4.371, 7.317 ______________________

c 5.968, 8.695, 5.698, 8.965, 6.895 ______________________

d 9.034, 4.093, 3.904, 9.34, 9.304 ______________________

e 5.243, 3.781, 9.501, 3.331, 9.529 ______________________

Write these decimals with place value words.

a 8.234 __ ones + __ tenths + __ hundredths __ + thousandths

b 5.40 ____________________

c 24.07 ____________________

d 51.78 ____________________

e 4.738 ____________________

f 24.825 ____________________

Write the decimal fraction.

a ______

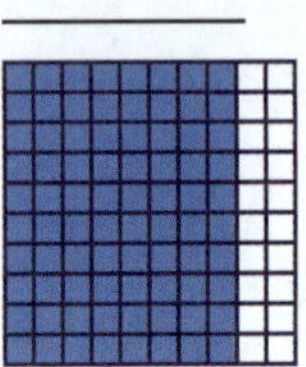

c ______

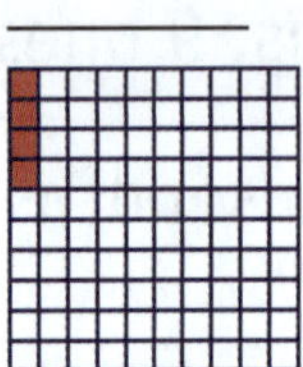

e ______

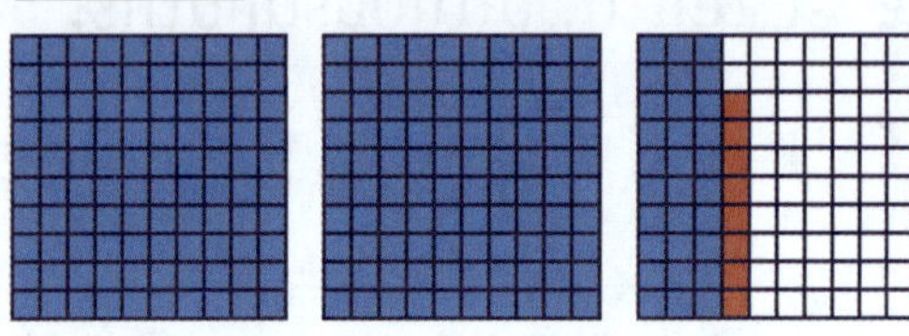

b ______

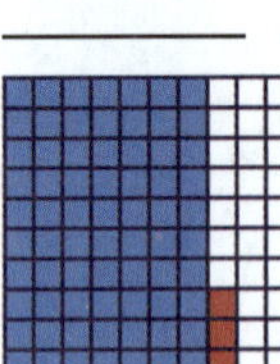

d ______

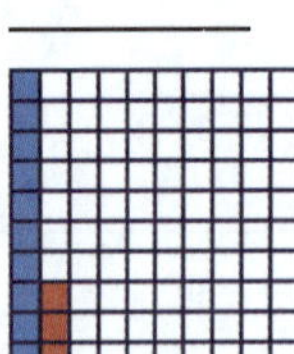

f ______

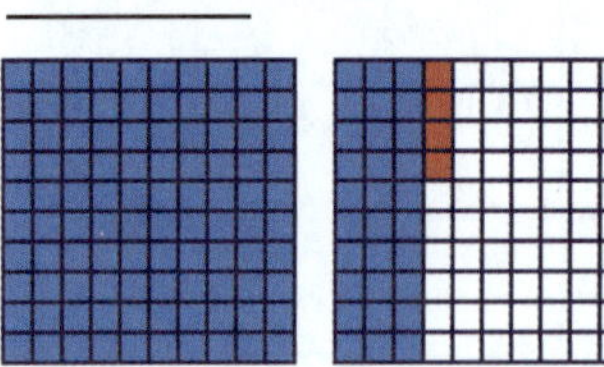

Show the decimal fractions. Use blue for tenths and red for hundredths.

a 0.51

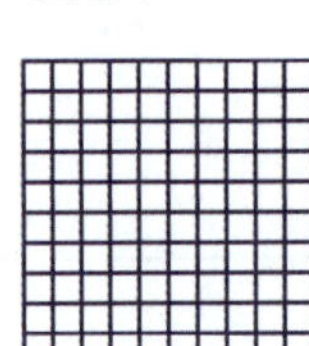

c 0.90

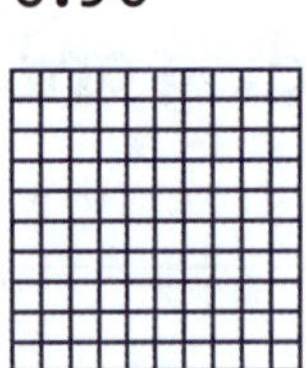

e 1.03

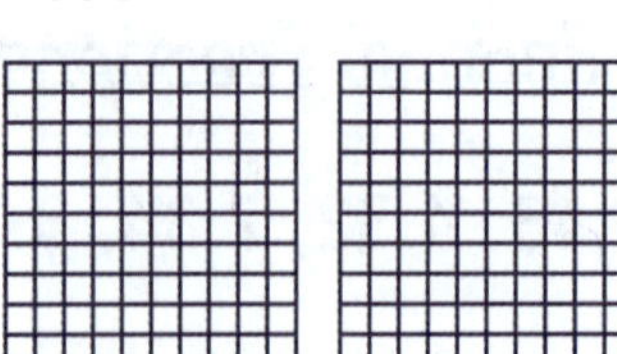

b 0.38

d 0.06

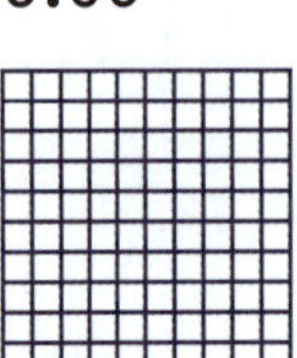

f 2.57

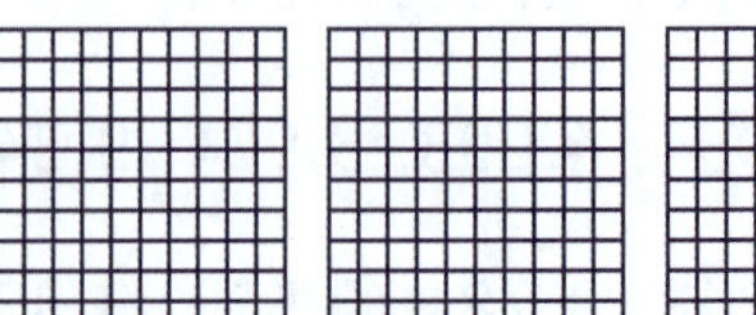

CATCH UP MATHS YEAR 5 BOOK A © PASCAL PRESS ISBN: 9781925726169

7 Fill in the table.

	Fraction	Fraction in words	Decimal fraction	Decimal in words
a	$\frac{29}{100}$			
b	$\frac{38}{100}$			
c		one hundred and ninety-seven hundredths		
d			2.54	
e				six point three four eight
f		two hundred and forty-one thousandths		
g			6.846	
h	$\frac{483}{100}$			

8 Write these decimal fractions in words.

a 6.593 ________ point ________ ________ ________

b 0.742 ________ point ________ ________ ________

c 3.561 ________ point ________ ________ ________

d 0.803 ________ point ________ ________ ________

e 2.41 ________ point ________ ________ ________

REVIEW

9 Write the digit in each place in 35.609.

a ones place __

b thousandths place __

c tens place __

d hundredths place __

e tenths place __

10 Write the digit in each place in 46.837.

a tens place __

b thousandths place __

c ones place __

d hundredths place __

e tenths place __

11 Write the digit in each place in 2.498.

a tenths place __

b ones place __

c hundredths place __

d thousandths place __

12 Write the numbers.

a 8 in the thousandths place,
3 in the ones place,
7 in the tenths place and
6 in the hundredths place

____

b 9 in the ones place,
7 in the hundredths place,
1 in the tenths place,
6 in the tens place,
8 in the thousandths place

____

c 3 in the hundredths place,
5 in the tens place,
2 in the ones place,
4 in the thousandths place and
8 in the tenths place

____

d 7 in the thousandths place,
8 in the tens place,
0 in the ones place,
9 in the tenths place
2 in the hundredths place

____

e 0 in the tenths place,
6 in the tens place,
4 in the ones place,
5 in the thousandths place
8 in the hundredths place

____

CATCH UP MATHS YEAR 5 BOOK A © PASCAL PRESS ISBN: 9781925726169

13 Write in decimal form.

a 5935 thousandths = ________

b 6710 thousandths = ________

c 8423 thousandths = ________

d 2872 thousandths = ________

e 4731 thousandths = ________

f 6481 thousandths = ________

g 3700 thousandths = ________

h 5490 thousandths = ________

i 4907 thousandths = ________

j 8034 thousandths = ________

14 Write in decimal form.

a two-tenths = ________

b 375 hundredths = ________

c 147 hundredths = ________

d nine-tenths = ________

e 733 hundredths = ________

f six-tenths = ________

g 843 hundredths = ________

h 125 hundredths = ________

15 Complete the table.

	Decimal	Tens	Ones	.	Tenths	Hundredths	Thousandths
a	15.723						
b	74.978						
c	73.098						
d	3.432						
e	5.24						
f	6.13						
g	8.354						
h	91.407						
i	70.320						
j	90.004						

REVIEW

Write the missing numbers.

a $2\frac{63}{100}$ = ___ + $\frac{}{10}$ + $\frac{}{100}$

b $7\frac{30}{100}$ = ___ + $\frac{}{10}$ + $\frac{}{100}$

c $1\frac{17}{100}$ = ___ + $\frac{}{10}$ + $\frac{}{100}$

d $5\frac{8}{100}$ = ___ + $\frac{}{10}$ + $\frac{}{100}$

Write the missing numbers.

a $6\frac{28}{100}$ = ___ + $\frac{}{100}$

b $8\frac{48}{100}$ = ___ + $\frac{}{100}$

c $4\frac{3}{100}$ = ___ + $\frac{}{100}$

d $3\frac{50}{100}$ = ___ + $\frac{}{100}$

Write the missing numbers.

a $2\frac{624}{1000}$ = 2 + $\frac{}{10}$ + $\frac{}{100}$ + $\frac{}{1000}$

b $8\frac{183}{1000}$ = 8 + $\frac{}{10}$ + $\frac{}{100}$ + $\frac{}{1000}$

c $9\frac{759}{1000}$ = ___ + $\frac{}{10}$ + $\frac{}{100}$ + $\frac{}{1000}$

d $4\frac{923}{1000}$ = ___ + $\frac{}{10}$ + $\frac{}{100}$ + $\frac{}{1000}$

e $15\frac{409}{1000}$ = ___ + $\frac{}{10}$ + $\frac{}{100}$ + $\frac{}{1000}$

f $10\frac{790}{1000}$ = ___ + $\frac{}{10}$ + $\frac{}{100}$ + $\frac{}{1000}$

19

Write the missing numbers.

a $3\frac{729}{1000}$ = ___ + $\frac{}{1000}$

b $8\frac{430}{1000}$ = ___ + $\frac{}{1000}$

c $5\frac{745}{1000}$ = ___ + $\frac{}{1000}$

d $14\frac{657}{1000}$ = ___ + $\frac{}{1000}$

e $20\frac{130}{1000}$ = ___ + $\frac{}{1000}$

f $39\frac{429}{1000}$ = ___ + $\frac{}{1000}$

Fill in the blank spaces.

a

3	2.	8	4	9	thousandths

= ____ + _____ thousandths = ___ + $\frac{}{1000}$

b

5	4.	9	8	hundredths	1	thousandths

= ____ + ___ hundredths + __ thousandth = ____ + $\frac{}{100}$ + $\frac{}{1000}$

CATCH UP MATHS YEAR 5 BOOK A © PASCAL PRESS ISBN: 9781925726169

c | 8 | 4 • 7 | tenths | 8 | hundredths | 1 | thousandths |

= ___ + __ tenths + __ hundredths + __ thousandth

= ___ + $\frac{\ }{10}$ + $\frac{\ }{100}$ + $\frac{\ }{1000}$

d | 4 | 3 • 0 | 0 | 5 | thousandths |

= ___ + _____ thousandths = ___ + $\frac{\ }{1000}$

Complete the table.

	Decimal	Mixed Number	Wholes	Tenths	Hundredths	Thousandths
a	5.632					
b			7	$\frac{3}{10}$	$\frac{6}{100}$	$\frac{5}{1000}$
c		$18\frac{792}{1000}$				
d			24	$\frac{8}{10}$	$\frac{3}{100}$	$\frac{9}{1000}$
e	16.516					
f			31	$\frac{1}{10}$	$\frac{2}{100}$	$\frac{6}{1000}$
g		$14\frac{203}{1000}$				
h			17	$\frac{0}{10}$	$\frac{4}{100}$	$\frac{1}{1000}$
i	5.278					
j			1	$\frac{4}{10}$	$\frac{3}{100}$	$\frac{5}{1000}$
k		$12\frac{349}{1000}$				
l			49	$\frac{6}{10}$	$\frac{3}{100}$	$\frac{6}{1000}$

NUMBER PATTERNS

A number pattern is a series of numbers that follows a rule. The rule tells you what to do with each number to make the pattern.

Example 1:
24, 26, 28, 30, 32, 34, 36 ...
Rule: Add 2

Example 2:
3, 12, 48, 192, 768, 3072 ...
Rule: Multiply by 4

Example 3:
97, 94, 91, 88, 85, 82, 79 ...
Rule: Subtract 3

Example 4:
128, 64, 32, 16, 8, 4, 2, 1
Rule: Divide by 2

Example 5:
73, 76, 79, 82, 85, 88, 91 ...
Rule: ____________

Example 6:
187, 181, 175, 169, 163, 157 ...
Rule: ____________

Use the rule to continue the pattern.

- 128, 132, 136, 140, 144 — Rule: + 4
- **a** 8, ____, ____, ____, ____ — Rule: × 2
- **b** 144, ____, ____, ____, ____ — Rule: Subtract 12
- **c** 96, ____, ____, ____, ____ — Rule: Divide by 2
- **d** 67, ____, ____, ____, ____ — Rule: Add 6
- **e** 6, ____, ____, ____, ____ — Rule: Multiply by 3

Check your answers
How many did you get correct?

CATCH UP MATHS YEAR 5 BOOK A © PASCAL PRESS ISBN: 9781925726169

PRACTICE

1 Complete the table.

	Rule	Number Pattern
●	+ 6	3, 9, 15, 21, 27, 33
a	÷ 2	4000, ____, ____, ____, ____, ____
b	+ 2	2000, ____, ____, ____, ____, ____
c	+ 3	185, ____, ____, ____, ____, ____
d	+ 8	1009, ____, ____, ____, ____, ____

2 Write the rule for each pattern.

● 8, 15, 22, 29, 36, 43 Rule: + 7

a 129, 126, 123, 120, 117 Rule: ____

b 7, 14, 28, 56, 112, 224 Rule: ____

c 5, 25, 125, 625, 3125 Rule: ____

d 10, 21, 32, 43, 54, 65 Rule: ____

e 6, 36, 216, 1296, 7776 Rule: ____

f 48, 24, 12, 6, 3 Rule: ____

3 Use the rule to continue these 2-step patterns.

● 10, 12, 18, 36, 90, 252, 738 Rule: − 6, × 3

a 5, 8, ____, ____, ____, ____, ____ Rule: − 1, × 2

b 4, 7, ____, ____, ____, ____, ____ Rule: × 2, − 1

c 260, 132, ____, ____, ____, ____, ____ Rule: ÷ 2, + 2

d 20, 26, ____, ____, ____, ____, ____ Rule: − 7, × 2

4 Continue the pattern and write the rule.

● 6, 12, 18, 24, 30, 36 Rule: + 6

a 81, 72, 63, ____, ____, ____ Rule: ____

b 4, 20, 100, ____, ____, ____ Rule: ____

c 73, 69, 65, ____, ____, ____ Rule: ____

5 Write the number sentence that describes the pattern.

● Rule: A + B = C

A	B	C
3	5	8
4	6	10
5	7	12
24	21	45

a Rule: ____________

●	■	▲
11	4	7
12	6	6
13	4	9
14	10	4

b Rule: ____________

D	E	F
2	3	6
4	2	8
5	3	15
9	6	54

c Rule: ____________

▲	■	★
10	2	5
3	1	3
8	4	2
24	3	8

d Rule: ____________

G	H	I
30	3	10
36	9	4
72	8	9
66	11	6

e Rule: ____________

X	⬢	A
6	15	9
12	27	15
15	29	14
20	52	32

CATCH UP MATHS YEAR 5 BOOK A © PASCAL PRESS ISBN: 9781925726169

PATTERN GRIDS

You can use tables to show patterns.
The rows of numbers are related by a rule.

The rule tells you how to work out the matching number in the other row.

Example 1:

Rule: ● = ■ + 6

■	2	3	4	5	6
●	8	9	10	11	12

● = ■ + 6
● = 2 + 6
● = 8

Example 2:

Rule: ▲ = ★ × 4

★	8	10	12	14	16
▲	32	40	48	56	64

▲ = ★ × 4
▲ = 8 × 4
▲ = 32

Example 3:

■ = ▲ ÷ 3

▲	36	33		27	
■	12		10		8

Example 4:

● = ⬢ − 5

⬢	93	95	97		101
●	88	90		94	

Check your answer on the video!

Your turn

Complete the patterns grids.

■ = ▲ × 3

▲	3	5	7	9	11
■	9	15	21	27	33

a ▲ = ● × 4

●	2	4	6	8	10
▲					

b ▲ = ⬢ − 5

⬢	5	10	15	20	25
▲					

c ● = ▲ ÷ 2

▲	6	12	18	24	30
●					

SELF CHECK Tick how you feel

Got it! Need help... I don't get it

Check your answers
How many did you get correct?

PRACTICE

1 Complete the pattern grids.

● B = A + 7

A	11	13	15	17	19	21	23
B	18	20	22	24	26	28	30

a D = 28 − C

C	16	14	12	10	8	4	2
D							

2 Write the rule for each pattern grid.

● Rule: ● = ▲ × 3

▲	3	5	7	9	11	13	15
●	9	15	21	27	33	39	45

a Rule: ______________

B	4	8	12	16	20	24	28
D	1	2	3	4	5	6	7

b Rule: ______________

C	1	2	3	4	5	6	7
K	11	22	33	44	55	66	77

c Rule: ______________

■	3	6	9	12	15	18	21
▲	1	2	3	4	5	6	7

3 Use the rule to continue these 2-step patterns.

● 3, 15, 63, 255, 1023, 4095 — Rule: Multiply by 4, plus 3

a 1, 16, ______, ______, ______, ______ — Rule: Add 7, multiply by 2

b 14, 16, ______, ______, ______, ______ — Rule: Subtract 6, multiply by 2

c 8, 20, ______, ______, ______, ______ — Rule: Minus 3, times 4

4 Complete the following.

●

Rule	3	4	7	8	12	11
+ 5	8	9	12	13	17	16
× 2	6	8	14	16	24	22

a

Rule	27	18	3	9	12	63
÷ 3						
+ 7						

b

Rule	25	5	65	10	60	20
÷ 5						
+ 8						

c

Rule	7	8	10	11	13	9
+ 12						
− 7						

CATCH UP MATHS YEAR 5 BOOK A © PASCAL PRESS ISBN: 9781925726169

EQUIVALENT NUMBER SENTENCES

Number sentences that are equal to each other are called equivalent number sentences. One side of the number sentence equals the other.

SCAN to watch video

Example 1:

34 + 2 = 20 + 16

Both sides equal 36.

Example 2:

38 − 22 = 20 − 4

Both sides equal 16.

Example 3:

3 × 4 = 12 × 1

Both sides equal 12.

Example 4:

20 − 13 = 21 ÷ 3

Both sides equal 7.

Example 5:

56 ÷ 7 = 2 × 4

Both sides equal ___.

Example 6:

2 × 9 = 10 + 8

Both sides equal ___.

Check your answer on the video!

Your turn

Complete these equivalent number sentences.

- 24 + [4] = 20 + 8
- **a** 8 + 4 = ☐ − 8
- **b** 7 × 3 = ☐ × 7
- **c** 50 − 25 = ☐ + 5
- **d** 36 ÷ 3 = ☐ × 3
- **e** 5 × 4 = ☐ ÷ 5
- **f** 35 − 5 = ☐ + 5
- **g** 36 + 4 = ☐ − 10
- **h** 4 × 4 = 2 × ☐
- **i** 40 + 2 = ☐ + 6

SELF CHECK Tick how you feel

Got it!	Need help...	I don't get it
 ☐	 ☐	  ☐

Check your answers

How many did you get correct? ☐

PRACTICE

1 Fill in the boxes to make equivalent number sentences.

● 7 × 8 = [60] – 4

a 120 ÷ 10 = [ ] × 3

b 8 × 6 = 32 + [ ]

c 64 ÷ 8 = 24 ÷ [ ]

d [ ] × 9 = 30 + 6

e [ ] + 10 = 4 × 11

f 56 ÷ [ ] = 20 – 12

g 132 ÷ 12 = 20 – [ ]

h 3 × [ ] = 80 – 50

i 5 + [ ] = 108 ÷ 12

j 144 ÷ 12 = [ ] – 8

k 169 ÷ 13 = 8 + [ ]

2 Write an equivalent number sentence.

● 24 ÷ 2 = 6 × 2

a 3 × 6 = ________

b 12 + 3 = ________

c 50 ÷ 5 = ________

d 8 × 8 = ________

e 20 – 13 = ________

f 72 ÷ 9 = ________

g 4 × 4 = ________

h 30 – 27 = ________

i 6 + 4 = ________

j 24 ÷ 3 = ________

k 5 × 5 = ________

3 Fill in the boxes to make equivalent number sentences.

● 3 × 6 | 20 – [2]

a 30 – [ ] | 9 × 3

b 2 × 9 | 40 – [ ]

c [ ] + 45 | 9 × 10

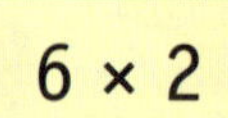

d 4 × [ ] | 6 × 2

e 110 ÷ 11 | 5 × [ ]

f 50 – 15 | 7 × [ ]

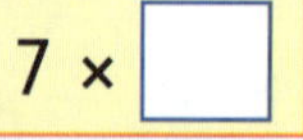

g 24 + 32 | 30 + [ ]

CATCH UP MATHS YEAR 5 BOOK A © PASCAL PRESS ISBN: 9781925726169

NUMBER SENTENCES AND PATTERNS WITH FRACTIONS AND DECIMALS

These number sentences and patterns look like other number sentences and patterns you have seen and solved before, but they have fractions and decimals in them.

SCAN to watch video

Number sentences with fractions

Example 1:

$\frac{1}{4} \times 12 = 3$

(Read $\frac{1}{4}$ of 12,

Solve $\frac{1}{4} \times 12$) (× top, ÷ bottom)

Number sentences with decimals

Example 2:

$7 \times 1.1 = 7.7$

Solve:

$$\begin{array}{r} 1.1 \\ \times \quad 7 \\ \hline 7.7 \end{array}$$

Patterns with fractions

Example 3:

$\frac{1}{4}, \frac{1}{2}, \frac{3}{4}, 1$

Rule: Add $\frac{1}{4}$

Patterns with decimals

Example 4:

2.2. 2.0, 1.8, 1.6, 1.4

Rule: Subtract 0.2

Example 5:

$\frac{3}{4} \times 20 =$ ____ (× top, ÷ bottom)

Example 6:

3.8, 4.1, 4.4, 4.7, 5.0

Rule: ____________

Complete the following.

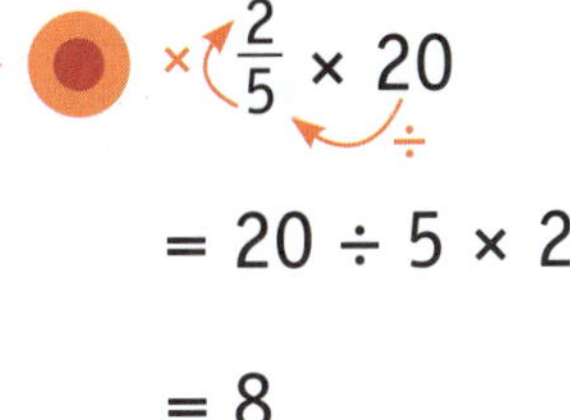

$\frac{2}{5} \times 20$ (× top, ÷ bottom)

$= 20 \div 5 \times 2$

$= 8$

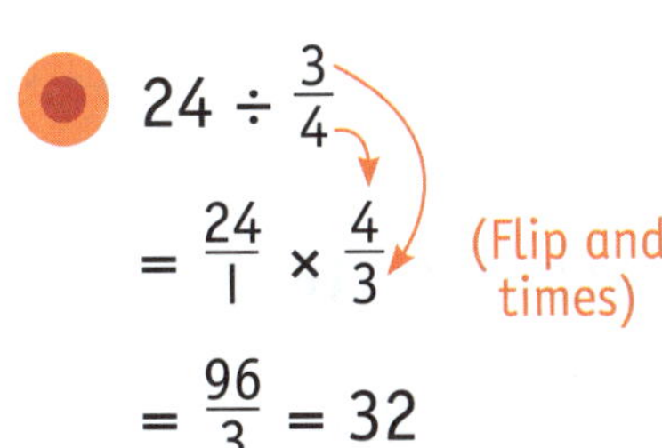

$24 \div \frac{3}{4}$

$= \frac{24}{1} \times \frac{4}{3}$ (Flip and times)

$= \frac{96}{3} = 32$

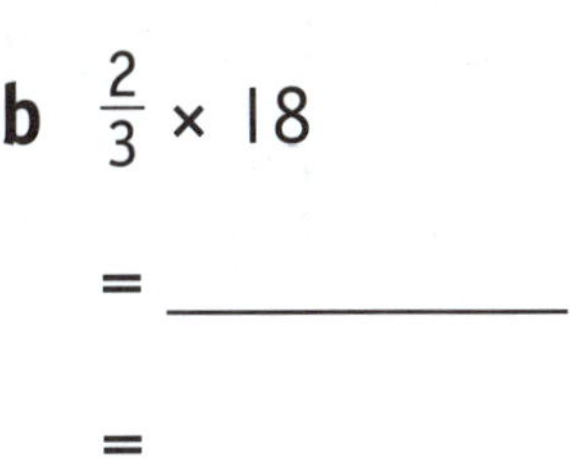

a $8 \times$ ____ $= 8.8$

b $\frac{2}{3} \times 18$

= ____________

= _____

c $12 \div \frac{3}{4}$

= ____________

= _____ = _____

d $1.2 \times 3 =$ _____

SELF CHECK Tick how you feel

Got it!	Need help...	I don't get it
 ☐	 ☐	☐

Check your answers

How many did you get correct? ☐

PRACTICE

1 Write the missing numbers.

- ● 1.2 × 3 = 6.6
- a ___ × 3 = 9.6
- b 4 × ___ = 4.8
- c 3.2 – 0.8 = ___
- d 7 × ___ = 7.7
- e 8.8 ÷ ___ = 2.2
- f 4.8 – ___ = 3.3
- g 3.4 × ___ = 6.8
- h 3 × ___ = 6.6
- i ___ × 1.1 = 3.3
- j 6 × 0.8 = ___
- k 0.9 × ___ = 7.2
- l 7 × ___ = 6.3
- m 4.1 × ___ = 8.2
- n 2 × ___ = 8.2

2 Complete these multiplications.

- ● $\frac{2}{3} \times 12 = 8$
- a $\frac{1}{4} \times 16 =$ ___
- b $\frac{2}{3} \times 24 =$ ___
- c $\frac{3}{8} \times 16 =$ ___
- d $\frac{2}{5} \times 15 =$ ___
- e $\frac{3}{8} \times 32 =$ ___
- f $\frac{2}{4} \times 24 =$ ___
- g $\frac{1}{8} \times 32 =$ ___
- h $\frac{1}{3}$ of 3 = ___
- i $\frac{2}{3} \times 27 =$ ___
- j $\frac{2}{9} \times 36 =$ ___
- k $\frac{3}{4}$ of 12 = ___

3 Complete these divisions.

- ● $8 \div \frac{1}{4} = \frac{8}{1} \times \frac{4}{1}$
  (Flip and times) $= \frac{32}{1} = 32$
- a $2 \div \frac{2}{3} =$ ________ = ___ = ___
- b $6 \div \frac{1}{2} =$ ________ = ___ = ___
- c $4 \div \frac{1}{5} =$ ________ = ___ = ___
- d $4 \div \frac{2}{3} =$ ________ = ___ = ___
- e $5 \div \frac{1}{2} =$ ________ = ___ = ___

4 Write the next three terms in these number patterns.

- ● 1.4, 1.5, 1.6, 1.7, 1.8, 1.9
- a 2.7, 2.4, 2.1, ___, ___, ___
- b 2.9, 3.1, 3.3, ___, ___, ___
- c 0.4, 0.6, 0.8, ___, ___, ___
- d 1.9, 2.1, 2.3, ___, ___, ___
- e 6.1, 5.7, 5.3, ___, ___, ___
- f 10.6, 10.2, 9.8, ___, ___, ___
- g 8.8, 8.5, 8.2, ___, ___, ___

TERMS IN SEQUENCES

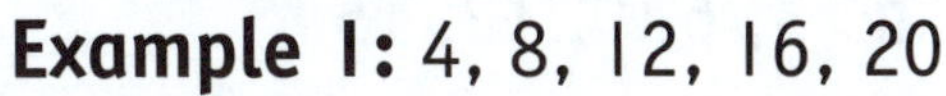

A term is one of the parts or members of a pattern.

Example 1: 4, 8, 12, 16, 20

This sequence has 5 terms.
The third term is 12.

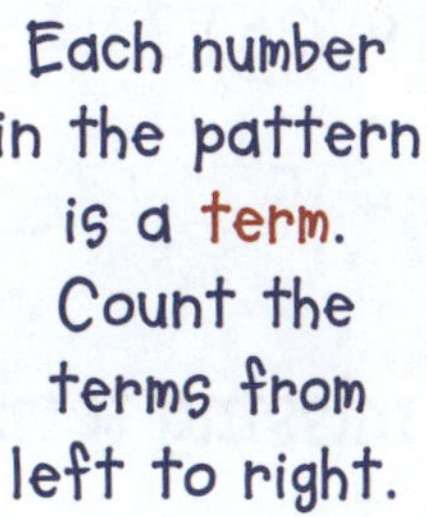

Example 2: 6, 12, 18, 24, 30, 36

This sequence has 6 terms.
The fifth term is 30.

Example 3: 24, 22, 20, 18, 16, 14, 12

This sequence has ___ terms. The third term is ___.

Example 4: 100, 103, 106, 109

This sequence has ___ terms. The first term is ___.

1 Circle the third term in each sequence.

- ● 2, 4, (6), 8, 10
- **a** 3, 6, 9, 12, 15
- **b** 20, 15, 10, 5, 0
- **c** 6, 12, 18, 24, 30
- **d** 100, 90, 80, 70, 60
- **e** 1.2, 2.2, 3.2, 4.2, 5.2

2 Write the second term in each sequence.

- ● 2, 4, 6, 8, 10 4
- **a** 4, 8, 12, 16, 20 ___
- **b** 5, 10, 15, 20, 25 ___
- **c** 24, 18, 12, 6, 0 ___
- **d** 7, 14, 21, 28, 35 ___
- **e** 64, 56, 48, 40, 32 ___

SELF CHECK Tick how you feel

Got it!	Need help...	I don't get it
☐	☐	☐

Check your answers
How many did you get correct? ☐

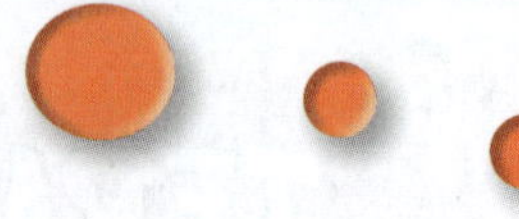

1 Which term is circled in each number sequence?

- ● 2, 4, 6, 8, 10 — 1st
- **a** 5, 12, 19, (26), 33 ____
- **b** 1, (2), 3, 4, 5 ____
- **c** 10, 8, 6, 4, (2) ____
- **d** 3, 6, (9), 12, 15 ____
- **e** 7, 14, 21, (28), 35 ____

2 Write the missing terms.

- ● 32, 40, 48, 56, 64
- **a** 12, 16, 20, ____, ____
- **b** 56, ____, ____, 77, 84
- **c** 109, 101, ____, ____, 77
- **d** 89, ____, 97, ____, 105
- **e** 28, ____, ____, 49, 56

3 Describe the pattern of the terms for each sequence in Question 2.

- ● The terms increase / ~~decrease~~ by 8
- **a** The terms increase / decrease by ____
- **b** The terms increase / decrease by ____
- **c** The terms increase / decrease by ____
- **d** The terms increase / decrease by ____
- **e** The terms increase / decrease by ____

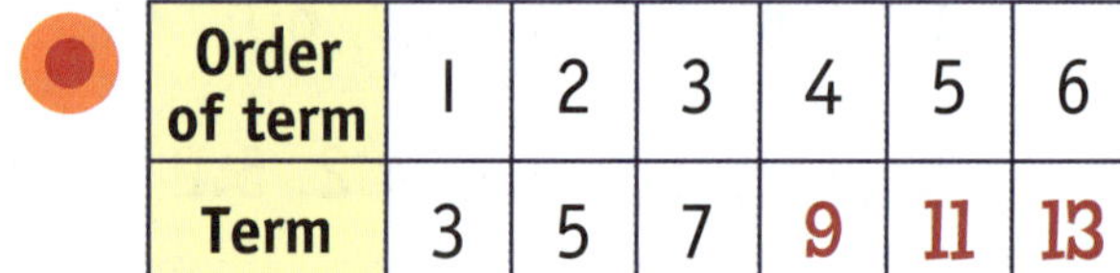

4 Complete the tables.

●

Order of term	1	2	3	4	5	6
Term	3	5	7	9	11	13

a

Order of term	1	2	3	4	5	6
Term	40	38	36			

b

Order of term	1	2	3	4	5	6
Term	2	4	6			

c

Order of term	1	2	3	4	5	6
Term	8	16	24			

5

Order of term	1	2	3	4	5	6	10	12
Term	1	4	9				?	?

- **a** Complete the table.
- **b** What is the 10th term? ____
- **c** What is the 12th term? ____

CATCH UP MATHS YEAR 5 BOOK A © PASCAL PRESS ISBN: 9781925726169

PATTERNS AND ALGEBRA REVIEW

1 Use the rule to continue the pattern.

a 6, _____, _____, _____, _____ Rule: × 3

b 102, _____, _____, _____, _____ Rule: + 2

c 500, _____, _____, _____, _____ Rule: ÷ 5

d 91, _____, _____, _____, _____ Rule: – 6

2 Write the rule for each pattern.

a 9, 16, 23, 30, 37 Rule: _____

b 128, 125, 122, 119, 116 Rule: _____

c 16, 32, 64, 128, 256 Rule: _____

d 68, 64, 60, 56, 52 Rule: _____

3 Use the rule to continue these 2-step patterns.

a 4, 6, _____, _____, _____, _____, _____ Rule: – 1, × 2

b 369, 126, _____, _____, _____, _____, _____ Rule: ÷ 3, + 3

c 1, 6, _____, _____, _____, _____, _____ Rule: × 2, + 4

4 Write the number sentence that describes the pattern.

a Rule: __________

A	B	C
7	4	28
5	9	45
10	3	30
24	2	48

b Rule: __________

▲	●	■
29	10	19
34	17	17
18	9	9
50	25	25

c Rule: __________

D	E	F
7	70	10
7	49	7
7	56	8
7	63	9

5 Complete the pattern grids.

a ⬢ = ▲ × 5

▲	9	7	5	3	1	0
⬢						

b B = A × 4

A	3	8	11	15	20	22
B						

c

Rule	60	50	25	10	15
÷ 5					
+ 9					

d

Rule	5	7	9	11	6
× 7					
– 4					

6 Complete these equivalent number sentences.

a $24 \div 4 = \square \times 2$

b $56 \div 8 = 20 - \square$

c $132 \div 11 = \square \times 1$

d $12 + 9 = 3 \times \square$

e $110 \div 11 = 50 - \square$

f $\square \times 3 = 36 - 12$

7 Write an equivalent number sentence.

a $24 \div 3 =$ ______

b $55 \div 5 =$ ______

c $30 - 13 =$ ______

d $72 \div 8 =$ ______

e $60 + 10 =$ ______

f $8 + 4 =$ ______

8 Complete the following.

a $\frac{3}{5} \times 20 =$ ______
= ______

b $\frac{2}{3} \times 24 =$ ______
= ______

c $\frac{4}{5} \times 25 =$ ______
= ______

d $12 \div \frac{1}{4} =$ ______
= ______

e $24 \div \frac{1}{4} =$ ______
= ______

f $2 \div \frac{2}{3} =$ ______
= ______

CATCH UP MATHS YEAR 5 BOOK A © PASCAL PRESS ISBN: 9781925726169

9 Fill in the missing numbers.

a 1.1 × ___ = 6.6

b 3 × ___ = 3.6

c 0.8 × ___ = 7.2

d 1.2 × ___ = 9.6

e ___ × 1.4 = 2.8

f 6 × 0.8 = ___

10 Write the next three terms in each number sequence.

a 1.9, 2.2, 2.5, ___, ___, ___

b 10.6, 10.2, 9.8, ___, ___, ___

c 3, 5.5, 8, ___, ___, ___

d 1.2, 2.2, 3.2, ___, ___, ___

e 4, 5.5, 7, ___, ___, ___

f 8.7, 7.6, 6.5, ___, ___, ___

11 What term is circled in each number sequence?

a 2, 4, 6, (8), 10, 12 ______

b 9, (18), 27, 36, 45 ______

c 1, 3, 5, 7, (9) ______

d (3.6), 3.3, 3.0, 2.7 ______

e 60, (54), 48, 42 ______

f 24, 27, 30, (33), 36 ______

12 Describe the pattern of the terms for each sequence in Question 11.

a The terms ______________________

b The terms ______________________

c The terms ______________________

d The terms ______________________

e The terms ______________________

f The terms ______________________

13

Order of term	1	2	3	4	5
Term	3	6	9		

a Complete the table.

b What would the 8th term be? _____

c What would the 10th term be? _____

d What would the 12th term be? _____

LIKELY AND UNLIKELY EVENTS

A **likely** event is something that would usually happen. It has a good chance of happening.

For example:
- You go to school.
- You play soccer.
- You eat breakfast.

An **unlikely** event is something that would not usually happen. It does not have a good chance of happening.

For example:
- You live in Hawaii for a year.
- You catch a taxi to school.
- The whole school has pizza.

Example 1: Your teacher sings in an opera. This event is **unlikely**.

Example 2: A dog barks. This event is **likely**.

Example 3: You have sport at school tomorrow.

This event is ____________.

Example 4: You fly on a plane for a school excursion.

This event is ____________.

Write likely or unlikely to describe the chance of the event happening.

- You go to sleep before midnight. likely
- **a** You eat lunch every day. ____________
- **b** You wear your school uniform on the weekend. ____________
- **c** You wear warm clothes in winter. ____________
- **d** You get a pony for your birthday. ____________

SELF CHECK Tick how you feel

Got it!	Need help...	I don't get it
☐	☐	☐

Check your answers

How many did you get correct? ☐

CATCH UP MATHS YEAR 5 BOOK A © PASCAL PRESS ISBN: 9781925726169

Match the event to the label that describes the chance of that event occurring.

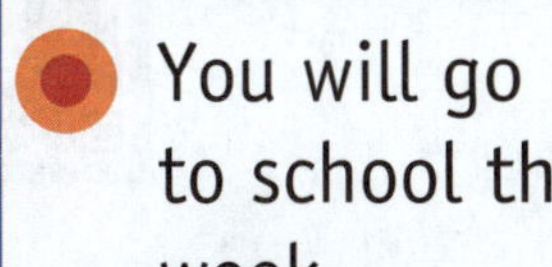

You will go to school this week.	**b** It will rain on a cloudy day.	**d** A hippopotamus will come to your football game.

Likely **Unlikely**

a If my dog is happy, she will wag her tail.	**c** I will read at school.	**e** I will eat pancakes for breakfast every day.

List five likely events and five unlikely events.

Likely events

Unlikely events

CERTAIN AND UNCERTAIN EVENTS

SCAN to watch video

A certain event means that the event will happen. There can be no other result.

Example 1:

The sun will rise in the morning.

This event is certain because the sun always rises in the morning.

An uncertain event means there is a different possible result.

Example 2:

My friend and I will play at the park after school.

There are other possible results:

- Your friend or you could get sick and not be able to play.
- There may be a storm and it would be dangerous to play outside.

Example 1: If this month is June, next month will be July.
This event is **certain**.

Example 2: New Year's Day will be sunny.
This event is **uncertain**.

Example 3: Christmas will be in December.

This event is ____________.

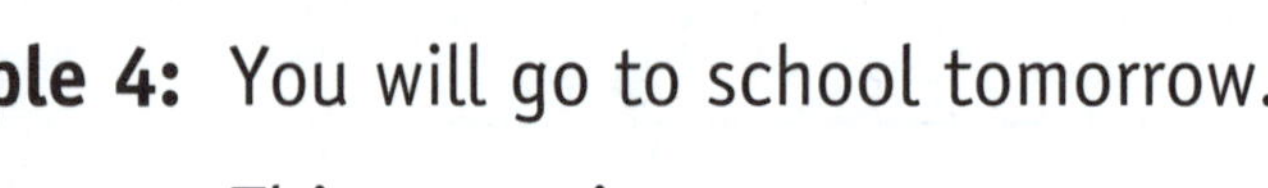

Example 4: You will go to school tomorrow.

This event is ____________.

Use green to circle the certain events and use red to circle the uncertain events.

a If today is Sunday, tomorrow will be Monday.

b We will eat tacos for dinner on Tuesday.

c The season after spring will be summer.

d It will be a cloudy day tomorrow.

SELF CHECK Tick how you feel		
Got it! ☐	Need help... ☐	I don't get it ☐

Check your answers

How many did you get correct? ☐

CATCH UP MATHS YEAR 5 BOOK A © PASCAL PRESS ISBN: 9781925726169

1 Join the events with the correct label.

- If you roll a dice, it will land on a number from 1 to 6. — Certain

a Your dad will cook a barbeque for dinner.

b If I choose a ball out of a bag that has blue balls and red balls, I will pick out a red ball.

c If I choose a jellybean from a jar of only red jellybeans, I will pick out a red jellybean.

d If you toss a coin, it will land on either heads or tails.

Certain

Uncertain

2 Write 'certain' or 'uncertain' to describe the chance of these events happening.

	Events	Certain or Uncertain?
•	Your teacher will wear a red shirt tomorrow.	uncertain
a	I will choose a red gummy bear out of a packet of coloured gummy bears.	
b	If I choose a card out of a deck of cards, it will be red.	
c	You will go to school next Monday.	
d	If I throw a dice, it will land on a 2.	
e	If there are only green balls in a box, you will pick out a green ball.	
f	The sun will set in the evening.	
g	You will go to the beach this weekend.	
h	If I flip a coin, it will land on heads.	
i	It will not snow tomorrow.	

PROBABILITY

Probability is the likelihood of an event happening.

Sometimes, if one event happens, another event cannot happen. Some events cannot happen at the same time.

Example 1:
If this event happens:
Lei has won every soccer game this season.

This event cannot happen:
Lei's team is in last place.

If Lei's team won every soccer game, they cannot be in last place.

Example 2:
If this event happens:
Rex won a race, beating five other children.

This event cannot happen:
Rex came third in the race.

If Rex won the race, he cannot be in third place.

Example 3:
If this event happens: Alyssa came third in a singing competition.

This event **can / cannot** happen: Alyssa won the singing competition.

If Alyssa came third, she **can / cannot** win the same competition.

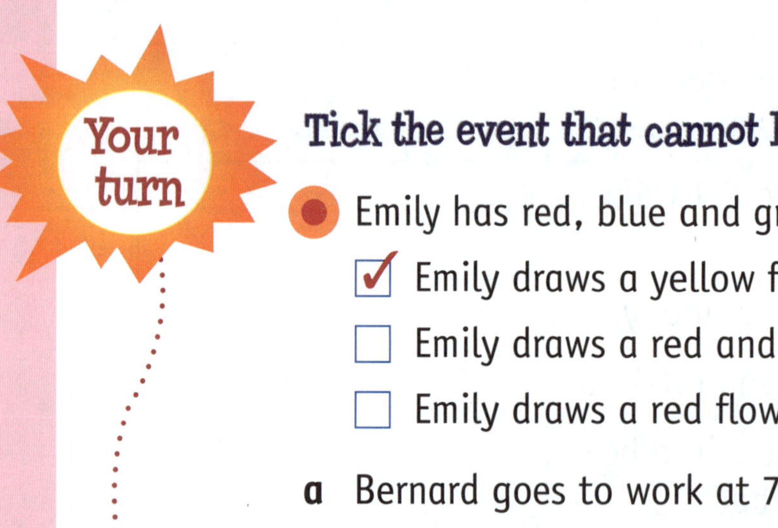

Tick the event that cannot happen.

- Emily has red, blue and green pencils.
  - ☑ Emily draws a yellow flower.
  - ☐ Emily draws a red and blue house.
  - ☐ Emily draws a red flower.

a Bernard goes to work at 7:00 am.
  - ☐ Bernard arrives at work at 6:00 am.
  - ☐ Bernard arrives at work at 7:30 am.
  - ☐ Bernard drops off his children on the way to work.

SELF CHECK Tick how you feel		
Got it! ☐	Need help... ☐	I don't get it ☐

Check your answers
How many did you get correct? ☐

CATCH UP MATHS YEAR 5 BOOK A © PASCAL PRESS ISBN: 9781925726169

1 **Tick the correct answer.**

- Phillip closes his store at 5 pm. So Phillip cannot:
  - [ ] leave the store at 6 pm.
  - [ ] serve a customer at 4 pm.
  - [x] serve a customer at 6 pm.

a If Zeb has yellow, red and white paint, Zeb cannot:
- [ ] paint a green car.
- [ ] paint a yellow car.
- [ ] paint a red and white car.

b Out of 12 balls in a box, there are 8 blue balls, 1 red ball and 3 pink balls. Jassin cannot:
- [ ] pick out a red ball.
- [ ] pick out a blue ball.
- [ ] pick out a black ball.

c A spinner has the numbers 1 to 20. When Annie spins it, she cannot:
- [ ] land on the number 16.
- [ ] land on the number 30.
- [ ] land on an even number.

d There are five jellybeans in a jar: one red, one blue, one black, one white and one yellow. Christine picks out the white jellybean. When Danny picks out a jellybean, he cannot:
- [ ] pick out a white jellybean.
- [ ] pick out a yellow jellybean.
- [ ] pick out a red jellybean.

e Today is Tiana's birthday, so it cannot be:
- [ ] her best friend's birthday today.
- [ ] Tiana's birthday tomorrow.
- [ ] her sister's birthday tomorrow.

RELATED EVENTS

Some events are related.
If one occurs, it is more likely another event will occur.

If you are learning to ride a bike, there is ~~is not~~ a chance you will fall off.	If you are learning to cook, you might ~~will not~~ burn your food.

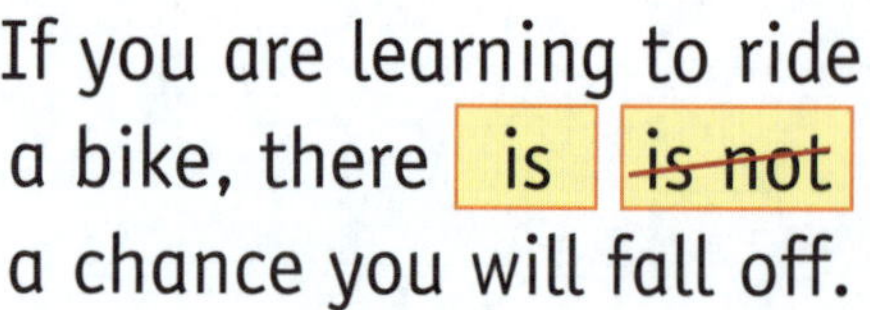

Sometimes the chance of something happening IS NOT affected by another event. The events are not related.

If Ju can swim, it ~~does~~ does not affect the chance of her learning to ride a bicycle.	If Ben loves to eat chicken, it ~~does~~ does not affect his chance of winning a race.

Example 1: If Allan can dance, it does/ does not affect the chance of him learning to ____________________.

Example 2: If Nani loves to fix cars, it does/ does not affect someone else's chance of ____________________.

Tick the event that is not related.

● Alexis is a champion swimmer.
- ☐ Alexis enjoys swimming.
- ☑ Alexis goes for a walk every day.

a Rosie is a master chef.
- ☐ Rosie eats healthy food.
- ☐ Rosie's best dish is curry.

SELF CHECK Tick how you feel

Got it!	Need help...	I don't get it
☐	☐	☐

Check your answers
How many did you get correct? ☐

CATCH UP MATHS YEAR 5 BOOK A © PASCAL PRESS ISBN: 9781925726169

PRACTICE

1 Write true or false beside each statement.

- Betty has a baby girl, so her next baby will be a boy. False

a If a coin is tossed and it lands on heads, the next time the coin is tossed it will land on tails. ________

b If you don't wear sunscreen on a hot summer's day, you will get sunburnt. ________

c If I buy a bag of potatoes and one is rotten, all the other potatoes in the bag will be rotten too. ________

2 Tick the event that is NOT related to the first event.

- I will eat lunch at noon today.
  - ☐ The cafe I go to for lunch opens at 12:30 pm.
  - ☑ I love eating cheese, lettuce and tomato sandwiches.

a Jason will go skiing tomorrow.
- ☐ Jason's friend loves skiing.
- ☐ There has not been much snowfall yet.

b Adrianna's bike tyre was flat.
- ☐ Adrianna's mum taught her how to pump up flat tyres.
- ☐ Adrianna's mum has a red bike.

c Roisin's oven is broken.
- ☐ Roisin will buy a new dress.
- ☐ Roisin wants to bake a cake.

d Jonah has been asked to sing in a concert.
- ☐ Jonah gets a sore throat.
- ☐ Jonah enjoys watching movies.

e Paul loves kayaking on the weekends.
- ☐ Paul lives near a river.
- ☐ Paul just learned a new skateboard trick.

OUTCOMES

An outcome is any possible result that can happen in a chance experiment.

Example 1:

A jar has the same number of red jellybeans and yellow jellybeans.

What are all the possible outcomes when three jellybeans are picked out of the jar?

Possible outcomes:

- 
- 
- 
- 

Choosing a blue jellybean is an example of an impossible outcome because there are no blue jellybeans in the jar.

Example 2:

List all the possible outcomes for one gummy bear and one jelly snake.

red bear + green snake	red bear + pink snake
orange bear + green snake	orange bear + pink snake
______________	______________
______________	______________

Check your answer on the video!

SCAN to watch video

Use the jar of marbles to answer the questions.

a List the possible outcomes when three marbles are picked out of this jar.

B B B ______________

______________ ______________

b It is possible / impossible to pick out a pink marble.

Check your answers
How many did you get correct?

  ISBN: 9781925726169

PRACTICE

1 List all the possible outcomes for the combination of a shirt and a pair of shorts.

- pink shirt + pink shorts
- pink shirt + orange shorts
- pink shirt + blue shorts
- __________
- __________
- __________
- __________
- __________
- __________

2 Alek's running club has a new uniform.
List all the possible uniform combinations Alek can make.

Hats are in three different colours: yellow, orange and red.
Shirts are in two different colours: yellow and red.
Shorts are in two different colours: orange and white.

1 yellow hat + yellow shirt + orange shorts
2 ________ hat + ________ shirt + ________ shorts
3 ________ hat + ________ shirt + ________ shorts
4 ________ hat + ________ shirt + ________ shorts
5 ________ hat + ________ shirt + ________ shorts
6 ________ hat + ________ shirt + ________ shorts
7 ________ hat + ________ shirt + ________ shorts
8 ________ hat + ________ shirt + ________ shorts
9 ________ hat + ________ shirt + ________ shorts
10 ________ hat + ________ shirt + ________ shorts
11 ________ hat + ________ shirt + ________ shorts
12 ________ hat + ________ shirt + ________ shorts

EVEN CHANCE EVENTS

SCAN to watch video

An even chance is when the possible outcomes are equally likely to happen.

When a coin is tossed, it lands on heads (H) or tails (T).
The chance of either outcome is even.
There are two possible outcomes.

There is a 1 in 2 ($\frac{1}{2}$) chance of getting heads.

There is a 1 in 2 ($\frac{1}{2}$) chance of getting tails.

Example 1:

A jar has 8 green jellybeans and 8 orange jellybeans.

When you pick out one jellybean, there is a 1 in 2 ($\frac{1}{2}$) chance of picking out a green jellybean and a 1 in 2 ($\frac{1}{2}$) chance of picking out an orange one.

Example 2:

Colour the marbles in each jar to show a 1 in 2 ($\frac{1}{2}$) chance.

Your turn

Is there an even chance of picking out a marble of each colour? Write yes or no.

● yes　　a ______　　b ______　　c ______

SELF CHECK　Tick how you feel

Got it!	Need help...	I don't get it
☐	☐	☐

Check your answers
How many did you get correct?

CATCH UP MATHS YEAR 5 BOOK A © PASCAL PRESS ISBN: 9781925726169

PRACTICE

1 Colour the spinners to show a one in two ($\frac{1}{2}$) chance outcome.

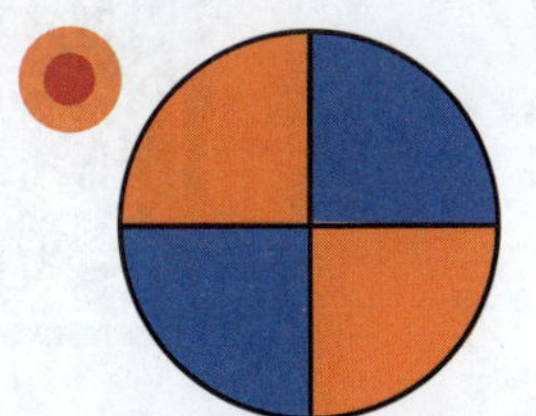

b 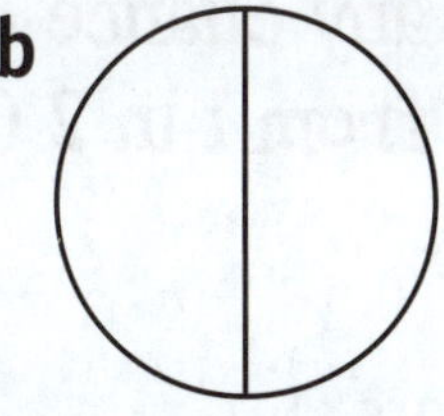

d 

f 

a 

c 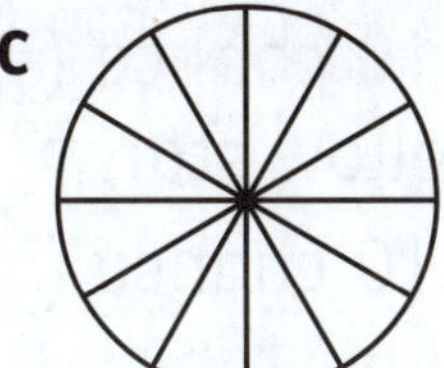

e 

g 

2 Draw extra marbles in each bag to show a one in two ($\frac{1}{2}$) chance outcome.

b 

d 

f 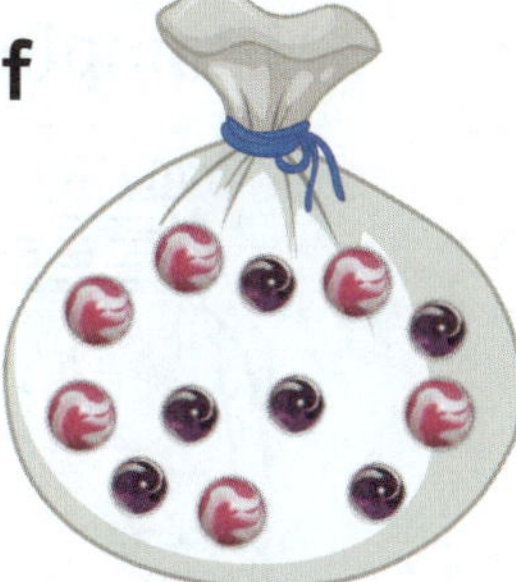

a 

c 

e 

g 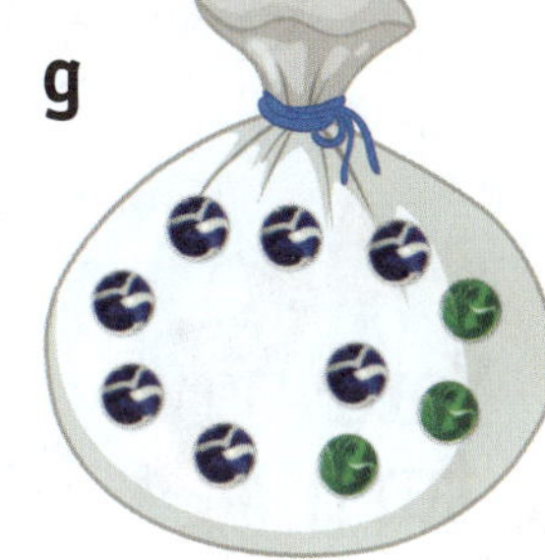

3 Tick the even chance events and cross the uneven chance events.

- ☐ I have three black counters and three green counters in a bag.
- a ☐ I have two red counters and six blue counters in a bag.
- b ☐ I flip a coin.
- c ☐ I have four yellow marbles and four green marbles in a jar.
- d ☐ I have two pink pegs and three white pegs in a jar.

UNEQUAL CHANCES

An unequal chance is any chance outcome that is different from 1 in 2 ($\frac{1}{2}$).

Example 1:

This jar has 10 stars.

I have a 6 out of 10 chance ($\frac{6}{10}$) of pulling out a yellow star.

I have a 4 out of 10 chance ($\frac{4}{10}$) of pulling out a blue star.

It is an unequal chance because there are more yellow stars than blue stars.

Example 2:

This jar has ___ gummy bears.

I have a ___ out of ___ chance ($\frac{}{}$) of pulling out a pink gummy bear.

I have a ___ out of ___ chance ($\frac{}{}$) of pulling out a green gummy bear.

It is an ___________ chance because there are more __________ gummy bears than __________ gummy bears.

Check your answer on the video!

Cross out the incorrect word to describe the chance.

● Equal ~~Unequal~~ chance

a Equal Unequal chance

b Equal Unequal chance

SELF CHECK Tick how you feel		
Got it!	Need help...	I don't get it

Check your answers

How many did you get correct?

CATCH UP MATHS YEAR 5 BOOK A © PASCAL PRESS ISBN: 9781925726169

PRACTICE

1 **Use the jar of 20 balls to answer the questions.**

- How many balls are green? **4** out of 20

a ☐ balls out of 20 are pink.

b ☐ balls out of 20 are blue.

c 3 balls out of ☐ are orange.

d 1 ball out of ☐ is yellow.

e The colour of ball with the greatest chance of being picked is ______________.

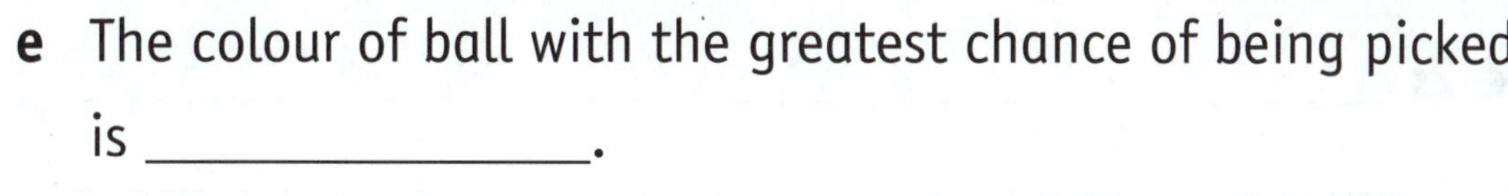

f The colour of ball with the least chance of being picked is ______________.

2 **Complete the missing information.**

- There are 10 jellybeans in the jar.

a __ in ___ chance ($\frac{\_}{10}$) of pulling out a blue jellybean.

b __ in ___ chance ($\frac{\_}{10}$) of pulling out a red jellybean.

c It is an ____________ chance.

3 **Draw balls in the jar below to show an unequal chance and then complete the missing information.**

a There are ___ balls in the jar.

b There is a __ in ___ (—) chance of pulling out a ____________ ball.

c There is a __ in ___ (—) chance of pulling out a ____________ ball.

PROBABILITY FROM 0 TO 1

The probability of an event happening is rated from 0 to 1.

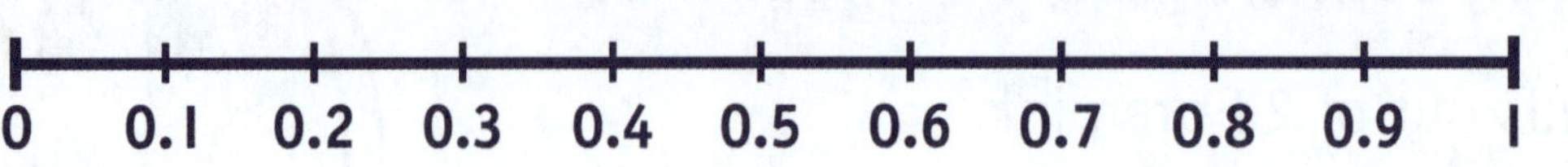

An impossible event has a probability of 0.

Even chance

A certain event has a probability of 1.

Impossible Events

Example 1: New Year's Day will be on 2 January.

Example 2: ____________________

____

Even Chance Events

Example 3: If there is a black ball and a white ball in a bag, there is an even chance of picking out the black ball.

Example 4: ____________________

____

Certain Events

Example 5: If today is Thursday, tomorrow will be Friday.

Example 6: ____________________

____

Check your answer on the video!

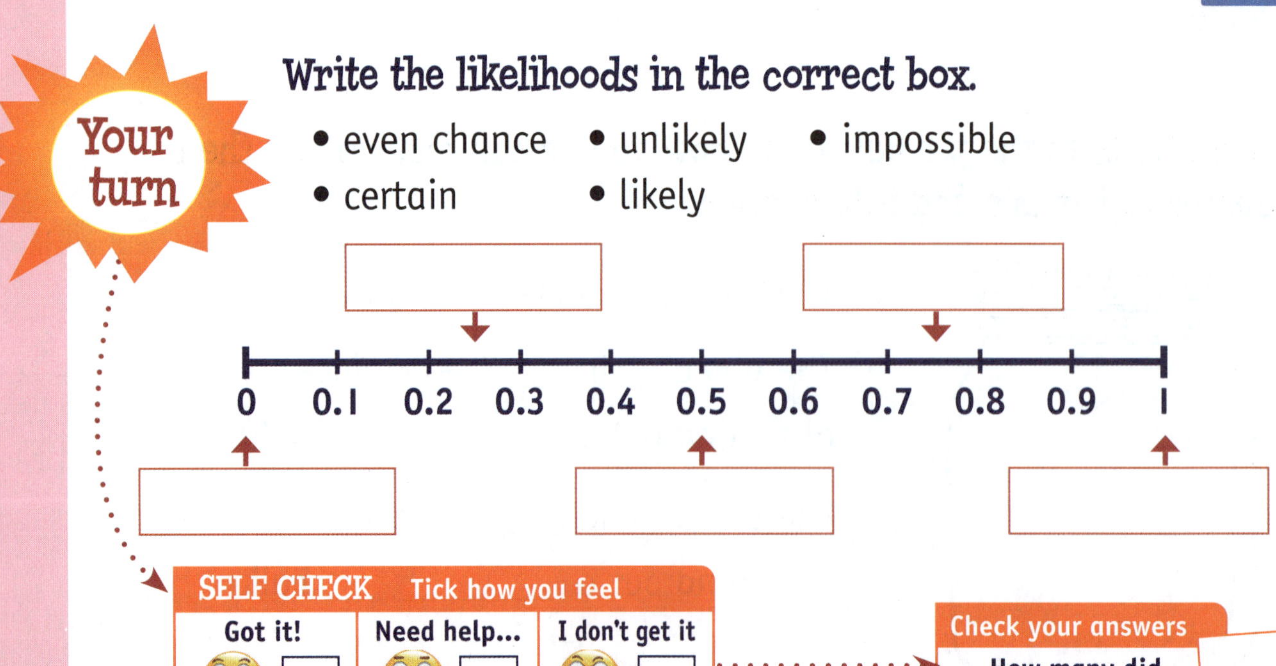

Write the likelihoods in the correct box.

- even chance
- unlikely
- impossible
- certain
- likely

SELF CHECK Tick how you feel

Got it!	Need help...	I don't get it
☐	☐	☐

Check your answers

How many did you get correct?

CATCH UP MATHS YEAR 5 BOOK A © PASCAL PRESS ISBN: 9781925726169

PRACTICE

1 Rate the probability of these events on a scale of 0 to 1.

	Rating	Event
●	0	You will grow 4 metres tall.
a		A tossed coin will land on heads.
b		The day after Wednesday will be Thursday.
c		I take a red ball out of a bag that has 8 red balls and 2 blue balls.
d		I take a red ball out of a bag that has 1 red ball and 5 blue balls.
e		January follows February.
f		This spinner will land on blue:

2 Describe an event that matches each probability.

	Rating	Event
●	0.3	I take a green ball out of a bag that has 3 green balls and 7 yellow balls.
a	0	
b	1	
c	0.5	
d	0.9	
e	0.6	
f	0.2	

CHANCE REVIEW

1 **Write likely or unlikely to describe the chance of the event happening.**

a You get a year off school. ________________

b You eat dinner at home. ________________

c You fly an aeroplane to see your grandparents. ________________

d The principal buys everyone a pet spider. ________________

e You play sport on the weekend. ________________

2 **Use green to circle the certain events and use red to circle the uncertain events.**

a If today is Tuesday, tomorrow will be Wednesday.

b I will visit my aunt tomorrow after school.

c I will go to my friend's birthday party on the weekend.

d After winter, spring will come.

3 **Explain why the two uncertain events in question 2 are uncertain.**

- ____

  ____

- ____

  ____

4 **Tick the correct answer.**

a Max's store opens at 7:00 am.

☐ Max cannot serve a customer at 7:30 am.

☐ Max cannot close the store at noon.

☐ Max cannot serve a customer at 6:00 am.

CATCH UP MATHS YEAR 5 BOOK A © PASCAL PRESS ISBN: 9781925726169

b Marissa has green, red and blue crayons.

- ☐ Marissa cannot draw a red boat.
- ☐ Marissa cannot draw an orange boat.
- ☐ Marissa cannot draw a blue and green boat.

c Matthew spins a spinner that has the numbers 1 to 15.

- ☐ Matthew cannot land on number 20.
- ☐ Matthew cannot land on a number from 1 to 10.
- ☐ Matthew cannot land on the number 15.

5 Write true or false.

a Nina tossed a coin and it landed on tails. The next time Nina tosses a coin, it will land on tails.

____

b Tony buys a bag of oranges and one is rotten, so all the oranges will be rotten.

____

c It is raining. If Mano does not use an umbrella, he will get wet.

____

d Maryrose has a deck of cards and she turns over one card. The card is red. This does not mean that the next card she turns over will be black.

____

6 Tick the event that cannot affect the chance of the first event happening.

a We will eat pasta for dinner.

- ☐ Dad will bring home takeaway.
- ☐ I love eating pasta for dinner.

b I will drive to work tomorrow.

- ☐ The car battery is flat.
- ☐ It is Tuesday tomorrow.

7 List the possible outcomes when three jellybeans are picked from this jar.

__________ __________ __________ __________

REVIEW

List all the possible outcomes for the combination of a shirt and a pair of shorts.

- ________________
- ________________
- ________________
- ________________
- ________________
- ________________
- ________________
- ________________
- ________________

Colour the spinners to show a 1 in 2 ($\frac{1}{2}$) chance outcome.

a 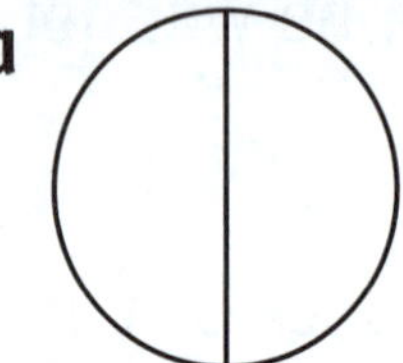 b 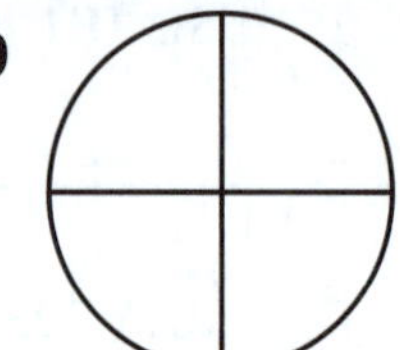 c 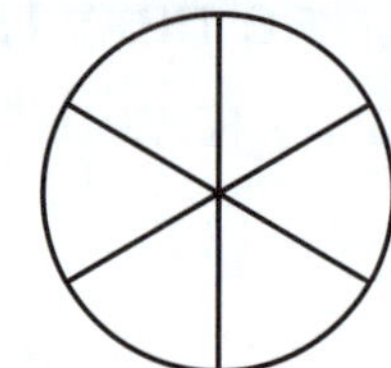 d 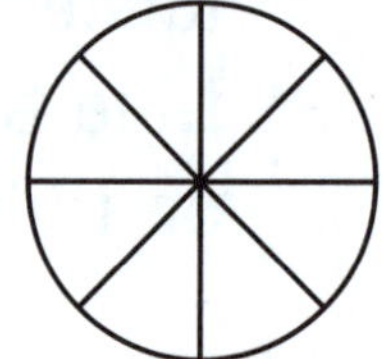

Draw extra marbles so that each bag has an even chance (0.5) outcome.

a  b 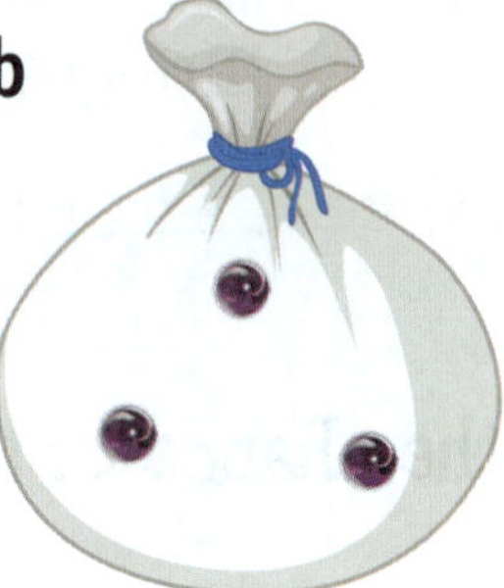 c  d 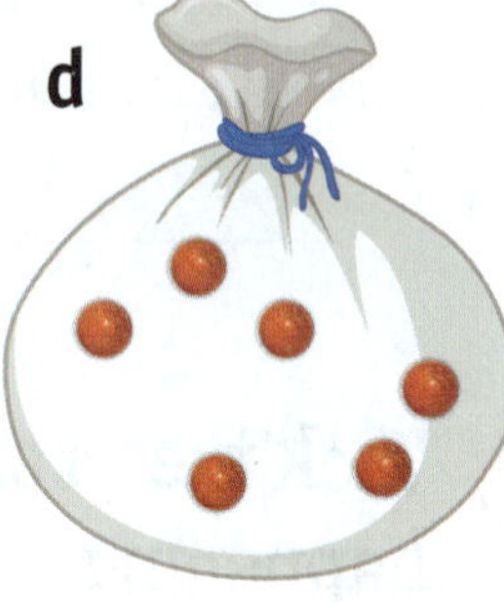

Complete the information.

a This jar has ___ balls. There are ___ green balls and ___ blue balls.
I have a ___ out of ___ chance (—) of taking out a green ball and a ___ out of ___ chance (—) of taking out a blue ball.
It is an equal / unequal chance.

CATCH UP MATHS YEAR 5 BOOK A © PASCAL PRESS ISBN: 9781925726169

b This jar has ___ balls. There are ___ pink balls and ___ green balls.

I have a ___ out of ___ chance (—) of taking out a pink ball and a ___ out of ___ chance (—) of taking out a green ball.

It is an equal / unequal chance.

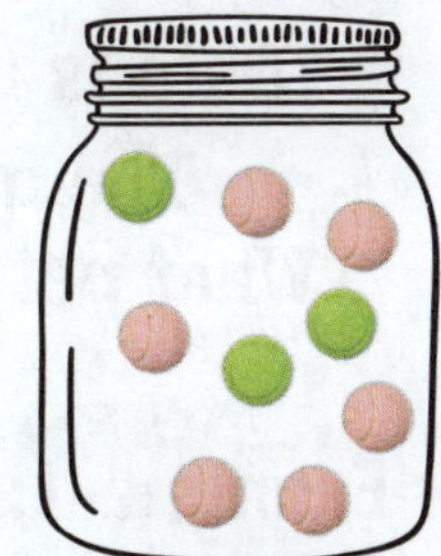

c This jar has ___ balls. There are ___ yellow balls and ___ orange balls.

I have a ___ out of ___ chance (—) of taking out an orange ball and a ___ out of ___ chance (—) of taking out a yellow ball.

It is an equal / unequal chance.

12 The scale below is a probability scale.

a Fill in the missing numbers on the probability scale.

b Write these labels on the probability scale:
impossible, even chance, certain, likely, unlikely

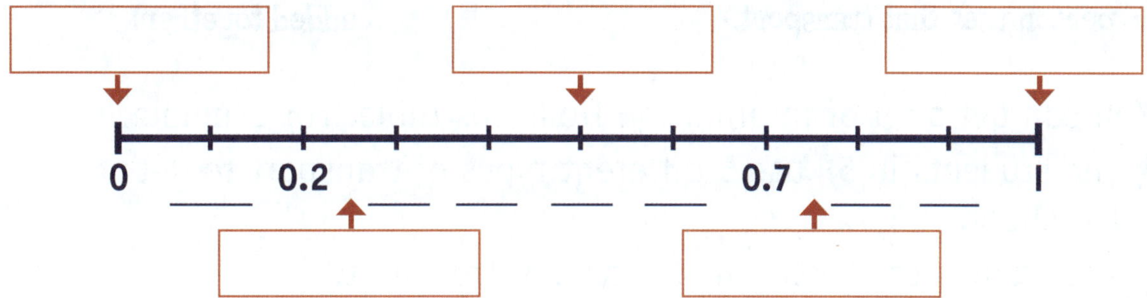

13 Rate the probability of these events happening on the scale of 0 to 1.

	Event	Rating
a	Ten students will be away from school.	
b	You will turn into a dinosaur.	
c	Christmas Day will be on 25 December.	
d	It will be hot during summer.	
e	I will take a red lolly from a bag that has 6 red lollies and 6 blue lollies.	

DATA TABLES

Data is information collected after a question is asked.

The question might be: How do you get to school? What pet(s) do you own? What is your favourite colour?

Data is often displayed in a table.

Example 1:

The table shows the data collected in answer to the question:

How do students in 5P get to school?

How 5P Gets To School ← This is the information being collected.

Transport	Tally	Total
Walk	卌 I	6
Bicycle	卌	5
Train	II	2
Car	III	3
Bus	卌 I	6
		22

These are the types of transport that data is collected about. →

After counting the tallies, write in the total. ←

The total number of students who provided data (all the totals added together). ←

Each line is a tally mark. One tally mark means one person uses that transport. ↑

You can get a lot of information from this table. For example:

- The students in 5P use 5 different types of transport to get to school.
- The train is the least popular type of transport.
- There are 22 students in 5P.

Use the table to answer the questions.

a How many students catch the bus to school? 6

b The form of transport used least by students in 5P to get to school is train.

c 5 students ride bikes to school.

d 6 students walk to school.

CATCH UP MATHS YEAR 5 BOOK A © PASCAL PRESS ISBN: 9781925726169

Example 2:
Count the tally marks and record the totals.

Colours of Cars Seen on Saturday

Colour	Tally	Total
Red	𝍸 𝍸	
Black	𝍸 𝍸 𝍸 |	
Silver	𝍸 𝍸 𝍸 |||	
White	𝍸 𝍸 |	
Other	𝍸 ||	

Your turn

Use this table to answer the questions.

Items Sold at Port Hacksville High

Item	Tally	Total
Nachos	𝍸 𝍸 𝍸	
Hot chips	𝍸 𝍸 𝍸 𝍸 |	
Hamburgers	𝍸 𝍸 𝍸 𝍸 𝍸 |||	
Hot dogs	𝍸 𝍸 𝍸 𝍸 𝍸 𝍸 ||	
Spring rolls	𝍸 𝍸 𝍸 𝍸 |	

a Complete the Total column.

b The item with the most sales is ________________.

c The items with the same number of sales are

________________ and ________________.

d The difference between the sales of nachos and hamburgers is ____.

e ____ hot dogs were sold.

SELF CHECK Tick how you feel

Got it!	Need help...	I don't get it
☐	☐	☐

Check your answers
How many did you get correct? ☐

PRACTICE

1 Fill in the missing information.

a 5T's Favourite Food

Food	Tally	Total
Pasta	𝍸 III	8
Pizza	𝍸 IIII	
Sushi	𝍸 II	
Tacos	III	
Hot dogs	II	

b 5M's Favourite Colour

Colour	Tally	Total
Red	IIII	
Blue		9
Green	𝍸	
Yellow		8

2 Taj surveyed 30 people about their favourite sport. He organised the data he collected in this table.

5P's Favourite Sport

Sport	Tally	Total
Ice hockey	𝍸 𝍸 II	12
Basketball	𝍸 III	8
Football	III	3
Volleyball	𝍸 II	7
		30

a What was the most popular sport?

____

b What was the least popular sport?

____

c What is the difference in votes between the most popular and least popular sports? _____

d How many people chose basketball? _____

e What is the difference in votes for volleyball and for football? _____

3 Organise the information in the table.

Colours of Tennis Balls

Colour	Tally	Total

CATCH UP MATHS YEAR 5 BOOK A © PASCAL PRESS ISBN: 9781925726169

PICTURE GRAPHS

A picture graph uses pictures to represent data.
The key tells you what the pictures mean.

SCAN to watch video

Example 1:

Use the graph to answer the following questions.

Cans collected in April

Kendra	
Jolie	
Billy	
Joe	

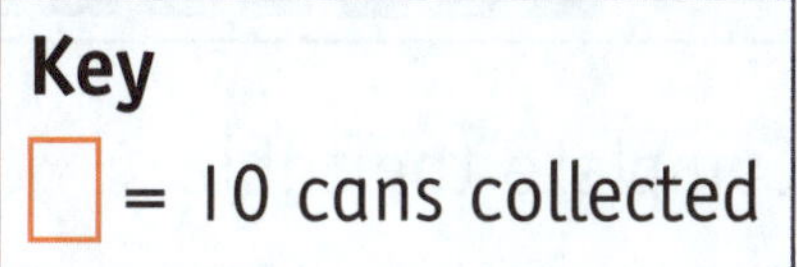

Here, = 10 cans collected so = 5 cans collected.

It is easy to see that Jolie collected the most cans and Billy collected the least.

a How many cans did Billy collect? 20

b How many cans did Jolie collect? 55

c Kendra collected 25 cans.

d Joe collected 30 cans.

e What is the difference between the number of cans collected by Joe and Kendra? 5

f What is the difference between the number of cans collected by Jolie and Billy? 35

g Jolie collected the most cans.

h Billy collected the least cans.

i 130 cans were collected altogether.

  ISBN: 9781925726169

Example 2: Use the graph to answer the following questions.

Year 5's Favourite Sport

Football	■■■■■■■■■
Basketball	■■■■■■■■■▌
Netball	■■■■■■▌
Tennis	■■■■■■■▌
Hockey	■■■■■■■■

Key

■ = 2 students

Year 5's Favourite Sport

Sport	Total
Football	
Basketball	
Netball	
Tennis	
Hockey	

a Complete the table.

b What was the most popular sport?

____

c What was the least popular sport?

____

d ___ students chose basketball.

e What is the difference between the numbers of votes for tennis and football? ____

f How many students were surveyed? ____

Your turn

Use the graph to complete the table.

Number of Ice Creams Sold in One Week

Sunday	●●●●●●
Monday	●●●●◖
Tuesday	●●●●
Wednesday	●●◖
Thursday	●●●●●
Friday	●●●●●●●
Saturday	●●●●●●●●◖

Key: ● = 4 ice creams

Number of Ice Creams Sold in One Week

Day	Total
Sunday	24
Monday	
Tuesday	
Wednesday	
Thursday	
Friday	
Saturday	

SELF CHECK Tick how you feel

Got it!	Need help...	I don't get it
☐	☐	☐

Check your answers

How many did you get correct? ☐

CATCH UP MATHS YEAR 5 BOOK A © PASCAL PRESS ISBN: 9781925726169

1 Use this picture graph to answer the following questions.

Number of Houses Built by Build-It Builders

Key: 🏠 = 20 houses built

- How many houses were built in 2018? 210

a How many houses were built in 2017? _____

b In what year were the most houses built? _____

c In what year were the least houses built? _____

d What is the difference between the numbers of houses built in 2020 and 2019? _____

e How many houses were built from 2017 to 2021? _____

2 5A made a graph of their favourite breakfast foods. Use it to answer the questions.

- Toast is the favourite breakfast of 1 girl.

a The most popular breakfast food for boys is _________________.

b The least popular breakfast food for girls is _________________.

c Cereal was chosen by ___ students.

d ___ girls chose eggs.

e ___ more boys than girls chose eggs.

f The difference between girls who chose waffles and girls who chose toast is ___.

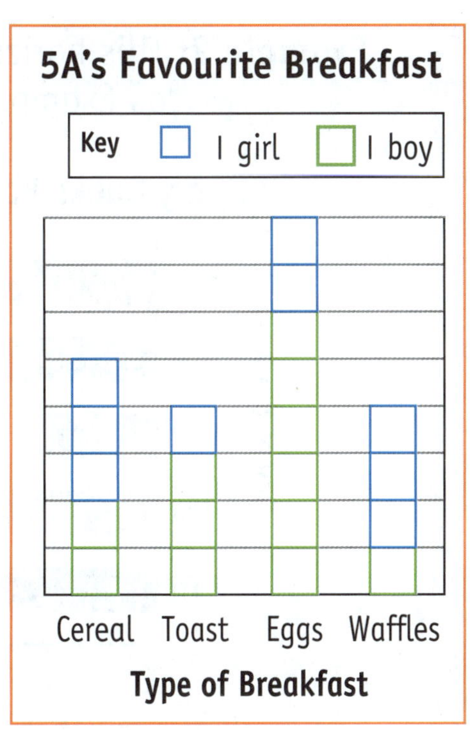

COLUMN GRAPHS

A column graph is a graph that shows the data in columns. The columns can be vertical or horizontal.

Example 1: This is a vertical column graph.
Use the graph to answer the following questions.

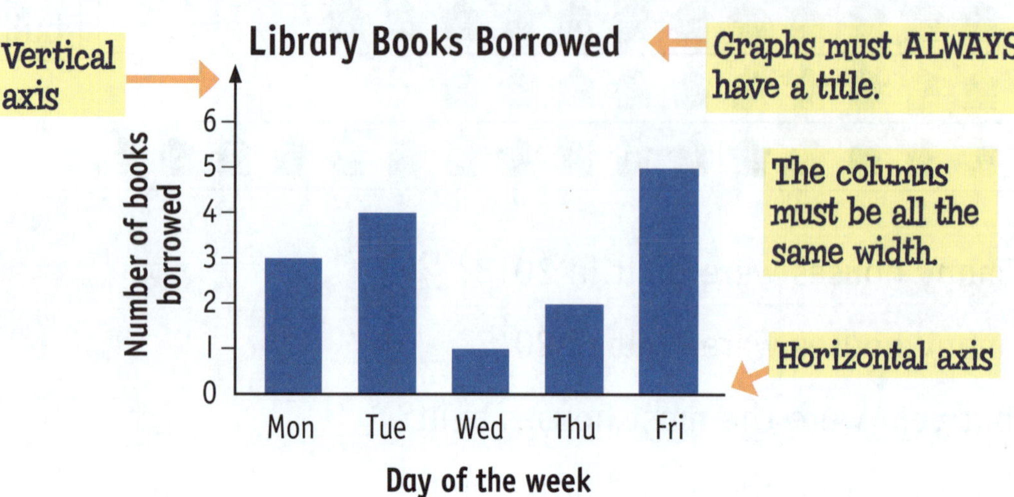

a How many books were borrowed on Tuesday? 4

b What day was the smallest number of books borrowed?

Wednesday

c How many books were borrowed in total? 15

Example 2: This horizontal column graph shows the same data as Example 1.

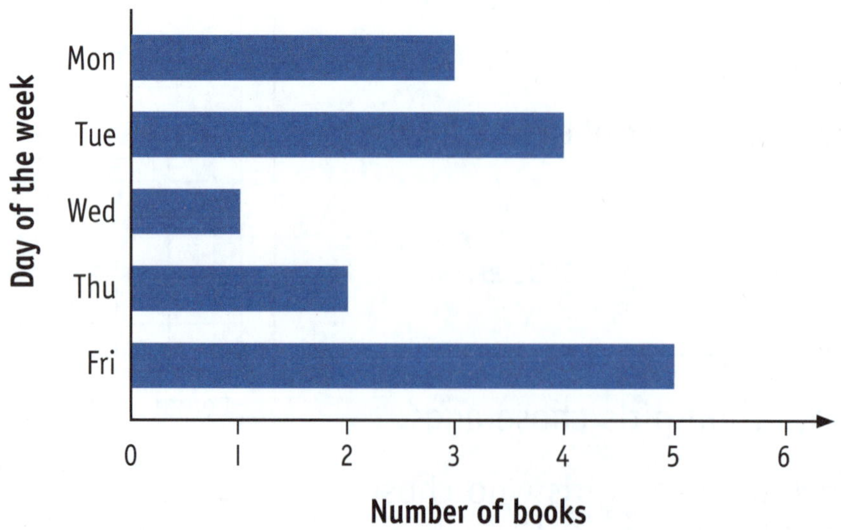

CATCH UP MATHS YEAR 5 BOOK A © PASCAL PRESS ISBN: 9781925726169

Example 3: Use this horizontal column graph to complete the table.

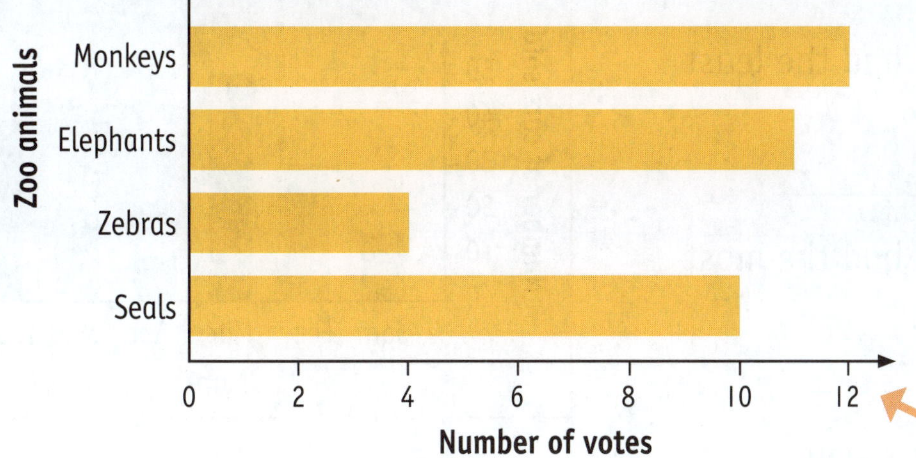

This scale increases by 2

Favourite Zoo Animal

Zoo Animal	Tally	Total
Monkeys		
Elephants		
Zebras		
Seals		

Use the graph to answer the questions.

What was the most popular animal? monkeys

a How many people chose elephants? ____

b ___ people chose zebras as their favourite zoo animal.

c The least popular zoo animal is ______________.

d The difference between the votes for monkeys and elephants is ____.

e The difference between the votes for the most popular and least popular zoo animal is ____.

SELF CHECK Tick how you feel

Got it!	Need help...	I don't get it
☐	☐	☐

Check your answers
How many did you get correct? ☐

PRACTICE

1 Use the graph to answer the questions.

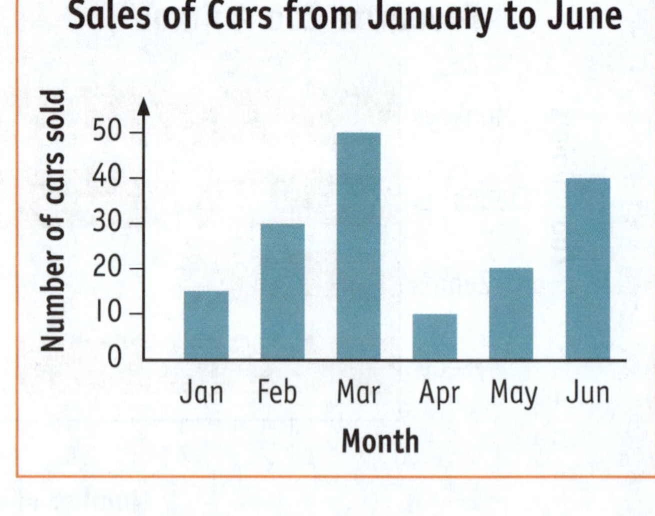

● Which month had the least sales?

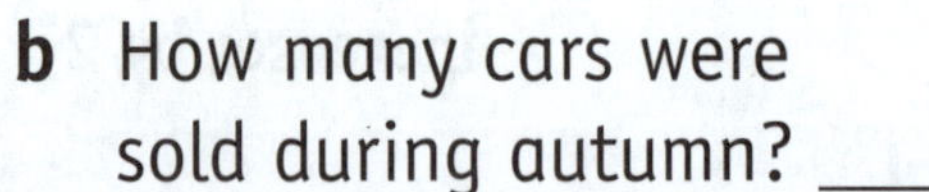

a Which month had the most sales? ____________

b How many cars were sold during autumn? ___

c What was the total number of cars sold in January and June? ___

d What was the total number of cars sold in February and May? ___

2 5W made a graph of the pets they have. Use their graph to answer the questions.

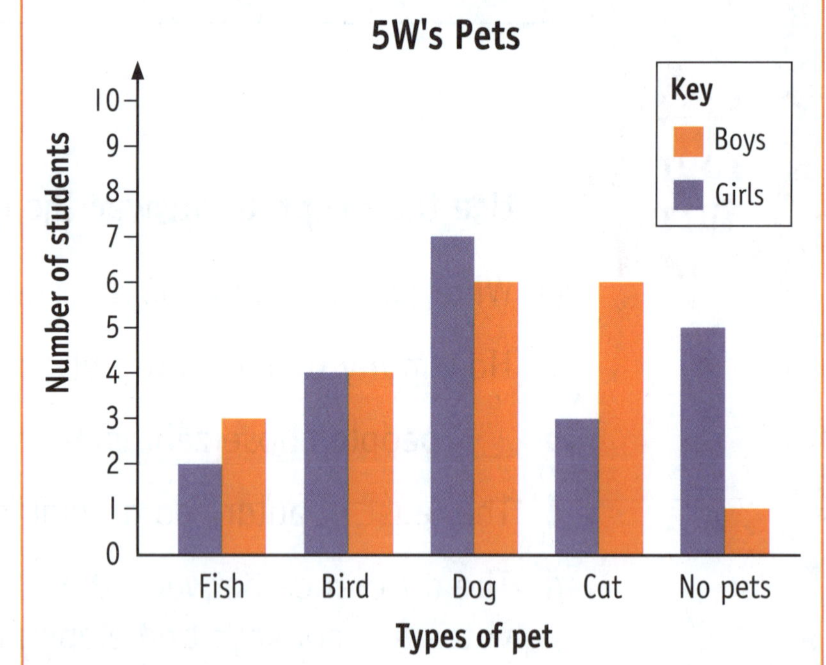

● How many boys have dogs? 6

a ___ girls have fish.

b ___ girls and ___ boys have no pets.

c ___ students own dogs.

d ___ students own cats.

e The difference between girls and boys who own birds is ___.

f The difference between girls and boys who own fish is ___.

CATCH UP MATHS YEAR 5 BOOK A © PASCAL PRESS ISBN: 9781925726169

BAR GRAPHS

In a bar graph, the bar is divided into different sections to show different values of data.

Example 1:

Favourite Season of Students in Year 5

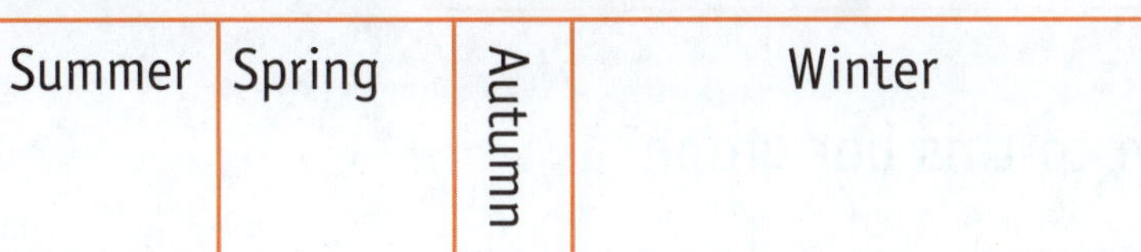

The Winter part of the bar graph is the longest and the Autumn part is the shortest.

It is easy to see that Winter is the most popular season and Autumn is the least popular season.

Example 2:

Use the following information to label the bar graph.

Favourite Subject of Students in Year 5

- Maths is the most popular and Art is the least popular.
- Sport and Science have the same number of votes.

Use the bar graph to answer the questions.

Favourite Fruit of Students in Year 5

Mango	Cherry	Strawberry	Pineapple

- The most popular fruit was strawberry.
- **a** The least popular fruit was ____________.
- **b** Strawberry was more/less popular than cherry.
- **c** Pineapple was more/less popular than mango.
- **d** Strawberry was more/less popular than pineapple.

SELF CHECK Tick how you feel

Got it!	Need help...	I don't get it
☐	☐	☐

Check your answers

How many did you get correct? ☐

PRACTICE

This bar graph shows the animals that were rehomed in November last year.

Birds	Fish	Cats	Dogs

a What title could be given to this bar graph?

____

b What animal was rehomed least in November? ________

c Were more or less birds rehomed than fish? ________

d If 100 animals were rehomed, about how many dogs is this? ____

e How many types of animals were rehomed in November? ___

f Were more or less cats rehomed than birds? ________

2 **Use the information to complete the following bar graph.**

- Money was spent five times on petrol: in January, February, March, April and May
- Half of the money was spent in January.
- One-quarter of the money was spent in May.
- The least amount of money was spent in March, followed by February.
- The amount of money spent in February was about the same as in April.

Money Spent on Petrol

CATCH UP MATHS YEAR 5 BOOK A © PASCAL PRESS ISBN: 9781925726169

DOT PLOTS

A dot plot is similar to a picture graph but uses dots instead of pictures. You can use a dot plot to show many-to-one data.

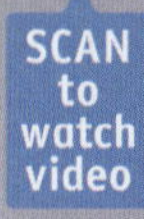

Many-to-one data means that one dot can represent 2 students, or 10 students. You don't need a dot for each student.

Example 1:

Use the dot plot to answer the questions.

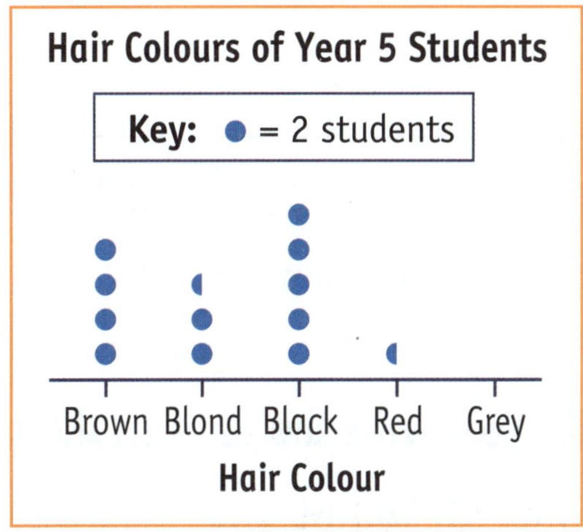

a Complete the table to show the values in the graph.

Hair Colour	Brown	Blond	Black	Red	Grey
Number of Students	8	5	10	1	0

b How many students were surveyed? 24

c What hair colour was most common? black

d Did more or less students have blond hair than black hair?

less

e Did more or less students have brown hair than red hair?

more

Example 2: Use the values in the table to complete the dot plot.

Number of People in Families of Year 5 Students

Number of Family Members	1	2	3	4	5	More than 5
Number of Students	0	3	5	10	4	2

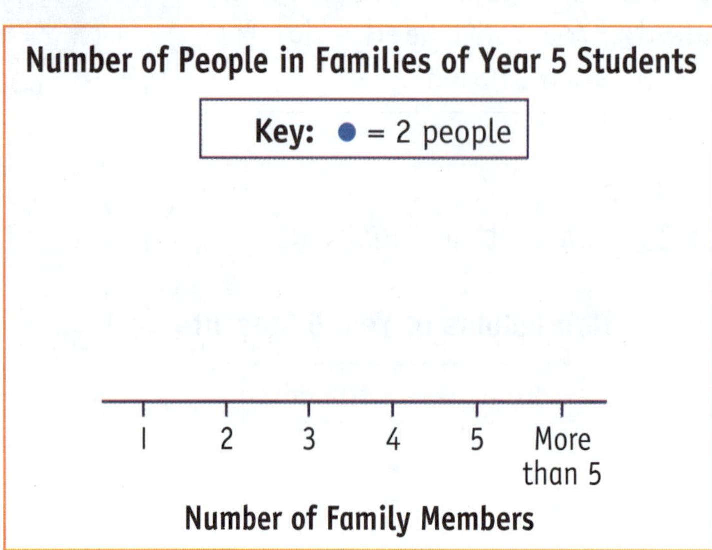

Your turn

Make a dot plot with the data in the table.

Soccer Goals Kicked by Zoe

Season	2017	2018	2019	2020	2021
Number of Goals	15	20	20	10	25

Check your answers

How many did you get correct?

CATCH UP MATHS YEAR 5 BOOK A © PASCAL PRESS ISBN: 9781925726169

1 Use the dot plot to answer the questions.

High Schools that 130 Students Attend

Key: ● = 10 students

Port High, Enda High, Woolly High, Kirra High, Sylvan High

Schools

- Which school has the most students attending? Port High

a How many students attend Sylvan High? ____

b How many students attend Woolly High? ____

c Which school has the least students attending? ____________________

d How many more students attend Kirra High than Enda High? ____

2 Use the dot plot to answer the questions.

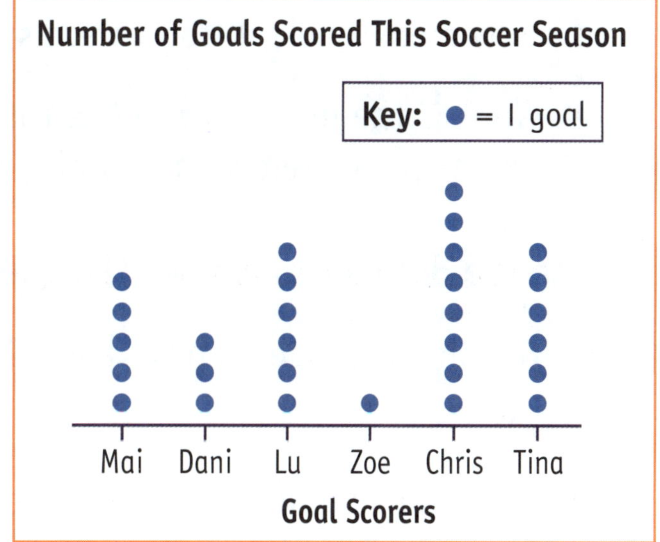

- How many goals did Tina score? 6

a Who scored the most goals? __________

b Who scored two more goals than Zoe? __________

c Who scored five goals? __________

d Who scored the least goals? __________

e How many more goals than Dani did Chris score? ____

f Who scored the same number of goals as Tina? _________

g What was the total number of goals scored? ____

3 Use the two dot plots to answer the following questions.

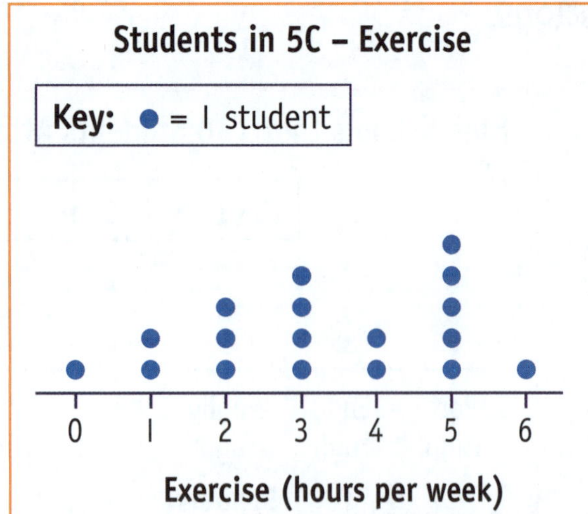

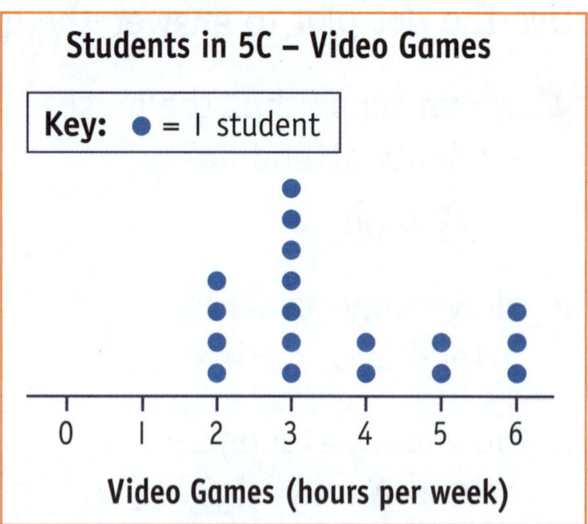

- How many students exercised 5 hours or more per week? 6

a How many students exercised 3 hours or less per week? ____

b How many students play video games for 4 hours per week? ____

c How many hours in total were spent exercising? ____

d How many hours in total were spent playing video games? ____

e What is the difference in the total numbers of hours spent exercising and playing video games? ____

4 Use the dot plot to answer the questions.

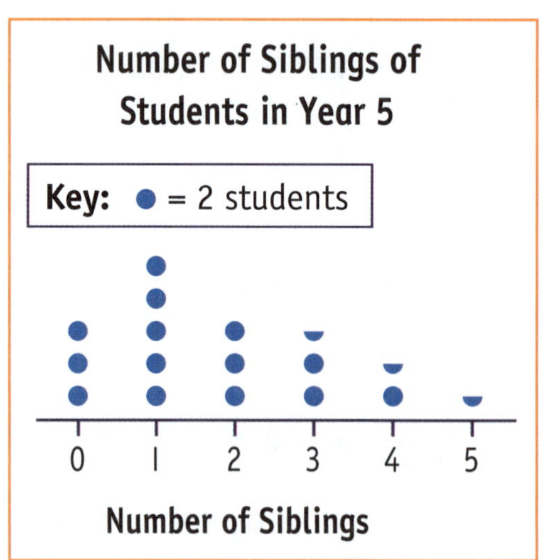

- How many students have two siblings? 6

a How many students are the only child? ____

b How many students have three or more siblings? ____

c ____ students have one sibling.

d How many students are there in Year 5? ____

e What is the difference between students with two siblings and students with four siblings? ____

CATCH UP MATHS YEAR 5 BOOK A © PASCAL PRESS ISBN: 9781925726169

5 **Twenty students were asked how many emails they sent on Monday. Their answers are in the box.**

Graph this data on the dot plot. The first one has been done for you. Then answer the following questions.

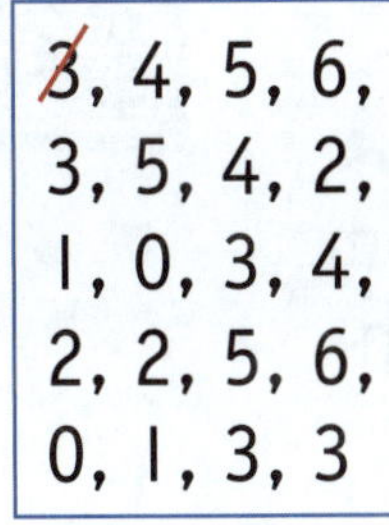

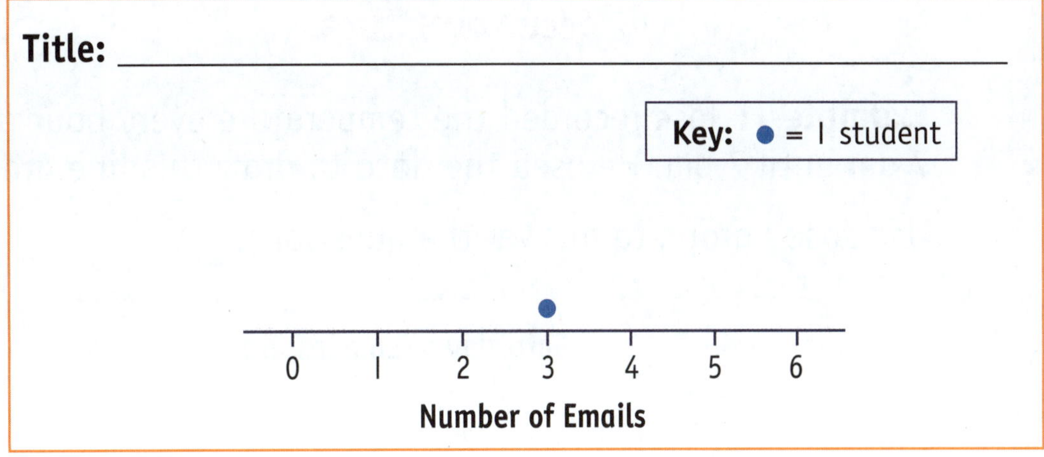

a What was the most common number of emails sent? ___

b How many people sent no emails? ___

c How many emails were sent altogether? ___

d How many people sent one email? ___

6 **Make a dot plot with the data in the table. Use a key of ● = 5 children.**

Children Attending Happy Days Primary

Year	Kinder	1	2	3	4	5	6
Number of Students	35	60	50	45	40	55	50

______

Key:

Kinder Year 1 Year 2 Year 3 Year 4 Year 5 Year 6

Year

LINE GRAPHS

A line graph is when data points on a set of axes are joined by a line. Line graphs are often used when small changes occur over time.

Example 1: Jose recorded the temperature every hour from 7 am until 7 pm. He used the data to draw this line graph.

Use Jose's graph to answer the questions.

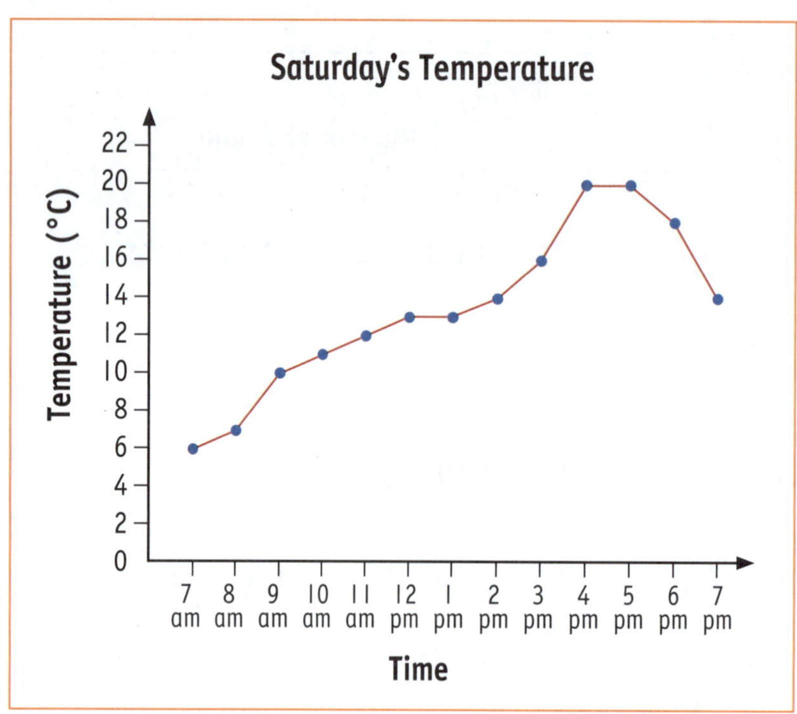

a What was the temperature at 7 am? 6 °C

b What was the lowest temperature Jose recorded? 6 °C

c The highest temperature recorded by Jose was 20 °C.

d At what time was 16°C recorded? 3 pm

e At what time was the temperature 11°C? 10 am

f What was the difference between the highest and lowest temperatures recorded by Jose? 14 °C

Example 2: This line graph shows the numbers of visitors to a park. There were 7 visitors at the park on Saturday, and 6 visitors on Sunday. Add this information to the graph.

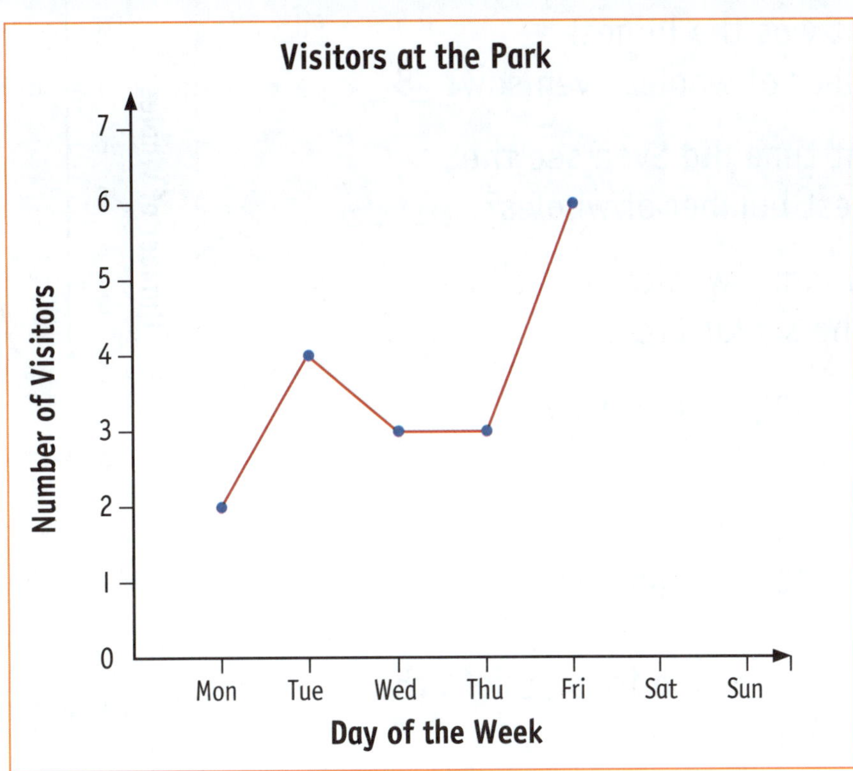

Check your answer on the video!

Your turn

Use the line graph above to answer the questions.

a Use the graph to complete the table.

Day of the Week	Mon	Tue	Wed	Thu	Fri	Sat	Sun
Number of Visitors	2						

b What day had the same number of visitors as Wednesday?

____

c How many people visited on Friday? ____

d What is the difference between the number of visitors at the park on Thursday and Monday? ____

e How many visitors were at the park on the weekend? ____

f How many visitors were at the park on weekdays? ____

SELF CHECK Tick how you feel

Got it!	Need help...	I don't get it
☐	☐	☐

Check your answers

How many did you get correct?

PRACTICE

1 **Sven recorded the number of whales he saw on a whale watching trip on 21 July.**

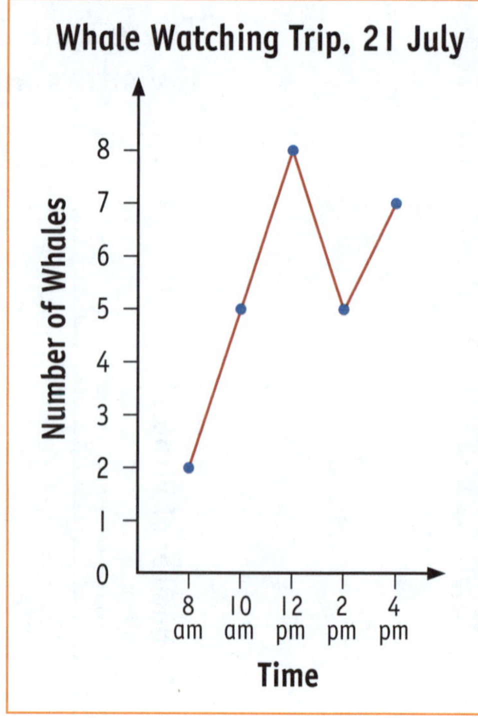

- What was the largest number of whales Sven saw? 8

a What time did Sven see the largest number of whales? ______

b How many whales did he see at 2 pm? ___

c How many whales were seen before 12 noon? ___

d How many whales were seen altogether? ___

2 **Use the line graph to complete the information.**

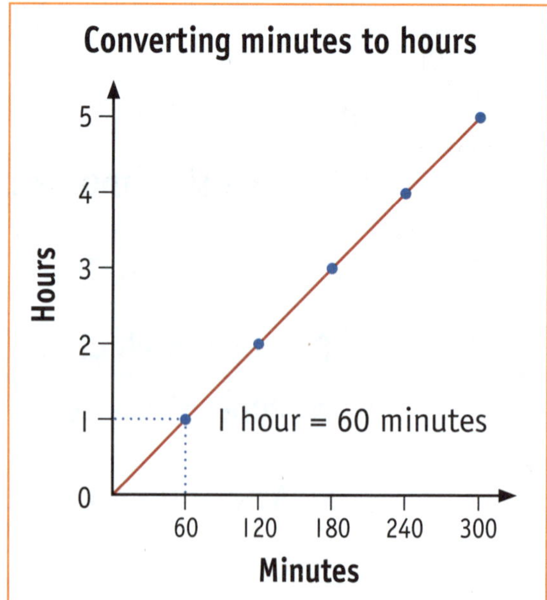

This line graph shows the relationship between ______ and minutes.

You can use this graph to convert hours to __________ or to convert minutes to hours.

The graph shows that:

__ hour = ___ minutes.

3 **Use the line graph to convert these hours into minutes.**

- 3 hours = 180 minutes

a 2 hours = ____________

b 5 hours = ____________

c 4 hours = ____________

4 **Use the line graph to convert these minutes into hours.**

- 90 minutes = 1.5 hours

a 270 minutes = __________

b 300 minutes = __________

c 150 minutes = __________

5 Use the line graph to answer the questions.

● After 1 minute, how many litres are in the bathtub? 10

a How long did it take to fill the bathtub with 40 litres? ________

b How many litres were in the bathtub after 5 min? ________

c What was the difference in litres between 4 min and 2 min?

____

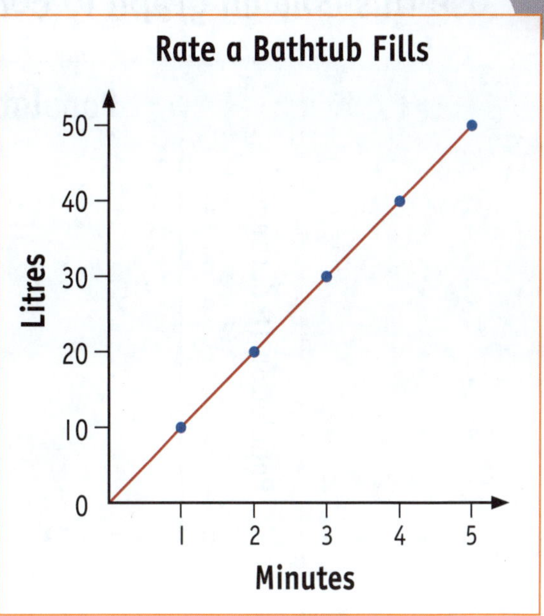

6 Noah's parents measured his height every year on his birthday. Here are their measurements:

Age (years)	0	1	2	3	4	5	6	7
Height (cm)	50	90	120	135	145	150	155	158

a Use the data in the table to complete the line graph.

b Between what ages did Noah grow the most?

____

c How much did Noah grow in 7 years?

____

d How much did Noah grow between 3 and 4 years old?

____

e How tall was Noah when he was born?

____

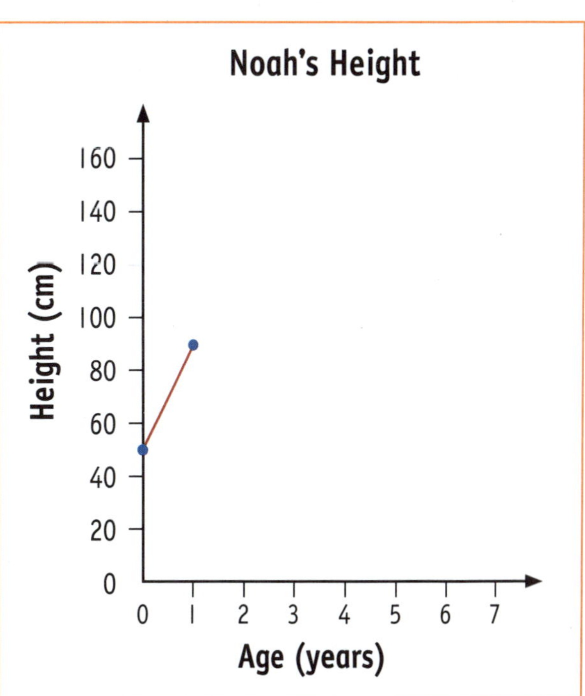

Use this column graph to complete the following questions.

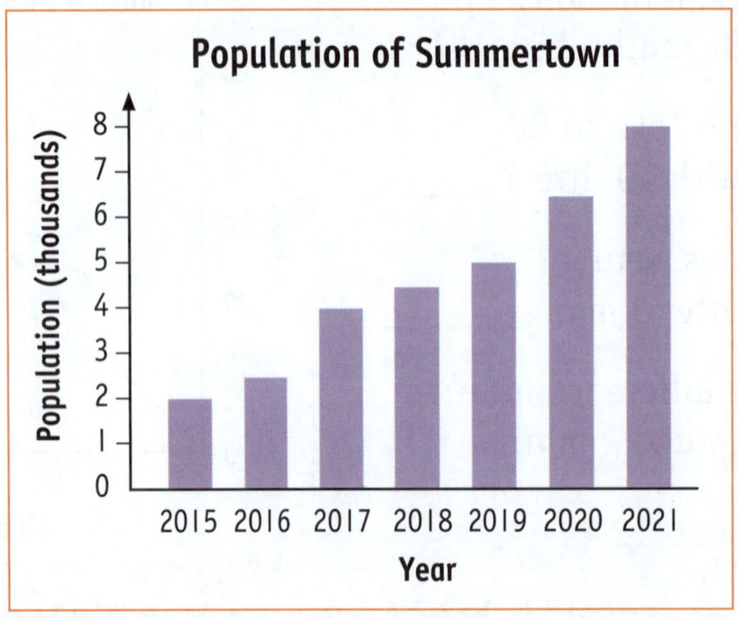

a Use the graph above to complete the table of values.

Population (thousands)	2						
Year	2015						

b Redraw the column graph above as a line graph.

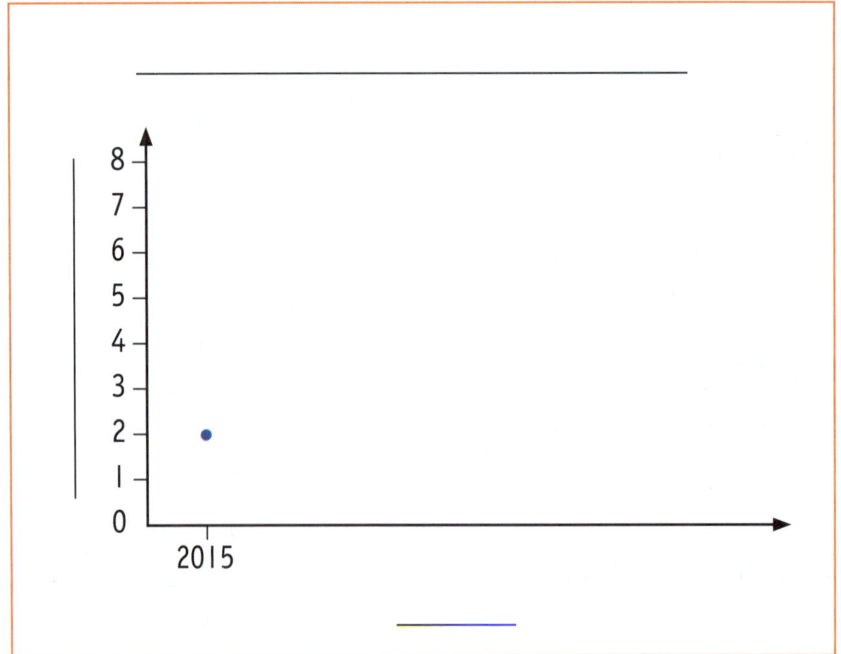

SPREADSHEETS

In a spreadsheet, data is organised into rows and columns.

Example 1: This spreadsheet shows Zac's savings.
Use the spreadsheet to answer the following questions.

	A	B	C	
1	**DATE**	**DEPOSIT**	**SUBTOTAL**	
2	31 JAN	\$2150.00	\$2150.00	= B2
3	28 FEB	\$200.00	\$2350.00	= C2 + B3
4	31 MAR	\$170.50	\$2520.50	= C3 + B4
5	30 APR	\$1500.00	\$4020.50	= C4 + B5
6	31 MAY	\$1000.00		= C5 + B6

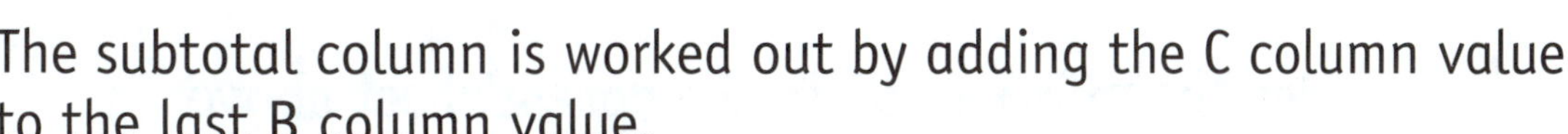

The subtotal column is worked out by adding the C column value to the last B column value.

For example, on 28 Feb, \$200 was deposited.
Add the C column (C2) to the deposit (B3).
\$2150.00 + \$200.00 = \$2350.00 (the value of C3).

a How much was deposited on 31 March? \$170.50

b What was the total of C4? \$2520.50

c When was \$1500.00 deposited? 30 April

d What amount of money was deposited on 31 May? \$1000.00

e What will the total be in C6? \$5020.50

f When was \$170.50 deposited? 31 March

g What date was the subtotal \$2350.00? 28 February

h Write the lowest subtotal. \$2150.00

i How much was deposited altogether? \$5020.50

Example 2:
This spreadsheet shows the deposits from Big Al's Pizza Bar. Complete the spreadsheet.

	A	B	C	
1	**DATE**	**DEPOSIT**	**SUBTOTAL**	
2	6 JUN	$2500.00	$2500.00	= B2
3	8 JUN	$1200.00		= C2 + B3
4	12 JUN	$3250.00		= C3 + B4
5	15 JUN	$1550.00		= C4 + B5
6	24 JUN	$675.00		= C5 + B6
7	30 JUN	$3600.00		= C6 + B7
8	6 JUL	$1850.00		= C7 + B8

Use the information in the spreadsheet above to answer the questions.

- What is the subtotal of C3? $3700.00

a What is the subtotal of C6? ___________

b How much money was deposited on 15 June? ___________

c When was $3600.00 deposited? ___________

d What amount of money was deposited on 6 July?

____

e How many deposits were made in June? ___

f How much money was deposited in total over 12 June and 15 June? ___________

g How much money was deposited altogether? ___________

SELF CHECK Tick how you feel

Got it!	Need help...	I don't get it
☐	☐	☐

Check your answers
How many did you get correct?

CATCH UP MATHS YEAR 5 BOOK A © PASCAL PRESS ISBN: 9781925726169

Use this spreadsheet to complete the following questions.

Hoop's Basketball Club Savings					
	A	B	C	D	
1	DATE	ITEM	DEPOSIT	SUBTOTAL	
2	1 Sep	Opening	$1400.00	$1400.00	= C2
3	5 Sep	Fundraiser	$2435.50	$3835.50	= D2 + C3
4	9 Sep	Donations	$1423.00		= D3 + C4
5	16 Sep	Fees received	$1320.00		= D4 + C5
6	24 Sep	Fundraiser	$ 600.00		= D5 + C6
7	28 Sep	Donations	$1255.00		= D6 + C7
8	30 Sep	Fees received	$1320.00		= D7 + C8

a Fill in the missing amounts in the spreadsheet.

b What was the largest deposit made? __________

c What date was the smallest deposit made? __________

d How much money was deposited on 28 September? __________

e What was the opening balance? __________

Add the following items to the Berries spreadsheet.

Berries Soccer Club Expenses					
	A	B	C	D	
1	DATE	ITEM	COST	BALANCE	
2	1 MAR	Opening	–	$9500.00	
3	5 MAR	Registrations	$1000.00	$8500.00	= D2 – C3
4	9 MAR	Bought bags	$500.00	$8000.00	= D3 – C4
5	15 MAR	Bought goals	$750.00	$7250.00	= D4 – C5
6	17 MAR	New balls	$150.00	$7100.00	= D5 – C6
7					= D6 – C7
8					= D7 – C8
9					= D8 – C9

- 17 March: new balls cost $150

a 21 March: trophies cost $295

b 28 March: awards cost $560

c 31 March: new shirts cost $1526

DATA REVIEW

Complete the table and then answer the questions.

Jellybean Colour

Colour	Tally	Total
Red	|||	
Blue	||||	
White		8
Green	~~||||~~ |	
Black		9

a How many jellybeans were there altogether? ____

b How many white jellybeans were there? ____

c What was the total of red and green jellybeans? ____

d What is the difference between the numbers of black and blue jellybeans?

____

e What colour had the smallest number of jellybeans? ________

This table shows how many books each class read during Book Week.

a Complete the total column in the table.

b Complete the picture graph to display the data.

Books Read in Book Week

Class	Tally	Total
5A	~~||||~~ ~~||||~~ ||	
5T	~~||||~~ ~~||||~~	
5P	~~||||~~ ||||	
5S	~~||||~~ ~~||||~~ |	
5D	~~||||~~ |||	

Number of Books Read in Book Week

Key ■ = 2 books

Classes	
5A	
5T	
5P	
5S	
5D	

CATCH UP MATHS YEAR 5 BOOK A © PASCAL PRESS ISBN: 9781925726169

Ava loves watching planes taking off at the airport.

She made this table from data she collected during one week of plane watching.

Number of Planes Taking Off

Day	Tally	Total
Monday	IIII	
Tuesday	~~IIII~~	
Wednesday	~~IIII~~ ~~IIII~~	
Thursday	IIII	
Friday	~~IIII~~ I	
Saturday	~~IIII~~ ~~IIII~~ II	
Sunday	~~IIII~~ ~~IIII~~ III	

a Fill in the missing information in the table.

b Write three questions you could ask and answer from the data in Ava's table.

- ______________________________
- ______________________________
- ______________________________

c Construct a vertical column graph and a horizontal column graph to display the data in Ava's table.

Vertical column graph

____

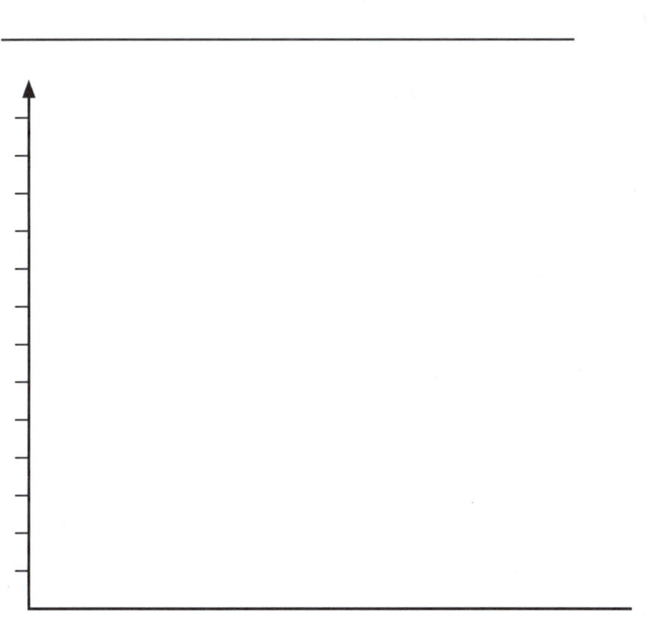

Horizontal column graph

____

REVIEW

4 Use this column graph to answer the following questions.

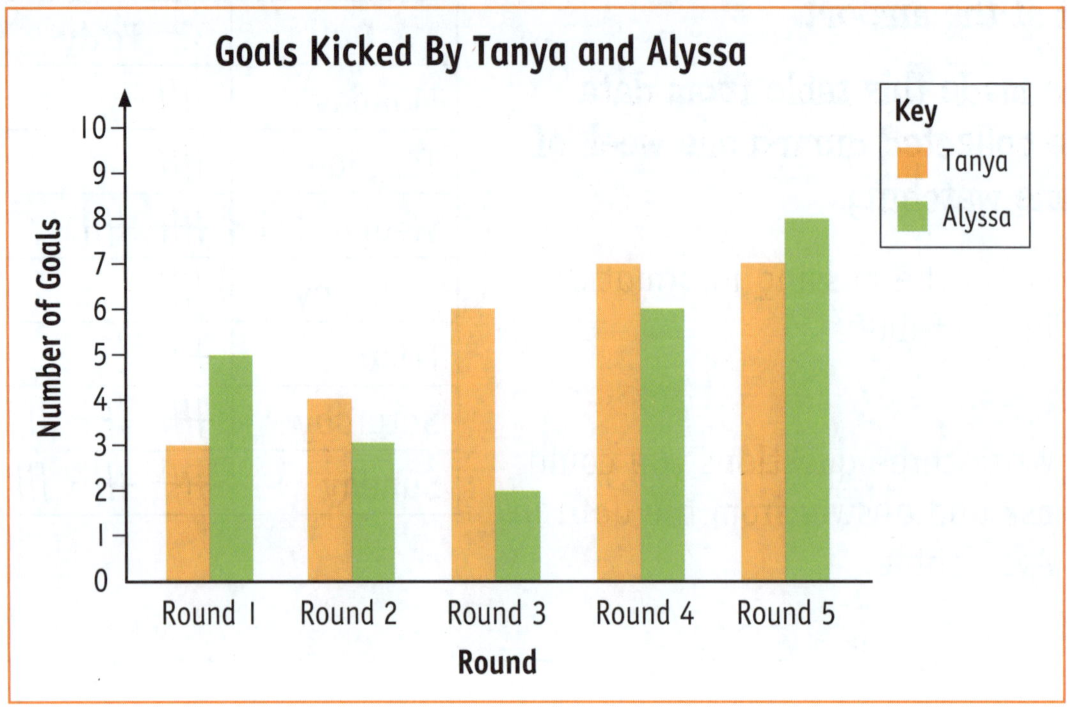

a How many goals altogether did Tanya and Alyssa score in Round 2?

____

b Tanya and Alyssa kicked 13 goals in Round 4. True or false? __________

c The most goals kicked was by ______________ in Round ___.

d Alyssa kicked ___ goals altogether.

e How many goals did Tanya and Alyssa kick altogether? ___

f What is the difference in the number of goals Tanya and Alyssa kicked in Round 3? ___

g What is the difference in the number of goals kicked by Tanya in Round 4 and Alyssa in Round 3? ___

h What was the least number of goals kicked? ___

i Who scored more goals in Round 1? ______________

j Who scored more goals altogether? ______________

CATCH UP MATHS YEAR 5 BOOK A © PASCAL PRESS ISBN: 9781925726169

5 Use this bar graph to answer the following questions.

Animals on Lina's Farm

Horses	Chickens	Goats	Sheep

a How many types of animals does Lina have on her farm? ___

b If Lina has 60 animals in total, how many horses does she have? ___

c The animal Lina has the smallest number of is ______________.

d The third-largest group of animals is ______________.

e If $\frac{1}{4}$ of the 60 animals are sheep, how many sheep does Lina have? ___

6 Construct a bar graph using the following information.

Of the people surveyed:

- $\frac{1}{2}$ chose chocolate
- $\frac{1}{4}$ chose potato chips
- $\frac{1}{8}$ chose lollies
- equal numbers of people chose pretzels and cake.

Favourite Snack Food

a If 100 people were surveyed, how many chose potato chips? ___

b If 120 people were surveyed, how many people chose lollies? ___

c Is chocolate the most or least popular snack food? _______

d How many snack foods were in the survey? ___

e What was the second most popular snack food? _________________

REVIEW

7 **Use this dot plot to answer the following questions.**

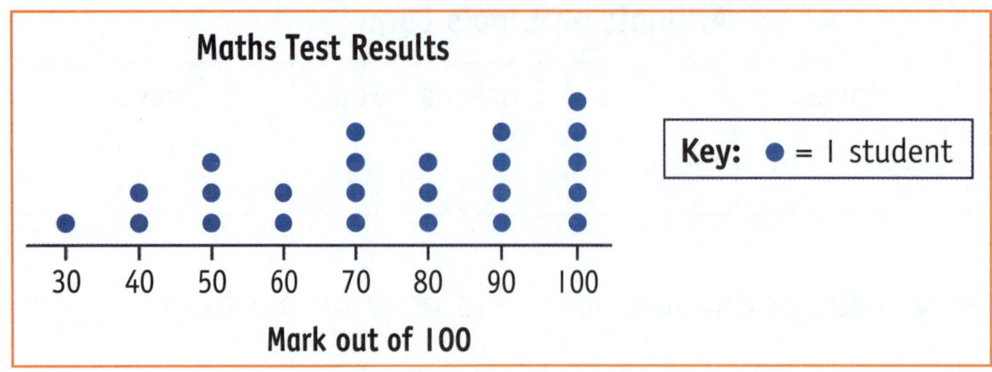

a How many students did the maths test? ____

b How many students scored less than 50? ____

c ____ students scored 90.

d ____ students scored 100.

e How many students scored 60 or more in the test? ____

f If 50 is a pass mark, how many students passed the exam? ____

g What was the most common score? ____

h What was the lowest score on the test? ____

8 **Make a dot plot of the data in the table. Use a key of ● = 2 students.**

Heights of Year 5 Students

Height (cm)	150	152	154	156	158	160	162
Number of Students	12	13	15	18	13	14	15

Key:

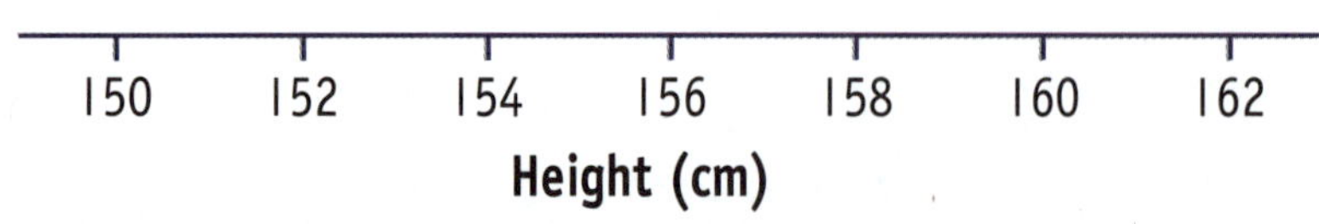

CATCH UP MATHS YEAR 5 BOOK A © PASCAL PRESS ISBN: 9781925726169

Use the data in the table to complete the activities.

Eli's Drive to Summertown

Time	Total km driven
10 am	0
11 pm	50
12 pm	150
1 pm	200
2 pm	300
3 pm	350

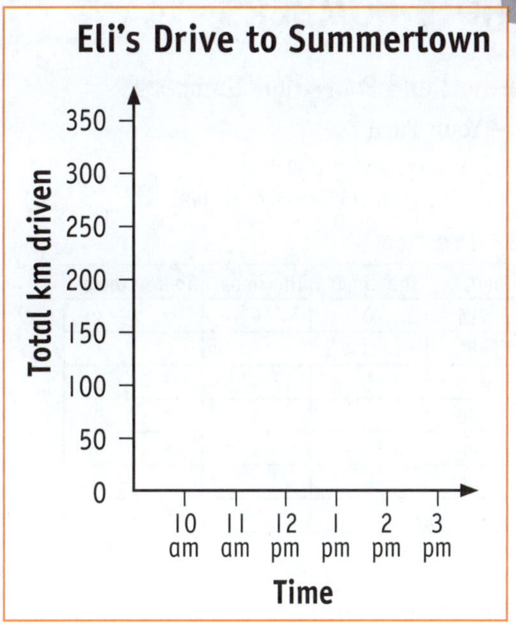

a Construct a line graph of Eli's drive.

b How many kilometres did Eli drive:

- 10 am–11 am? _____
- 11 am–12 noon? _____
- 2 pm–3 pm? _____
- altogether? _______

Use the spreadsheet to answer the following questions.

Antonia's Savings Spreadsheet					
	A	B	C	D	
1	**DATE**	**DEPOSIT**	**WITHDRAWAL**	**BALANCE**	
2	1 FEB	—	—	$220.00	
3	5 FEB	$450.00	—		= D2 + B3
4	9 FEB		$100.00		= D3 – C4
5	12 FEB		$200.00		= D4 – C5
6	14 FEB	$570.00			= D5 + B6
7	18 FEB	$175.00			= D6 + B7
8	28 FEB		$500.00		= D7 – C8

a Complete the Balance column.

b How much money did Antonia deposit in February? _________

c How much money did Antonia withdraw in February? _________

d What was the largest deposit in February? _________

e How much did Antonia withdraw on 12 February? _________

f When did Antonia deposit $450.00? _________

1 WHOLE NUMBERS

Three-digit and Four-digit Numbers

Page 1 – Your Turn

Circled: 361, 544, 471, 289, 107, 929
Underlined: 2593, 6382, 1139, 5945, 4134, 1495

Page 2 – Practice

1

	Number	Thousands	Hundreds	Tens	Ones
a	924	0	9	2	4
b	561	0	5	6	1
c	426	0	4	2	6
d	840	0	8	4	0
e	1384	1	3	8	4
f	8972	8	9	7	2
g	6259	6	2	5	9

2 a seven hundred and eighty-three
b two hundred and forty-eight
c five hundred and ten
d one thousand, nine hundred and sixty-three
e two thousand, four hundred and ninety
f five thousand, seven hundred and twenty-one

3 3 digits: 643, 108, 923 4 digits: 5960, 7463, 4129

Place Value THREE-DIGIT AND FOUR-DIGIT NUMBERS

Page 3 – Your Turn

1 a ones b thousands c tens d tens e ones

2 a 5921 b 253 c 320 d 4815

Page 4 – Practice

1 a hundreds b thousands c ones d tens e hundreds f tens g hundreds h ones i thousands j tens k hundreds l tens m ones

2 a 3458 b 900 c 756 d 1834 e 235

3 a 8435 b 4756 c 665 d 756 e 837 f 4587 g 926 h 8170 i 6109 j 219 k 2013 l 324 m 596 n 4923

Value THREE-DIGIT AND FOUR-DIGIT NUMBERS

Page 5 – Your Turn

Circle 634, 8231, 432

Page 6 – Practice

1 a 4 b 400 c 4 d 40 e 40 f 400 g 4000 h 400 i 400 j 4000 k 40 l 4000 m 40 n 40

2 Adult to check

3 a 530 b 1037 c 265 d 2589 e 9803 f 1009 g 8632 h 582 i 470 j 4936 k 5419 l 873 m 684 n 3781 o 9275

4 a 439, 447, 1472, 415
b 6358, 6426, 6156
c 1297, 407, 1357
d 735, 32, 4534, 937

Number Expanders THREE-DIGIT AND FOUR-DIGIT NUMBERS

Page 7 – Your Turn

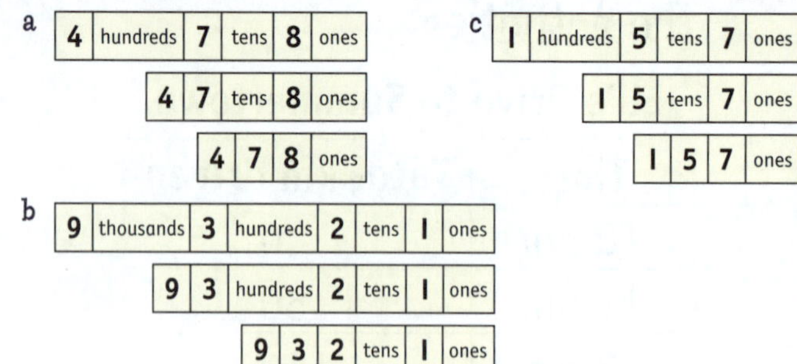

Page 8 – Practice

1 a 7143 b 672 c 251 d 5749 e 3486 f 3259

2 Incorrect number expanders that should be crossed out:

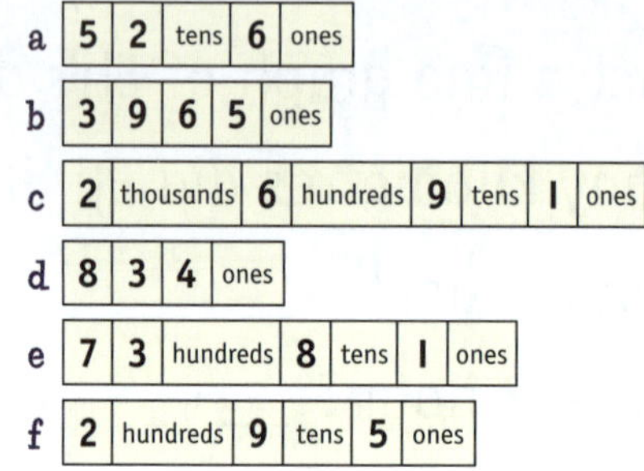

Expanded Numbers THREE-DIGIT AND FOUR-DIGIT NUMBERS

Page 9 – Your Turn

a 400 + 30 + 2
= 4 hundreds + 3 tens + 2 ones
b 3000 + 60 + 5
= 3 thousands + 6 tens + 5 ones

Page 10 – Practice

1 a 800 + 50 + 7 b 300 + 20 + 9 c 1000 + 90 + 3 d 1000 + 800 + 30 + 4 e 500 + 6 f 4000 + 300 + 70

2 a 300 + 2 b 1000 + 600 + 40 + 2 c 2000 + 30 + 7 d 4000 + 60 e 800 + 50

3 a 1000 + 400 + 60 + 3 b 700 + 7 c 1000 + 400 + 90 d 500 + 40 e 6000 + 800 + 9 f 400 + 20 + 4

Modelling Numbers THREE-DIGIT AND FOUR-DIGIT NUMBERS

Page 11 – Your Turn

a 1224 b 361

Page 12 – Practice

1 a 0, 7, 3, 2 b 0, 4, 1, 0 c 0, 5, 0, 3 d 1, 0 ,8, 3 e 4, 8, 6, 0 f 3, 6, 0, 9 g 5, 9, 1, 7

2 a b c

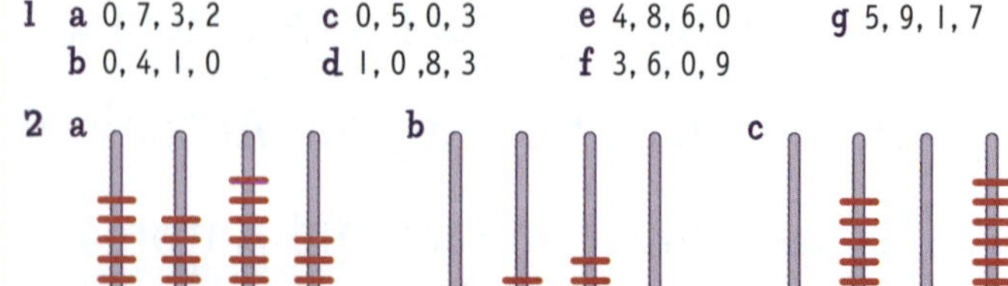

CATCH UP MATHS YEAR 5 BOOK A © PASCAL PRESS ISBN: 9781925726169

d e

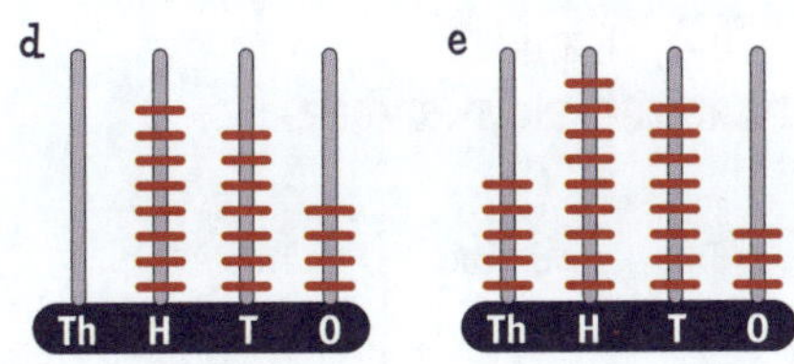

Ordering Numbers THREE-DIGIT AND FOUR-DIGIT NUMBERS

Page 13 – Your Turn

a A b A c D d A e D

Page 14 – Practice

1 a 101, 217, 437, 536
b 2417, 2560, 2597, 2650
c 243, 342, 423, 432
d 1179, 1197, 1719, 1791
e 5060, 5660, 6560, 6650

2 a 110, 620, 635, 712, 743
b 2795, 2714, 2605, 2597, 2560
c 603, 423, 362, 342, 204
d 3752, 3725, 3572, 3275, 3257
e 4911, 4901, 4191, 4119, 4109

3 a ascending
b descending
c ascending
d ascending
e descending

Five-digit Numbers

Page 15 – Your Turn

1 37 651, 10 350, 29 734, 13 659, 42 715

2 a forty thousand, three hundred and ninety-five
b sixty-two thousand, five hundred and thirty-six

Page 16 – Practice

1 a fifty thousand, two hundred and eighty-six
b ninety-three thousand, two hundred and seventy-five
c eighty-seven thousand, three hundred and forty
d sixty-seven thousand and thirty-eight
e twenty-four thousand, one hundred
f seventy-one thousand, eight-hundred and fifty-four
g thirty-six thousand and nine
h forty-two thousand and ten

2 a seventy-four thousand and fifty-six
b eighty thousand and ninety
c thirty-two thousand, eight hundred and forty
d sixty-two thousand and eight

3 68 560, 22 870, 32 590, 28 490, 30 003, 45 310

Place Value FIVE-DIGIT NUMBERS

Page 17 – Your Turn

43 240, 329, 7267, 927, 1352, 672, 93, 3289, 14, 63 053

Page 18 – Practice

1 a 42 560
b 95 390
c 59 936
d 99 363
e 38 342
f 67 429
g 84 132
h 53 931
i 19 150
j 10 000
k 35 003
l 62 415
m 71 307
n 59 020

2

	Number	Ten Thousands	Thousands	Hundreds	Tens	Ones
a	63 000	6	3	0	0	0
b	42 618	4	2	6	1	8
c	86 254	8	6	2	5	4
d	38 913	3	8	9	1	3
e	71 352	7	1	3	5	2

3 a ten thousands
b ones
c thousands
d tens
e hundreds
f ten thousands
g thousands
h hundreds
i tens
j ones
k thousands
l hundreds

Value FIVE-DIGIT NUMBERS

Page 19 – Your Turn

1 Adult to check

2 54, 64 357, 3156

Page 20 – Practice

1 a 30 000
b 7
c 50
d 0
e 10 000
f 2000
g 900
h 10 000
i 40
j 500
k 5000

2 Adult to check. Sample answers:
a 94 258
b 28 945
c 24 958
d 85 249
e 28 495
f 89 425
g 52 948
h 45 894

3 a 28, 0, 8, 3428
b 0, 320, 6320, 0

Number Expanders FIVE-DIGIT NUMBERS

Page 21 – Your Turn

5 3 thousands 2 hundreds 4 tens 7 ones
5 3 2 hundreds 4 tens 7 ones
5 3 2 4 tens 7 ones
5 3 2 4 7 ones

Page 22 – Practice

1 a 50 165
b 70 800
c 47 294
d 96 428
e 59 603
f 84 059

2 a 4 ten thousands 7 thousands 3 hundreds 2 tens 6 ones
b 8 ten thousands 1 thousands 0 hundreds 1 tens 6 ones
c 9 ten thousands 2 thousands 3 hundreds 0 tens 4 ones

3 a 2 7 3 4 tens 9 ones
b 9 1 4 6 2 ones
c 8 2 9 hundreds 5 tens 4 ones
d 9 0 3 6 tens 5 ones
e 5 7 9 hundreds 5 tens 2 ones

Expanded Numbers FIVE-DIGIT NUMBERS

Page 23 – Your Turn

a 300 b 1000 c 700 d 30

Page 24 – Practice

1 a 10 000 + 5000 + 400 + 70 + 1
b 3000 + 500 + 80 + 4
c 20 000 + 9000 + 700 + 60 + 2
d 30 000 + 5000 + 800 + 40 + 7
e 50 + 7
f 100 + 90 + 3
g 5000 + 600 + 40 + 9

2 a 20 000 + 2000 + 300 + 80 + 7
b 90 000 + 1000 + 400 + 70 + 6
c 80 000 + 9000 + 200 + 90 + 5
d 70 000 + 3000 + 40 + 9

3 a 70 000, 800, 60
b 1000, 60
c 10 000, 30, 6
d 300, 8
e 20 000, 500, 9
f 200, 7

Modelling Numbers FIVE-DIGIT NUMBERS

Page 25 – Your Turn

a 16 209 2 a 35 235 b 51 037

1 WHOLE NUMBERS CONTINUED

Page 26 – Practice

1 a

TT Th H T O

37, 6, 4, 7

d

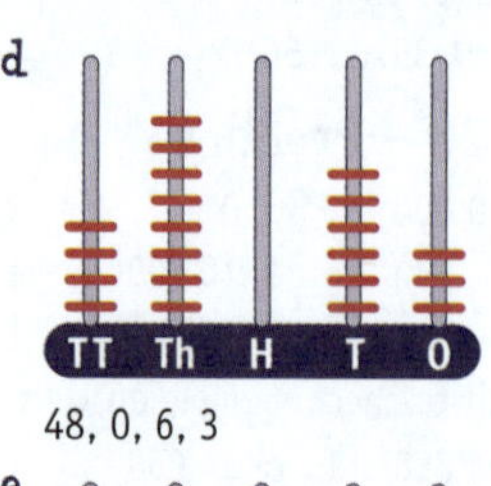

48, 0, 6, 3

b

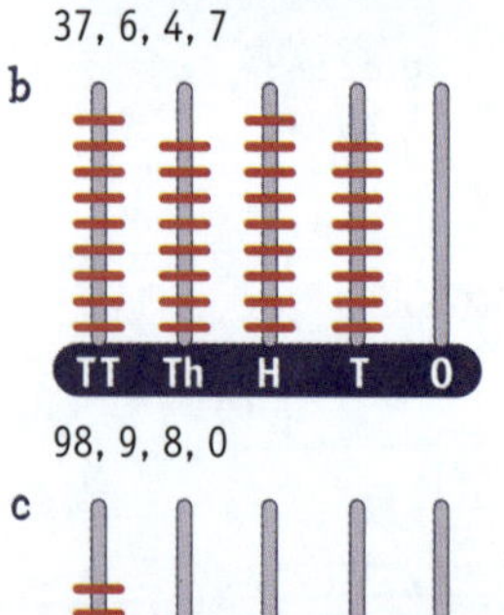

98, 9, 8, 0

e

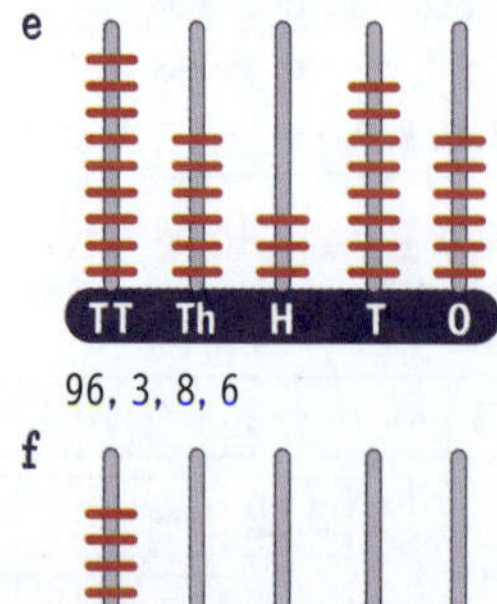

96, 3, 8, 6

c

TT Th H T O

74, 3, 4, 2

f

TT Th H T O

82, 1, 0, 3

Ordering Numbers FIVE-DIGIT NUMBERS

Page 27 – Your Turn

a ✓ b ✗ c ✗ d ✓

Page 28 – Practice

1 a Ascending order 1797, 2849, 3974, 4937, 9745
Descending order 9745, 4937, 3974, 2849, 1791
b Ascending order 26 395, 29 356, 59 263, 69 352, 96 532
Descending order 96 532, 69 352, 59 263, 29 356, 26 395
c Ascending order 13 657, 16 375, 57 136, 67 531, 75 631
Descending order 75 631, 67 531, 57 136, 16 375, 13 657
d Ascending order 23 468, 32 648, 48 632, 68 324, 86 423
Descending order 86 423, 68 324, 48 632, 32 648, 23 468

2

Prize Money	Order
$57 293	5
$75 216	2
$52 999	7
$10 195	10
$12 639	9

Prize Money	Order
$63 147	3
$24 652	8
$55 250	6
$98 343	1
$60 249	4

Odd and Even Numbers

THREE-DIGIT, FOUR-DIGIT AND FIVE-DIGIT NUMBERS

Page 29 – Your Turn

Circle red: 389, 1367, 13 247, 415, 133, 71 529, 9009
Circle green: 432, 2468, 3572, 25 836, 146, 56 930

Page 30 – Practice

1 a 477, 479, 481
b 1824, 1826, 1828
c 900, 902, 904
d 12 442, 12 444, 12 446
e 98 324, 98 326, 98 328
f 65 232, 65 230, 65 228
g 19 081, 19 079, 19 077
h 2398, 2396, 2394
i 869, 867, 865
j 24 940, 24 942, 24 944

2 Odd Numbers: 235, 57 975, 767, 1351, 13 571, 98 999, 3593
Even Numbers: 42 436, 2492, 148, 25 920, 646, 4244

Greater Than, Less Than, Equal To

THREE-DIGIT, FOUR-DIGIT AND FIVE-DIGIT NUMBERS

Page 31 – Your Turn

a False b True c True d True e True

Page 32 – Practice

1 a <, less than
b >, greater than
c =, equal to
d >, greater than
e <, less than

2 a > b < c < d = e < f < g > h > i < j = k = l < m > n >

3 a ✓ b ✓ c ✗ d ✗ e ✗ f ✓ g ✓ h ✓ i ✗ j ✗ k ✗

Rounding to the Nearest 100, 1000 and 10 000

Page 33 – Your Turn

a 5800 b 700

Page 34 – Practice

1 a 700 b 1500 c 9900 d 15 500 e 59 800 f 64 300 g 700 h 8900
2 a 2000 b 3000 c 7000 d 16 000 e 41 000
3 a 50 000 b 50 000 c 60 000 d 60 000 e 10 000
4

	Nearest 100	Nearest 1000	Nearest 10 000
a 68 437	68 400	68 000	70 000
b 40 305	40 300	40 000	40 000
c 23 036	23 000	23 000	20 000

Largest and Smallest Numbers

THREE-DIGIT, FOUR-DIGIT AND FIVE-DIGIT NUMBERS

Page 35 – Your Turn

1 a 3557 b 246 c 10 279
2 a 8643 b 921 c 99 840

Page 36 – Practice

1 a 707 b 309 c 236 d 145 e 238 f 155 g 609 h 235
2 a 8543 b 9852 c 3331 d 6330 e 5400 f 9531 g 9860 h 8700
3 a 22 345, 54 322
b 34 568, 86 543
c 34 679, 97 643
d 20 577, 77 520
e 11 688, 88 611
f 70 099, 99 700
g 10 289, 98 210

Six-digit Numbers

Page 37 – Your Turn

	Number	Hundred Thousands	Ten Thousands	Thousands	Hundreds	Tens	Ones
a	137 397	1	3	7	3	9	7
b	950 133	9	5	0	1	3	3

Page 38 – Practice

1 a five hundred and twenty-four thousand, three hundred and seventy-eight
b nine hundred and fifty-four thousand and forty-eight
c one hundred and twenty-five thousand, two hundred and ninety-five
d one hundred and eighty thousand and thirty-two
e nine hundred and eighty-nine thousand, six hundred and twenty-four

  ISBN: 9781925726169

2

	Number	Hundred Thousands	Ten Thousands	Thousands	Hundreds	Tens	Ones
a	102 252	1	0	2	2	5	2
b	679 035	6	7	9	0	3	5
c	100 324	1	0	0	3	2	4
d	959 615	9	5	9	6	1	5
e	980 711	9	8	0	7	1	1
f	897 241	8	9	7	2	4	1
g	948 111	9	4	8	1	1	1
h	468 037	4	6	8	0	3	7
i	975 152	9	7	5	1	5	2

Value SIX-DIGIT NUMBERS

Page 39 – Your Turn

a 600 000 c 2000 e 90
b 70 000 d 400 f 5

Page 40 – Practice

1 a 200 000 d 0 g 6000 j 1
b 40 000 e 3000 h 50 000 k 0
c 300 f 300 000 i 10

2 a 2 d 2 g 2000 j 20
b 200 000 e 20 h 200 000 k 20
c 2000 f 200 i 20 000

3 Adult to check. Sample responses:
a 269 010 b 492 864 c 908 586 d 183 622 e 786 554

4 a 40 000 b 40 c 400 d 4 e 400 000

Number Expanders SIX-DIGIT NUMBERS

Page 41 – Your Turn

a 279 113 b 713 490

Page 42 – Practice

1 a 438 536 c 546 005 e 873 290 g 538 178
b 444 823 d 553 429 f 437 302

2
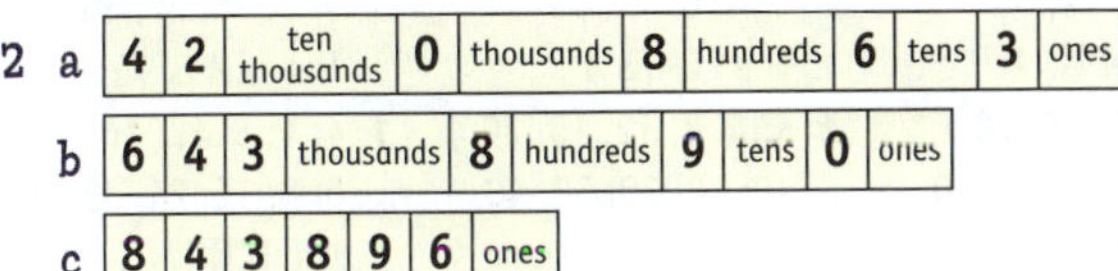
a 4 2 ten thousands 0 thousands 8 hundreds 6 tens 3 ones
b 6 4 3 thousands 8 hundreds 9 tens 0 ones
c 8 4 3 8 9 6 ones
d 7 3 6 2 8 tens 3 ones

3 a 869 375 b 135 350 c 240 593

Expanded Numbers SIX-DIGIT NUMBERS

Page 43 – Your Turn

a 100 + 20 + 6
b 2000 + 500 + 90 + 3
c 30 000 + 2000 + 800 + 40 + 6
d 700 000 + 40 000 + 3000 + 200 + 90 + 1

Page 44 – Practice

1 a 20 + 9
b 6000 + 300 + 80 + 2
c 700 + 60 + 3
d 20 000 + 4000 + 600 + 30 + 9
e 600 000 + 10 000 + 7000 + 300 + 40 + 5
f 700 000 + 20 000 + 1000 + 500 + 9
g 800 000 + 9000 + 3000 + 10 + 2
h 900 000 + 30 000 + 600 + 30
i 600 000 + 80 000 + 2000 + 900
j 400 000 + 300 + 20 + 6

2 a 200 000 + 40 000 + 9000 + 200 + 5
b 600 000 + 2000 + 900 + 50
c 800 000 + 70 000 + 300 + 20 + 3
d 900 000 + 30 000 + 6000 + 5
e 900 000 + 30 000 + 8000 + 500 + 5
f 300 000 + 20 000 + 6000 + 10
g 400 000 + 20 000 + 1000 + 6
h 700 000 + 20 000 + 5000 + 400
i 400 000 + 10 000 + 2000 + 500 + 10 + 6

Ordering Numbers SIX-DIGIT NUMBERS

Page 45 – Your Turn

a 540 480 (854 584) d (197 913) 157 593
b 952 437 (954 273) e (395 482) 386 529
c 257 984 (752 948)

Page 46 – Practice

1 a (985 673) 589 736 d (568 924) 245 896
b (473 921) 129 734 e (908 250) 508 290
c (438 293) 335 274

2 a 1, 3, 2, 4, 5 b 5, 2, 4, 3, 1 c 4, 1, 3, 5, 2
3 a 4, 2, 3, 5, 1 b 2, 4, 5, 3, 1 c 4, 1, 5, 3, 2

Odd and Even Numbers SIX-DIGIT NUMBERS

Page 47 – Your Turn

Circle green: 631 514, 483 160, 989 336
Circle red: 326 513, 537 611, 534 419

Page 48 – Practice

1 Odd Numbers: 625 437, 542 837, 351 255, 491 311, 931 391, 282 533
Even Numbers: 756 284, 982 534, 843 712, 736 526, 118 318, 149 032

2 Odd numbers to cross out:
a 456 983 b 427 289 c 203 859 d 362 487 e 508 963

3 a 273 512, 273 514, 273 516 d 709 500, 709 502, 709 504
b 896 540, 896 542, 896 544 e 431 506, 431 508, 431 510
c 911 088, 911 090, 911 092

Greater Than, Less Than, Equal To SIX-DIGIT NUMBERS

Page 49 – Your Turn

a True b False c True d True e True

Page 50 – Practice

1 a is less than c is greater than e is less than
b is equal to d is less than

2 a > c < e > g = i > k <
b > d = f > h < j =

3 a ✗ c ✓ e ✗ g ✗ i ✓ k ✓ m ✓ o ✓
b ✗ d ✓ f ✗ h ✓ j ✗ l ✓ n ✓

Rounding to 10 000 and 100 000

Page 51 – Your Turn

a 30 000 b 900 000

Page 52 – Practice

1 a 40 000 b 60 000 c 20 000 d 40 000 e 80 000
2 a 800 000 b 400 000 c 400 000 d 800 000 e 700 000
3 a 660 000, 700 000 e 170 000, 200 000 i 730 000, 700 000
b 470 000, 500 000 f 840 000, 800 000 j 640 000, 600 000
c 520 000, 500 000 g 250 000, 300 000
d 460 000, 500 000 h 490 000, 500 000

1 WHOLE NUMBERS CONTINUED

Largest and Smallest Numbers SIX-DIGIT NUMBERS

Page 53 – Your Turn

a Smallest: 236 789 Largest: 987 632
b Smallest: 234 467 Largest: 764 432
c Smallest: 406 789 Largest: 987 640

Page 54 – Practice

1 Smallest, Largest:
a 223 457, 745 322
b 304 789, 987 430
c 223 569, 965 322
d 134 899, 998 431
e 457 889, 988 754
f 122 455, 554 221
g 234 679, 976 432
h 112 228, 822 211

2 Numbers to cross out:
a 205 388
b 300 576
c 132 567
d 230 977
e 774 302
f 164 000
g 775 301
h 526 000
i 023 569

The Role of Zero

Page 55 – Your Turn

a tens
b thousands

Page 56 – Practice

1 a tens
b ones
c hundreds
d hundreds
e ones
f ten thousands
g tens
h tens
i ones
j ones
k hundreds
l ones
m thousands
n ten thousands
o thousands
p tens
q tens

2 Adult to check

Abbreviations of Large Numbers

Page 57 – Your Turn

1 a 72K b 146K
2 a 138 thousand
3 a 916 thous. b 57 thous.

Page 58 – Practice

1 a 246 thous.
b 810K
c 900 thous.
d 5K
e 25 thousand
f 43K
g 217 thousand
h 951 thous.
i 1K
j 14K

2 a 315 thous.
b 16 thous.
c 400 thous.
d 814 thous.
e 2 thous.
f 8 thous.
g 97 thous.

3 a 73K
b 3K
c 52K
d 15K
e 376K
f 988K
g 1K

Factors

Page 59 – Your Turn

a 1, 3, 5, 15
b 1, 3, 9
c 1, 7

Page 60 – Practice

1 a 1, 2, 7, 14
b 1, 2, 4, 8, 16
c 1, 2, 3, 5, 6, 10, 15, 30
d 1, 2, 4, 8, 16, 32
e 1, 2, 3, 4, 6, 9, 12, 18, 36
f 1, 2, 4, 5, 10, 20
g 1, 2, 11, 22
h 1, 2, 3, 6

2 Numbers to cross out:
a 6
b 3
c 6
d 4
e 7
f 2
g 3
h 2
i 6
j 4
k 10

Highest Common Factor (HCF)

Page 61 – Your Turn

a 4 b 3 c 5

Page 62 – Practice

1 a 14: 1, 2, 7, 14
21: 1, 3, 7, 21
7 is the HCF

b 21: 1, 3, 7, 21
24: 1, 2, 3, 4, 6, 8, 12, 24
3 is the HCF

c 18: 1, 2, 3, 6, 9, 18
36: 1, 2, 3, 4, 6, 9, 12, 18, 36
18 is the HCF

d 10: 1, 2, 5, 10
24: 1, 2, 3, 4, 6, 8, 12, 24
2 is the HCF

e 16: 1, 2, 4, 8, 16
20: 1, 2, 4, 5, 10, 20
4 is the HCF

f 22: 1, 2, 11, 22
33: 1, 3, 11, 33
11 is the HCF

g 10: 1, 2, 5, 10
30: 1, 2, 3, 5, 6, 10, 15, 30
10 is the HCF

h 9: 1, 3, 9
27: 1, 3, 9, 27
9 is the HCF

i 12: 1, 2, 3, 4, 6, 12
18: 1, 2, 3, 6, 9, 18
6 is the HCF

Multiples

Page 63 – Your Turn

a 3, 6, 9
b 4, 8, 12
c 12, 24, 36

Page 64 – Practice

1 a 2, 4, 6, 8, 10, 12, 14, 16
b 4, 8, 12, 16, 20, 24, 28, 32
c 5, 10, 15, 20, 25, 30, 35, 40
d 7, 14, 21, 28, 35, 42, 49, 56
e 9, 18, 27, 36, 45, 54, 63, 72
f 10, 20, 30, 40, 50, 60, 70, 80
g 11, 22, 33, 44, 55, 66, 77, 88

2 Numbers to cross out:
a 115
b 28
c 20
d 46
e 28
f 23
g 46
h 35
i 37
j 56

Lowest Common Multiples (LCM)

Page 65 – Your Turn

a 12 b 6 c 6

Page 66 – Practice

1 a 3: 3, 6, 9, 12, 15
4: 4, 8, 12, 16
The LCM is 12

b 5: 5, 10, 15, 20
10: 10, 20, 30, 40
The LCM is 10

c 2: 2, 4, 6, 8, 10
6: 6, 12, 18, 24, 30
The LCM is 6

d 3: 3, 6, 9, 12
6: 6, 12, 18, 24
The LCM is 6

e 1: 1, 2, 3, 4
4: 4, 8, 12, 16
The LCM is 4

f 4: 4, 8, 12, 16, 20
10: 10, 20, 30, 40
The LCM is 20

g 3: 3, 6, 9, 12, 15
5: 5, 10, 15, 20
The LCM is 15

h 2: 2, 4, 6, 8, 10
10: 10, 20, 30, 40
The LCM is 10

i 3: 3, 6, 9, 12, 15
9: 9, 18, 27, 36
The LCM is 9

j 4: 4, 8, 12, 16, 20
8: 8, 16, 24, 32
The LCM is 8

Whole Numbers Review Page 67

1 a sixty-two
b one hundred and thirty-seven
c five hundred and sixty-three
d one thousand, two hundred and fifty
e two thousand and thirty
f forty-five thousand, eight hundred and ninety-three
g eighty-six thousand, seven hundred and twenty-four
h two hundred and fifty-one thousand, three hundred and eighty-six

CATCH UP MATHS YEAR 5 BOOK A © PASCAL PRESS ISBN: 9781925726169

2

	Number	Hundred Thousands	Ten Thousands	Thousands	Hundreds	Tens	Ones
a	56	0	0	0	0	5	6
b	250	0	0	0	2	5	0
c	1346	0	0	1	3	4	6
d	8007	0	0	8	0	0	7
e	32 430	0	3	2	4	3	0
f	40 003	0	4	0	0	0	3
g	100 200	1	0	0	2	0	0
h	840 937	8	4	0	9	3	7
i	647 300	6	4	7	3	0	0
j	420 030	4	2	0	0	3	0

3 a 70 b 7 c 700 d 7000 e 700 f 700 000 g 7000 h 70 000 i 7 j 7 k 7000 l 70

4 a 800 b 20 c 20 d 4 e 20 000 f 6000 g 100 000 h 50 000 i 7 j 500 k 80 l 700 000 m 6 n 90

5 a 328 569 b 674 805 c 635 003 d 270 407

6 a 7, 34 674
73, 4674
734, 674
7346, 74
73 467, 4
734 674, 0

b 5, 83 652
58, 3652
583, 652
5836, 52
58 365, 2
583 652, 0

7 a 109 580 b 961 c 7620 d 87 314 e 89 f 73 554 g 346 h 406 789

8 Adult to check

9 Adult to check

10 236 400, 23 436, 437, 983 402, 127 483, 1438

11 a 4 3 6 9 hundreds 3 tens 2 ones

b 5 1 ten thousands 9 thousands 0 hundreds 6 tens 3 ones

c 3 hundred thousands 2 ten thousands 8 thousands 4 hundreds 3 tens 1 ones

d 7 5 4 9 0 2 ones

e 1 4 7 3 8 tens 9 ones

f 6 9 5 3 hundreds 6 tens 2 ones

12 a 5 hundred thousands 7 ten thousands 3 thousands 9 hundreds 8 tens 6 ones
5 7 ten thousands 3 thousands 9 hundreds 8 tens 6 ones
5 7 3 thousands 9 hundreds 8 tens 6 ones
5 7 3 9 hundreds 8 tens 6 ones
5 7 3 9 8 tens 6 ones
5 7 3 9 8 6 ones

b 6 hundred thousands 4 ten thousands 0 thousands 3 hundreds 7 tens 2 ones
6 4 ten thousands 0 thousands 3 hundreds 7 tens 2 ones
6 4 0 thousands 3 hundreds 7 tens 2 ones
6 4 0 3 hundreds 7 tens 2 ones
6 4 0 3 7 tens 2 ones
6 4 0 3 7 2 ones

13 a 700 + 20 + 3
b 1000 + 400 + 30 + 9
c 500 + 30 + 3
d 20 000 + 5000 + 200 + 90 + 5
e 300 000 + 80 000 + 4000 + 600 + 20 + 9
f 70 000 + 3000 + 800 + 60
g 500 000 + 9000 + 300 + 50
h 70 000 + 6000 + 30 + 9
i 600 000 + 30 000 + 200 + 80 + 4

14

	Number	Th	H	T	O
a	536	0	5	3	6
b	782	0	7	8	2
c	1597	1	5	9	7
d	7846	7	8	4	6
e	39 873	39	8	7	3
f	50 437	50	4	3	7
g	142 583	142	5	8	3
h	235 870	235	8	7	0
i	643 209	643	2	0	9

15

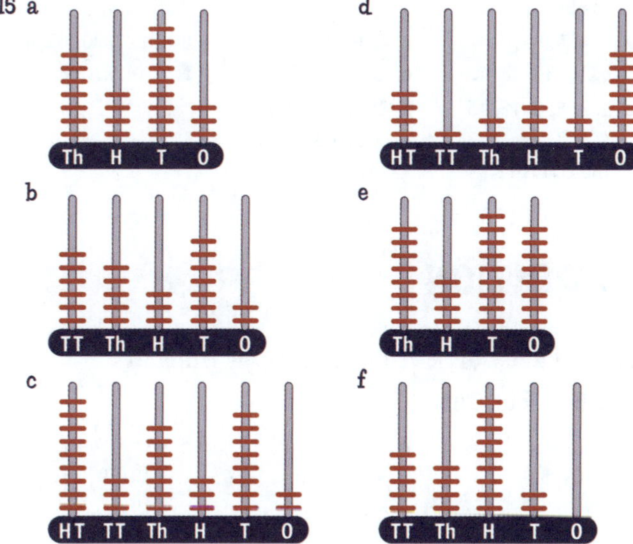

16 a 3526, 5236, 5623, 5632, 6523
b 14 759, 41 579, 54 791, 75 154, 91 547
c 531 682, 582 316, 613 285, 631 582, 815 632
d 290 543, 390 245, 453 029, 904 532, 920 354

17 a 9524, 9254, 5429, 4529, 2945
b 97 330, 97 033, 90 337, 73 903, 37 390
c 97 316, 79 613, 69 137, 67 139, 61 379
d 942 355, 495 234, 459 352, 395 542, 253 954
e 864 023, 682 403, 420 368, 403 862, 320 468

18 Circle red: 27, 53 847, 3423, 656 521, 439
Circle green: 16, 138, 849 342, 1526, 16 430

19 a > b < c < d > e > f < g < h = i > j >

20

Round to	a 532 487	b 643 981	c 857 603
nearest 10	532 490	643 980	857 600
nearest 100	532 500	644 000	857 600
nearest 1000	532 000	644 000	858 000
nearest 10 000	530 000	640 000	860 000
nearest 100 000	500 000	600 000	900 000

1 WHOLE NUMBERS CONTINUED

21 Smallest, Largest
- **a** 156, 651
- **b** 3579, 9753
- **c** 24 578, 87 542
- **d** 13 469, 96 431
- **e** 145 599, 995 541

22
- **a** tens
- **b** ones
- **c** hundreds
- **d** thousands
- **e** hundreds
- **f** ten thousands
- **g** tens
- **h** ten thousands

23 Adult to check

24 Adult to check

25 Adult to check

26 **a** 350K **b** 1K **c** 291K **d** 739K **e** 400K **f** 395K

27
- **a** 635 thous.
- **b** 5 thous.
- **c** 173 thous.
- **d** 849 thous.
- **e** 500 thous.
- **f** 572 thous.

28
- **a** 1, 2, 3, 4, 6, 12
- **b** 1, 2, 4, 8, 16
- **c** 1, 2, 3, 6, 9, 18
- **d** 1, 2, 3, 4, 6, 9, 12, 18, 36

29
- **a** 8: 1, 2, 4, 8
  12: 1, 2, 3, 4, 6, 12
  The HCF is 4.
- **b** 20: 1, 2, 4, 5, 10, 20
  24: 1, 2, 3, 4, 6, 8, 12, 24
  The HCF is 4.
- **c** 15: 1, 3, 5, 15
  18: 1, 2, 3, 6, 9, 18
  The HCF is 3.

30
- **a** 2, 4, 6, 8, 10
- **b** 8, 16, 24, 32, 40
- **c** 5, 10, 15, 20, 25
- **d** 9, 18, 27, 36, 45
- **e** 10, 20, 30, 40, 50
- **f** 7, 14, 21, 28, 35

31
- **a** 3: 3, 6, 9, 12, 15
  4: 4, 8, 12, 16, 24
  The LCM is 12.
- **b** 2: 2, 4, 6, 8, 10
  3: 3, 6, 9, 12, 15
  The LCM is 6.

2 ADDITION

Adding Three or More Single-Digit Numbers

Page 76 – Your Turn

- **a** 2 + 8 + 5 = 10 + 5 = 15
- **b** 3 + 7 + 5 + 2 = 10 + 7 = 17
- **c** 4 + 6 + 7 + 2 = 10 + 9 = 19

Page 77 – Practice

1
- **a** 3 + 7 + 8 = 10 + 8 = 18
- **b** 5 + 5 + 9 = 10 + 9 = 19
- **c** 1 + 9 + 7 = 10 + 7 = 17
- **d** 6 + 4 + 5 = 10 + 5 = 15
- **e** 2 + 8 + 1 = 10 + 1 = 11

2
- **a** 6 + 4 + 3 + 4 = 10 + 7 = 17
- **b** 2 + 8 + 9 + 3 = 10 + 12 = 22
- **c** 5 + 5 + 6 + 7 = 10 + 13 = 23
- **d** 4 + 6 + 3 + 9 = 10 + 12 = 22
- **e** 4 + 6 + 5 + 1 = 10 + 6 = 16

Sum

Page 78 – Your Turn

- **a** 6 + 3 + 1 = 10
- **b** 5 + 9 + 2 = 16
- **c** 24 + 10 + 26 = 60

Page 79 – Practice

1
- **a** The sum is 24.
- **b** The sum is 100.
- **c** The sum is 67.
- **d** The sum is 88.
- **e** The sum is 82.
- **f** The sum is 130.
- **g** The sum is 198.

2
- **a** 93
- **b** 89
- **c** 320
- **d** 193
- **e** 90
- **f** 351
- **g** 469
- **h** 1180
- **i** 872

Relating Addition and Subtraction

Page 80 – Your Turn

- **a** 73 + 15 = 88
  15 + 73 = 88
  88 − 15 = 73
  88 − 73 = 15
- **b** 15 + 32 = 47
  32 + 15 = 47
  47 − 32 = 15
  47 − 15 = 32

Page 81 – Practice

1 **a** 51, 51, 51 **b** 83, 98, 83

2
- **a** 24 + 11 = 35
  11 + 24 = 35
  35 − 11 = 24
  35 − 24 = 11
- **b** 25 + 17 = 42
  17 + 25 = 42
  42 − 17 = 25
  42 − 25 = 17
- **c** 34 + 42 = 76
  42 + 34 = 76
  76 − 42 = 34
  76 − 34 = 42
- **d** 15 + 19 = 34
  19 + 15 = 34
  34 − 19 = 15
  34 − 15 = 19
- **e** 23 + 44 = 67
  44 + 23 = 67
  67 − 44 = 23
  67 − 23 = 44

3
- **a** 45 + 32 = 77
  32 + 45 = 77
  77 − 32 = 45
  77 − 45 = 32
- **b** 9 + 124 = 133
  124 + 9 = 133
  133 − 124 = 9
  133 − 9 = 124

Addition Without Trading

TWO-DIGIT AND THREE-DIGIT NUMBERS

Page 82 – Your Turn

a 57 **b** 177 **c** 379 **d** 869 **e** 997

Page 83 – Practice

1 **a** 66 **b** 104 **c** 59

2 **a** 105 **b** 114 **c** 47

3 **a** 259 **b** 447 **c** 387

4 **a** 596 **b** 399 **c** 795

5
- **a** 477
- **b** 299
- **c** 778
- **d** 949
- **e** 688
- **f** 1030
- **g** 935

Adding With Trading TWO-DIGIT AND THREE-DIGIT NUMBERS

Page 84 – Your Turn

a 145 **b** 71 **c** 542 **d** 663 **e** 1151

Page 85 – Practice

1 **a** 85 **b** 92 **c** 110

2 **a** 90 **b** 73 **c** 111

3 **a** 542 **b** 457 **c** 558

4 **a** 342 **b** 917 **c** 902

5
- **a** 590
- **b** 904
- **c** 1009
- **d** 1409
- **e** 1407
- **f** 704
- **g** 869

Adding With and Without Trading FOUR-DIGIT NUMBERS

Page 86 – Your Turn

a 2367 **b** 6699 **c** 12 319 **d** 1622 **e** 8146

Page 87 – Practice

1 **a** 6019 **b** 5668

2 **a** 8997 **b** 6310 **c** 8828

CATCH UP MATHS YEAR 5 BOOK A © PASCAL PRESS ISBN: 9781925726169

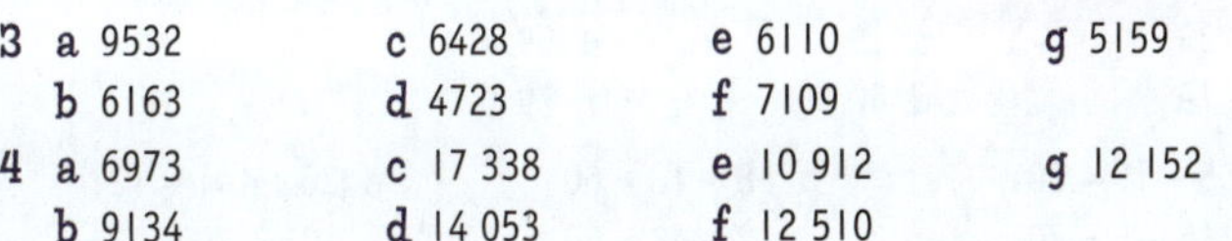

3 a 9532 b 6163 c 6428 d 4723 e 6110 f 7109 g 5159

4 a 6973 b 9134 c 17 338 d 14 053 e 10 912 f 12 510 g 12 152

Rounding to Estimate Addition Answers

Page 88 – Your Turn

a 200, 200, 400 b 550, 100, 650

Page 89 – Practice

1 a 60, 50, 110 b 70, 10, 80 c 280, 40, 320 d 30, 580, 610 e 30, 440, 470

2 a 500, 600, 1100 b 600, 200, 800 c 900, 400, 1300 d 400, 800, 1200

Using the Jump Strategy to Solve Addition

TWO-DIGIT AND THREE-DIGIT NUMBERS

Page 90 – Your Turn

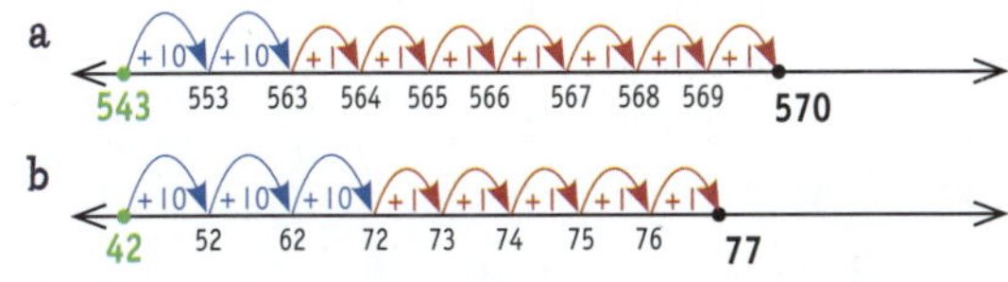

Page 91 – Practice

a 78 b 77 c 469 d 275 e 855 f 745 g 536 h 280

Using the Jump Strategy

THREE-DIGIT AND FOUR-DIGIT NUMBERS

Page 92 – Your Turn

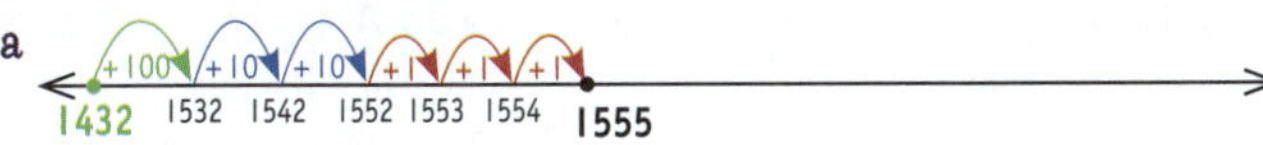

Page 93 – Practice

a 648 b 427 c 692 d 530 e 1758 f 2052 g 3868 h 10 877

Using the Split Strategy to Solve Addition

TWO-DIGIT AND THREE-DIGIT NUMBERS

Page 94 – Your Turn

a 400
60 + 30 = 90
2 + 5 = 7
400 + 90 + 7 = 497

b 500 + 100 = 600
20 + 30 = 50
4 + 3 = 7
600 + 50 + 7 = 657

c 600 + 300 = 900
40 + 30 = 70
2 + 6 = 8
900 + 70 + 8
= 978

Page 95 – Practice

1 a 30 + 20 = 50
6 + 3 = 9
50 + 9 = 59

b 500 + 200 = 700
30 + 30 = 60
4 + 5 = 9
700 + 60 + 9 = 769

c 100 + 0 = 100
30 + 40 = 70
2 + 3 = 5
100 + 70 + 5 = 175

d 700 + 800 = 1500
40 + 30 = 70
2 + 5 = 7
1500 + 70 + 7 = 1577

e 800 + 0 = 800
50 + 40 = 90
8 + 1 = 9
800 + 90 + 9 = 899

f 500 + 400 = 900
50 + 20 = 70
2 + 5 = 7
900 + 70 + 7 = 977

g 200 + 0 = 200
80 + 10 = 90
3 + 4 = 7
200 + 90 + 7 = 297

h 600 + 200 = 800
0 + 50 = 50
7 + 4 = 11
800 + 50 + 11 = 861

Using the Split Strategy

THREE-DIGIT AND FOUR-DIGIT NUMBERS

Page 96 – Your Turn

a 2000 + 4000 = 6000
500 + 100 = 600
20 + 30 = 50
5 + 2 = 7
6000 + 600 + 50 + 7
= 6657

Page 97 – Practice

1 a 500 + 200 = 700
60 + 10 = 70
1 + 5 = 6
700 + 70 + 6 = 776

b 3000 + 4000 = 7000
200 + 300 = 500
50 + 20 = 70
1 + 3 = 4
7000 + 500 + 70 + 4
= 7574

c 8000 + 0 = 8000
700 + 100 = 800
30 + 40 = 70
2 + 6 = 8
8000 + 800 + 70 + 8
= 8878

d 700 + 100 = 800
50 + 20 = 70
7 + 1 = 8
800 + 70 + 8 = 878

e 7000 + 1000 = 8000
200 + 300 = 500
60 + 20 = 80
7 + 1 = 8
8000 + 500 + 80 + 8
= 8588

Rounding Money

Page 98 – Your Turn

a $8.55 b $14.35 c $15.25 d $20.65 e $84.95

Page 99 – Practice

1 a $1.60 b $0.95 c $3.40 d $5.55 e $11.30 f $13.65 g $23.95 h $27.80 i $7.30 j $41.85 k $0.10 l $33.90 m $31.05 n $9.85 o $101.95 p $35.50 q $123.55 r $25.65 s $15.65 t $0.85

2 a $2.45 b $3.75 c $24.15 d $41.60 e $157.55 f $276.00 g $200.00

Working Out Change

Page 100 – Your Turn

a $2.25 b $8.45

Page 101 – Practice

1 a $1.65 b $2.90 c $1.05 d $3.20 e $7.45 f $4.80 g $1.70 h $7.50 i $11.25 j $26.45

2 ADDITION CONTINUED

Money Problems – Spending and Saving

Page 102 – Your Turn

a $37.30

Page 103 – Practice

1 a Danny $246.05, Kai $237.35
b $348.25
c $316.40
d Danny
e Danny

2 a

Deposit	Withdraw	Balance
		$ 0.00
$ 348.25		$ 348.25
	$ 20.00	$ 328.25

b

Deposit	Withdraw	Balance
		$ 0.00
$ 316.40		$ 316.40
	$ 20.00	$ 296.40

3 a $100.00
b $72.00
c $1007
d $150.00
e $287.00
f $815.00
g $715.00
h 23/3
i 1/4
j $815.00
k $520.00
l $20.00
m 23/3
n 15/3

4

Date	Deposit	Withdraw	Balance
5/5	$20.00		$20.00
7/5	$30.00		$50.00
9/5	$120.00		$170.00
15/5		$50.00	$120.00
21/5		$20.00	$100.00
30/5	$60.00		$160.00
1/6		$10.00	$150.00
6/6	$250.00		$400.00

a $50.00
b $120.00
c $160.00
d $400.00
e $250.00
f 5
g 3
h $480.00
i $80.00
j 30/5
k $230.00
l $70.00

5

Earnings	
Item	**Amount**
Pay	$979.80
Pay for chores	$65.50
Pay for walking dogs	$75.00
Total	$1120.30

Expenses	
Item	**Amount**
Soccer fees	$120.00
Clothes	$135.50
Guitar lesson	$55.00
Food	$127.50
Movies	$55.00
Total	$493.00

a $493.00
b $140.50
c $1120.30
d $627.30
e Clothes
f $777.80

6

Necessity	Amount
Clothes	$55.00
Medicine	$16.90
Milk	$4.20
Fruit and veg	$29.50
Bread	$5.00
Total	$110.60

Luxury	Amount
Soft drink	$4.20
Toy car	$15.20
Flowers	$22.50
Takeaway	$42.00
Chocolate	$5.50
Total	$89.40

a Answers will vary.
b $90.40
c Answers will vary.

Addition Review Page 108

1 a 6 + 4 + 7
= 10 + 7
= 17
b 5 + 5 = 3
= 10 + 3
= 13
c 3 + 7 + 8
= 10 + 8
= 18
d 8 + 2 + 1 + 5
= 10 + 6
= 16
e 1 + 9 + 8 + 2
= 10 + 10
= 20
f 4 + 6 + 3 + 7
= 10 + 10
= 20

2 a 23
b 19
c 19
d 40
e 59
f 79

3 a 5 + 15 = 20
15 + 5 = 20
20 − 15 = 5
20 − 5 = 15
b 48 + 12 = 60
12 + 48 = 60
60 − 12 = 48
60 − 48 = 12
c 20 + 101 = 121
101 + 20 = 121
121 − 101 = 20
121 − 20 = 101

4 a 24 + 35 = 59
35 + 24 = 59
59 − 35 = 24
59 − 24 = 35
b 15 + 47 = 62
47 + 15 = 62
62 − 47 = 15
62 − 15 = 47
c 53 + 31 = 84
31 + 53 = 84
84 − 31 = 53
84 − 53 = 31

5 a 88 b 79 c 87 d 68 e 108

6 a 856
b 485
c 947
d 874
e 2793
f 3788
g 4765
h 5247

7 a 113
b 115
c 121
d 93
e 165
f 564
g 1412
h 1020
i 1608
j 1210
k 1574
l 8256
m 12 890
n 14 755
o 14 878
p 10 918

8 a 30, 30, 60
b 50, 60, 110
c 140, 60, 200
d 280, 30, 310

9 a 300, 100, 400
b 500, 300, 800
c 900, 300, 1200
d 400, 700, 1100

10 a 62
b 110
c 485
d 253
e 6123
f 6268
g 4575
h 11 653

11 a 30 + 50 = 80
5 + 2 = 7
80 + 7 = 87
b 60 + 70 = 130
2 + 1 = 3
130 + 3 = 133
c 500 + 0 = 500
40 + 50 = 90
2 + 5 = 7
500 + 90 + 7 = 597
d 700 + 0 = 700
30 + 20 = 50
4 + 3 = 7
700 + 50 + 7 = 757
e 500 + 200 = 700
30 + 30 = 60
5 + 4 = 9
700 + 60 + 9 = 769
f 600 + 100 = 700
50 + 40 = 90
2 + 3 = 5
700 + 90 + 5 = 795
g 200 + 300 = 500
50 + 40 = 90
1 + 7 = 8
500 + 90 + 8 = 598
h 5000 + 1000 = 6000
700 + 200 = 900
80 + 10 = 90
4 + 3 = 7
6000 + 900 + 90 + 7 = 6997
i 2000 + 5000 = 7000
600 + 300 = 900
10 + 10 = 20
3 + 1 = 4
7000 + 900 + 20 + 4 = 7924

12 a $1.25
b $2.55
c $8.50
d $14.15
e $25.40
f $72.55
g $102.50
h $251.40

13 a $1.35
b $3.45
c $2.65
d $6.30
e $3.55
f $4.80
g $12.35
h $46.60
i $21.90
j $35.25

14 $315.20

15 a Lina $336.00, Ali $230
b Ali
c Lina
d Lina
e Ali
f $660.50

CATCH UP MATHS YEAR 5 BOOK A © PASCAL PRESS ISBN: 9781925726169

3 SUBTRACTION

Using the Jump Strategy to Solve Subtraction

TWO-DIGIT AND THREE-DIGIT NUMBERS

Page 114 – Your Turn

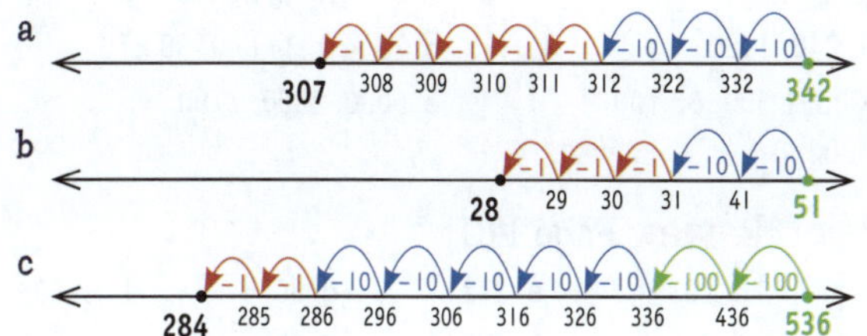

Page 115 – Practice

1 a 14 c 28 e 492 g 672
b 37 d 491 f 408 h 289

Using the Jump Strategy

THREE-DIGIT AND FOUR-DIGIT NUMBERS

Page 116 – Your Turn

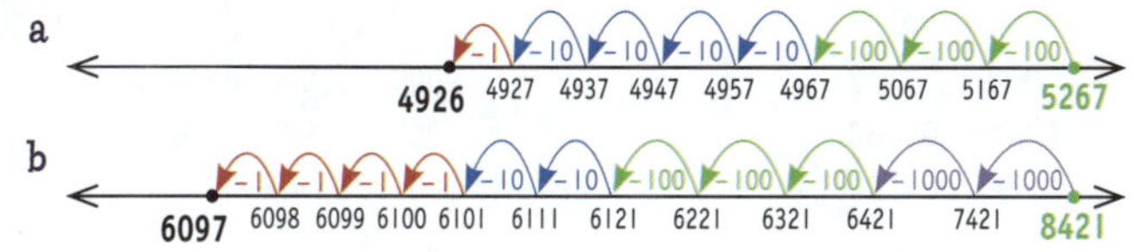

Page 117 – Practice

1 a 5845 c 9732 e 5027 g 401
b 1951 d 2184 f 4397 h 1759

Using the Split Strategy to Solve Subtraction

TWO-DIGIT AND THREE-DIGIT NUMBERS

Page 118 – Your Turn

a 300 – 0 = 300
60 – 50 = 10
2 – 1 = 1
300 + 10 + 1 = 311

Page 119 – Practice

1 a 500 – 300 = 200
80 – 70 = 10
9 – 8 = 1
200 + 10 + 1 = 211

b 80 – 40 = 40
2 – 1 = 1
40 + 1 = 41

c 90 – 20 = 70
9 – 7 = 2
70 + 2 = 72

d 300 – 0 = 300
70 – 20 = 50
3 – 2 = 1
300 + 50 + 1 = 351

e 800 – 0 = 800
30 – 20 = 10
4 – 1 = 3
800 + 10 + 3 = 813

f 800 – 200 = 600
50 – 10 = 40
4 – 3 = 1
600 + 40 + 1 = 641

g 700 – 500 = 200
80 – 20 = 60
9 – 6 = 3
200 + 60 + 3 = 263

h 800 – 400 = 400
40 – 30 = 10
2 – 0 = 2
400 + 10 + 2 = 412

Using the Split Strategy

THREE-DIGIT AND FOUR-DIGIT NUMBERS

Page 120 – Your Turn

a 4000 – 2000 = 2000
700 – 100 = 600
30 – 30 = 0
2 – 2 = 0
2000 + 600 + 0 + 0 = 2600

Page 121 – Practice

1 a 7000 – 2000 = 5000
200 – 100 = 100
30 – 0 = 30
8 – 3 = 5
5000 + 100 + 30 + 5 = 5135

b 6000 – 0 = 6000
900 – 500 = 400
40 – 40 = 0
8 – 3 = 5
6000 + 400 + 0 + 5 = 6405

c 4000 – 0 = 4000
700 – 600 = 100
50 – 10 = 40
2 – 1 = 1
4000 + 100 + 40 + 1 = 4141

d 8000 – 5000 = 3000
900 – 200 = 700
30 – 10 = 20
9 – 3 = 6
3000 + 700 + 20 + 6 = 3726

e 3000 – 0 = 3000
800 – 700 = 100
50 – 20 = 30
6 – 4 = 2
3000 + 100 + 30 + 2 = 3132

Subtraction Without Trading

TWO-DIGIT AND THREE-DIGIT NUMBERS

Page 122 – Your Turn

a 531 b 812 c 36 d 771 e 205

Page 123 – Practice

1 a 541 b 33 c 603

2 a 163 d 521 g 100 j 534
b 831 e 30 h 14 k 12
c 621 f 202 i 671

3 a 63 – 41 = 22
b 792 – 342 = 450
c 535 – 124 = 411
d 768 – 320 = 448
e 996– 772 = 224
f 838 – 527 = 311
g 492 – 362 = 130

Subtraction Without Trading

THREE-DIGIT AND FOUR-DIGIT NUMBERS

Page 124 – Your Turn

a 4621 b 7260 c 8181 d 1310 e 201

Page 125 – Practice

1 a 8035 b 5141 c 6201
2 a 7121 b 2351 c 5111
3 a 4431 c 4110 e 1012 g 9211
b 9262 d 2420 f 8221

Subtraction Without Trading

FOUR-DIGIT AND FIVE-DIGIT NUMBERS

Page 126 – Your Turn

a 41 611 b 72 155

Page 127 – Practice

1 a 11 611 b 72 630
2 a 54 510 b 32 012 c 71 612 d 36 122 e 33 300
3 a 70 291 b 15 132 c 21 111 d 32 321 e 15 010

ANSWERS

3 SUBTRACTION CONTINUED

Subtraction With Trading
TWO-DIGIT AND THREE-DIGIT NUMBERS

Page 128 – Your Turn
a 64 b 539 c 19 d 216 e 28

Page 129 – Practice
1 a 209 b 62 c 654
2 a 46 b 644 c 66 d 44 e 477 f 668 g 29
3 a 74 – 25 = 49
b 55 – 36 = 19
c 88 – 69 = 19
d 91 – 78 = 13
e 63 – 48 = 15
f 42 – 17 = 25
g 42 – 19 = 23
h 326 – 133 = 193
i 524 – 239 = 285
j 837 – 243 = 594
k 903 – 342 = 561
l 741 – 236 = 505
m 847 – 208 = 639
n 424 – 136 = 288
o 625 – 285 = 340

Subtraction With Trading
THREE-DIGIT AND FOUR-DIGIT NUMBERS

Page 130 – Your Turn
a 5881 b 2911 c 4141

Page 131 – Practice
1 a 2519 b 6879
2 a 6918 b 2941 c 4921 d 5299 e 4488
3 a 3831 b 8109 c 4949 d 6619 e 4427

Subtraction With Trading
FIVE-DIGIT NUMBERS

Page 132 – Your Turn
a 68 496 b 20 862 c 46 106

Page 133 – Practice
1 a 70 996 b 48 817
2 a 11 078 b 48 948 c 18 916 d 78 585 e 34 099
3 a 34 097 b 68 379
4 a 58 273 – 41 382 = 16 891
b 52 493 – 14 275 = 38 218
c 63 425 – 44 284 = 19 141
d 82 310 – 43 413 = 38 897
e 91 472 – 43 860 = 47 612

Trading From Larger Place Values

Page 134 – Your Turn
a 16 580 b 866 c 268

Page 135 – Practice
1 a 67 427 b 4853 c 6177 d 19 698 e 125 997 f 24 564 g 44 787 h 5747 i 268 j 48 528 k 18 996
2 a \$36.38 b \$6.50 c \$44.83 d \$47.50 e \$44.75 f \$205.05 g \$12.54 h \$102.43

What's the Difference?

Page 136 – Your Turn
a 3986 b 460 514

Page 137 – Practice
1 a \$354 599 b \$9442 c \$204 243
2 a \$346 758 b \$243 080 c \$24 001

Rounding to Estimate Subtraction Answers

Page 138 – Your Turn
1 a 610, 420, 190 2 a 1600, 200, 1400

Page 139 – Practice
1 a 740, 260, 480
b 1360, 240, 1120
c 43 520, 36 420, 7100
d 73 430, 34 560, 38 870
2 a 73 600, 21 500, 52 100
b 300, 300, 0
c 6000, 3300, 2700

Subtraction Review Page 140
1 a 31 b 38 c 203 d 372 e 232 f 1083 g 8075 h 4550 i 4350
2 a 50 – 40 = 10
8 – 2 = 6
10 + 6 = 16
b 70 – 50 = 20
3 – 0 = 3
20 + 3 = 23
c 500 – 0 = 500
60 – 40 = 20
3 – 1 = 2
500 + 20 + 2 = 522
d 800 – 0 = 800
60 – 40 = 20
2 – 2 = 0
800 + 20 + 0 = 820
e 400 – 100 = 300
20 – 0 = 20
7 – 2 = 5
300 + 20 + 5 = 325
f 500 – 200 = 300
90 – 40 = 50
3 – 1 = 2
300 + 50 + 2 = 352
g 5000 – 0 = 5000
900 – 200 = 700
40 – 30 = 10
3 – 1 = 2
5000 + 700 + 10 + 2 = 5712
h 9000 – 0 = 9000
300 – 200 = 100
40 – 10 = 30
2 – 2 = 0
9000 + 100 + 30 = 9130
i 7000 – 4000 = 3000
500 – 300 = 200
30 – 10 = 20
6 – 2 = 4
3000 + 200 + 20 + 4 = 3224
j 9000 – 3000 = 6000
300 – 0 = 300
40 – 20 = 20
2 – 1 = 1
6000 + 300 + 20 + 1 = 6321
3 a 32 b 21 c 46 d 37 e 38 f 17 g 552 h 542 i 530 j 541 k 605 l 861
4 a 5111 b 3211 c 8632 d 1441 e 6387 f 4076 g 13 135 h 18 612 i 81 220 j 21 142 k 26 169 l 30 589
5 a 14 b 115 c 190 d 1117 e 3242 f 3134 g 5053 h 54 960 i 92 217 j 22 741 k 33 122 l 68 951
6 a 368 b 457 c 32 d 4847 e 6417 f 5579 g 625 h 4970 i 2693 j 26 476 k 49 848 l 19 927 m 64 648 n 52 947 o 41 679 p 8650 q 24 710 r 44 696
7 a 36 b 475 c 549 d 8684 e 2093 f 33 488
8 a 340, 170, 170
b 3580, 430, 3150
c 17 440, 1240, 16 200
9 a 400, 100, 300
b 53 400, 5400, 48 000

4 MULTIPLICATION

Skip Counting

Page 146 – Your Turn

1 a 10, 15, 20, 25, 30
b 8, 12, 16, 20, 24

2 a 30, 24, 18, 12, 6
b 80, 72, 64, 56, 48

Page 147 – Practice

1 a 8, 16, 24, 32, 40, 48, 56, 64
b 9, 18, 27, 36, 45, 54, 63, 72
c 5, 10, 15, 20, 25, 30, 35, 40
d 1, 2, 3, 4, 5, 6, 7, 8
e 12, 24, 36, 48, 60, 72, 84, 96
f 24, 22, 20, 18, 16, 14, 12, 10
g 84, 77, 70, 63, 56, 49, 42, 35
h 66, 60, 54, 48, 42, 36, 30, 24
i 40, 36, 32, 28, 24, 20, 16, 12
j 70, 60, 50, 40, 30, 20, 10, 0

2 a 7, 6
b 36, 48
c 35, 49
d 50, 35
e 24, 36
f 16, 20
g 91, 87
h 27, 28
i 18, 27
j 117, 127
k 77, 44
l 80, 60
m 107, 125

Product, Factors and Multiples

Page 148 – Your Turn

1 a 2 b 3
2 a 42 b 23 c 1

Page 149 – Practice

1 a 18
1 2 3 6 9 18
1 × 18 = 18
2 × 9 = 18
3 × 6 = 18
b 7
1 7
1 × 7 = 7
c 24
1 2 3 4 6 8 12 24
1 × 24 = 24
2 × 12 = 24
3 × 8 = 24
4 × 6 = 24
d 16
1 2 4 8 16
1 × 16 = 16
2 × 8 = 16
4 × 4 = 16

2 a 20 b 56 c 108 d 132 e 72

3 a 42, 48, 54, 60, 66 b 28, 35, 42, 49, 56 c 36, 45, 54, 63, 72

4 a 4
b 8
c 60
d 11
e 11
f 8
g 6
h 0

Multiplying 2-digit by 1-digit Numbers

Page 150 – Your Turn

Known facts
30 × 8 = 240
240 + 8 + 8 = 256

Multiply tens then ones
8 tens × 3 + 8 twos
= 240 + 16 = 256

Area model

	30	2
8	240	16

= 240 + 16 = 256

Page 151 – Practice

1 a Known facts
30 × 5 = 150
150 + 5 + 5 + 5 + 5 = 170
Multiply tens then ones
3 tens × 5 + 5 fours
= 150 + 20 = 170
Area model

	30	4
5	150	20

150 + 20 = 170

b Known facts
60 × 8 = 480
480 + 8 + 8 + 8 + 8 + 8 + 8 + 8
= 536
Multiply tens then ones
6 tens × 8 + 8 sevens
= 480 + 56 = 536
Area model

	60	7
8	480	56

480 + 56 = 536

c Known facts
40 × 3 = 120
120 + 3 = 123
Multiply tens then ones
4 tens × 3 + 3 ones
= 120 + 3 = 123
Area model

	40	1
3	120	3

120 + 3 = 123

Formal Algorithms

Page 152 – Your Turn

1 a

	H	T	O
		[1]3	5
×			2
		7	0

b

	H	T	O
		6	1
×			8
	4	8	8

c

	H	T	O
		7	2
×			3
	2	1	6

d

	H	T	O
		[2]2	3
×			7
	1	6	1

e

	H	T	O
		[5]8	9
×			6
	5	3	4

Page 153 – Practice

1 a 128
b 290
c 384
d 729
e 736
f 280
g 110

2 Score: 5 out of 8
a correct
b correct
c incorrect: 360
d correct
e correct
f incorrect: 140
g correct
h incorrect: 488

3 a 1088 b 1645 c 2821 d 3096 e 7136

Multiplying 3-digit and 4-digit Numbers by 1-digit Numbers

Page 155 – Your Turn

a 631 × 4
Multiply thousands, hundreds, tens, ones
= (600 × 4) + (30 × 4) + (1 × 4)
= 2400 + 120 + 4
= 2524
Formal algorithm

	[1]6	3	1
×			4
2	5	2	4

Area model

	600	30	1
4	2400	120	4

= 2400 + 120 + 4
= 2524

b 742 × 8
Multiply thousands, hundreds, tens, ones
= (700 × 8) + (40 × 8) + (2 × 8)
= 5600 + 320 + 16
= 5936
Formal algorithm

	[3]7	[1]4	2
×			8
5	9	3	6

Area model

	700	40	2
8	5600	320	16

= 5600 + 320 + 16
= 5936

Page 156 – Practice

1 a 1956 b 7630 c 3073 d 47 704

2 a 2460 b 9192 c 5075 d 39 488 e 3521

3 a 2928
b 29 640
c 14 675
d 3352
e 3542
f 2481
g 13 230
h 44 970
i 39 746

4 MULTIPLICATION CONTINUED

Multiplying 2-digit and 3-digit Numbers by 2-digit Numbers

Page 159 – Your Turn

a

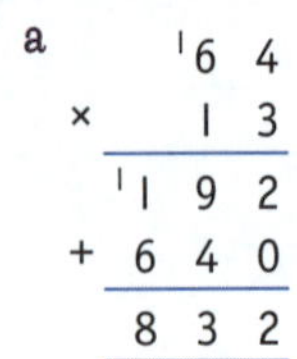

	60	4
10	600	40
3	180	12

640
192
832

b

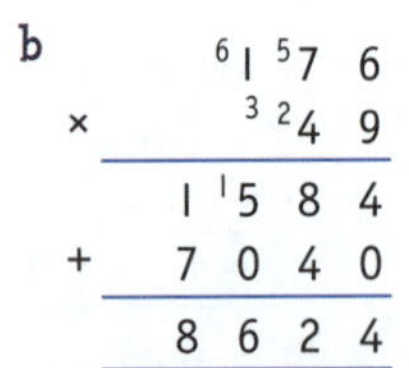

	100	70	6
40	4000	2800	240
9	900	630	54

7040
1584
8624

c

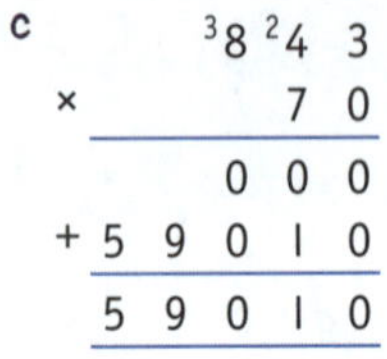

	800	40	3
70	56 000	2800	210
0	0	0	0

59 010
0
59 010

Page 160 – Practice

1 a 6360 b 1863
2 a 2772 b 6264 c 16 320 d 3784 e 47 192 f 12 474 g 33 271 h 60 444
3 a 1323 b 5628 c 49 248 d 32 340 e 6882

Multiplication Review Page 162

1 a 15, 18, 21, 24, 27, 30 b 24, 30, 36, 42, 48, 54 c 60, 70, 80, 90, 100, 110 d 36, 45, 54, 63, 72, 81
2 a 28, 24, 20, 16, 12, 8 b 49, 42, 35, 28, 21, 14 c 64, 56, 48, 40, 32, 24 d 120, 108, 96, 84, 72, 60
3 a 54 b 49 c 48 d 48 e 72 f 30
4 a 6, 12, 18, 24, 30, 36 b 8, 16, 24, 32, 40, 48 c 9, 18, 27, 36, 45, 54 d 4, 8, 12, 16, 20, 24 e 7, 14, 21, 28, 35, 42
5 a 1, 2, 3, 4, 6, 9, 12, 18, 36 b 1, 2, 3, 4, 6, 8, 12, 16, 24, 48 c 1, 3, 5, 15 d 1, 2, 3, 4, 6, 12
6 a 10 b 4 c 9 d 4 e 56 f 5 g 9 h 6 i 81 j 5 k 9 l 4
7 a 511 b 147 c 738 d 316 e 222 f 357 g 861 h 1126 i 4543 j 4968 k 771 l 2015
8 a 144 b 256 c 301
9 a 702 b 567 c 291
10 a 84 b 294 c 378
11 a 1448 b 29 070 c 2865
12 a 2608 b 23 612 c 4092
13 a 2692 b 7650 c 4465 d 30 737 e 1791 f 52 227
14 a 1288 b 5166 c 1050 d 4416 e 10 738 f 25 757 g 44 992 h 52 080 i 47 304
15 a 2772 b 3610 c 32 606 d 5248 e 1008 f 34 839 g 1480 h 46 480 i 15 210
16 a correct b incorrect: 252 c correct d correct e incorrect: 1540 f correct g incorrect: 7050

5 DIVISION

Groups and Equal Rows

Page 168 – Your Turn

1 a 5
2 a ●●●●●●●●● / ●●●●●●●●● b ●●● / ●●● / ●●●

Page 169 – Practice

1 a 2, 2 b 9, 9 c 4, 4 d 9, 9 e 2, 2
2 a 3, 3 b 8, 8 c 1, 1 d 2, 2 e 8, 8

Relating × to ÷

Page 170 – Your Turn

a 6, 6 b 21, 21 c 100, 100 d 11, 11 e 7, 7

Page 171 – Practice

1 a 45 ÷ 5 = 9 b 2 ÷ 1 = 2 c 64 ÷ 8 = 8 d 42 ÷ 7 = 6
2 a 2 × 6 = 12, 6 × 2 = 12, 12 ÷ 6 = 2, 12 ÷ 2 = 6
b 2 × 9 = 18, 9 × 2 = 18, 18 ÷ 9 = 2, 18 ÷ 2 = 9
c 4 × 5 = 20, 5 × 4 = 20, 20 ÷ 5 = 4, 20 ÷ 4 = 5
d 9 × 8 = 72, 8 × 9 = 72, 72 ÷ 8 = 9, 72 ÷ 9 = 8
e 12 × 10 = 120, 10 × 12 = 120, 120 ÷ 10 = 12, 120 ÷ 12 = 10
3 a 9 × 10 b 5 × 9 c 8 × 5 d 7 × 4 e 6 × 5

Quotient, Divisor and Dividend

Page 172 – Your Turn

1 Adult to check
2 a 48, 6, 8 b 72, 8, 9 c 10 ÷ 5 = 2 d 20 ÷ 10 = 2

Page 173 – Practice

1 a 7 b 2 c 10 d 4 e 7
2 24 ÷ 6, 16 ÷ 4, 28 ÷ 7
3 The first number should be circled in red, and the second number should be circled in blue.
a 3 b 6 c 9 d 4 e 5 f 4 g 9 h 8
4 a 3 b 3 c 7 d 6 e 7 f 11 g 7 h 9 i 5 j 2 k 1 l 11 m 11 n 5 o 10 p 6 q 3 r 5 s 10 t 9 u 12

Formal Division

Page 174 – Your Turn

a 2 b 3 c 6 d 8 e 3 f 4 g 6 h 1

CATCH UP MATHS YEAR 5 BOOK A © PASCAL PRESS ISBN: 9781925726169

Page 175 – Practice

1 a 4, 4 × 4 = 16
b 9, 2 × 9 = 18
c 4, 5 × 4 = 20
d 4, 6 × 4 = 24
e 7, 4 × 7 = 28

2 a 8 b 44 c 8 d 8 e 10 f 48 g 8 h 6

3
a 8 ÷ 8 = 1, $8\overline{)8}$ quotient 1
b 20 ÷ 5 = 4, $5\overline{)20}$ quotient 4
c 63 ÷ 7 = 9, $7\overline{)63}$ quotient 9
d 120 ÷ 10 = 12, $10\overline{)120}$ quotient 12
e 10 ÷ 2 = 5, $2\overline{)10}$ quotient 5
f 33 ÷ 11 = 3, $11\overline{)33}$ quotient 3
g 70 ÷ 10 = 7, $10\overline{)70}$ quotient 7
h 72 ÷ 8 = 9, $8\overline{)72}$ quotient 9
i 42 ÷ 7 = 6, $7\overline{)42}$ quotient 6

Different Ways to Write Division

Page 176 – Your Turn

a $\frac{43}{2}$ b $\frac{74}{3}$ c $4\overline{)52}$

Page 177 – Practice

1 a $6\overline{)42}$ b $10\overline{)93}$ c $9\overline{)81}$ d $7\overline{)63}$ e $25\overline{)75}$ f $4\overline{)230}$ g $11\overline{)121}$ h $5\overline{)182}$

2 a $\frac{53}{4}$ b $\frac{72}{3}$ c $\frac{37}{4}$ d $\frac{82}{3}$ e $\frac{51}{2}$ f $\frac{95}{3}$ g $\frac{64}{7}$ h $\frac{42}{6}$

3 a 27 ÷ 3 b 47 ÷ 4 c 64 ÷ 5 d 58 ÷ 4 e 83 ÷ 7 f 62 ÷ 3 g 73 ÷ 4 h 97 ÷ 8

4

	Fraction	$\overline{)\ }$	÷
a	$\frac{61}{4}$	$4\overline{)61}$	61 ÷ 4
b	$\frac{25}{3}$	$3\overline{)25}$	25 ÷ 3
c	$\frac{72}{6}$	$6\overline{)72}$	72 ÷ 6
d	$\frac{49}{7}$	$7\overline{)49}$	49 ÷ 7
e	$\frac{81}{9}$	$9\overline{)81}$	81 ÷ 9
f	$\frac{74}{5}$	$5\overline{)74}$	74 ÷ 5
g	$\frac{69}{9}$	$9\overline{)69}$	69 ÷ 9
h	$\frac{22}{3}$	$3\overline{)22}$	22 ÷ 3
i	$\frac{16}{4}$	$4\overline{)16}$	16 ÷ 4

Division With Remainders

Page 178 – Your Turn

a 3 remainder 3 b 8 remainder 4 c 6 remainder 1

Page 179 – Practice

1 a 7 r 7 b 6 r 3 c 8 r 3 d 7 r 6 e 8 r 1 f 9 r 2 g 7 r 2 h 9 r 4 i 5 r 7

2 a 4 r 1 b 7 r 2 c 5 r 3 d 9 r 5

3 a 25 ÷ 4 = 6 r 1
b 70 ÷ 6 = 11 r 4
c 84 ÷ 9 = 9 r 3
d 148 ÷ 12 = 12 r 4
e 106 ÷ 10 = 10 r 6
f 62 ÷ 5 = 12 r 2
g 29 ÷ 6 = 4 r 5
h 78 ÷ 8 = 9 r 6

Division of 2-digit Numbers

Page 180 – Your Turn

a 12 b 21 c 11 d 22 e 11

Page 181 – Practice

1 a 11 b 13 c 11 d 11 e 11 f 12 g 13

2 a 17 b 14 c 17 d 12 e 17 f 13 g 29 h 15 i 28 j 14 k 13 l 18 m 16 n 27 o 18

3 a 19 b 26 c 37 d 15 e 32 f 13 g 11

Division of 3-digit Numbers

Page 182 – Your Turn

a 212 b 242 c 111 d 211 e 341 f 120 g 120

Page 183 – Practice

1 a 135 b 124 c 159 d 321 e 142 f 98 g 84 h 27 i 56 j 98 k 65

2 a 117 r 1 b 188 c 150 r 3

3 a 90, 100, 120, 40, 300
b 90, 20, 80, 70, 60
c 70, 110, 30, 60, 50

Recording Remainders as Fractions and Decimals

Page 184 – Your Turn

a 74 r 1 = $74\frac{1}{2}$ = 74.5
b 45 r 1 = $45\frac{1}{3}$ = 45.33
c 129 r 2 = $129\frac{2}{5}$ = 129.4

Page 185 – Practice

1 a 49 r 1 = $49\frac{1}{3}$
b 111 r 3 = $111\frac{3}{4}$
c 173 r 1 = $173\frac{1}{2}$
d 63 r 2 = $63\frac{1}{2}$
e 171 r 2 = $171\frac{2}{3}$

2 a 54 r 3 = 54.75
b 62 r 2 = 62.4
c 43 r 5 = 43.625

3

	Dividend	Divisor	Quotient	Remainder	Quotient and Remainder as a Fraction	Quotient and Remainder as a Decimal
a	15	2	7	1	$7\frac{1}{2}$	7.5
b	32	5	6	2	$6\frac{2}{5}$	6.4
c	57	8	7	1	$7\frac{1}{8}$	7.125
d	38	3	12	2	$12\frac{2}{3}$	12.67

5 DIVISION CONTINUED

Division Review Page 185

1 a 5, 5 b 4, 16, 4 c 3, 24, 3
2 a 7, 7 b 5, 5 c 6, 6
3 a 48 ÷ 6 = 8 b 72 ÷ 8 = 9 c 10 ÷ 2 = 5 d 28 ÷ 7 = 4 e 12 × 10 = 120 f 3 × 12 = 36 g 8 × 8 = 64 h 4 × 12 = 48 i 9 × 7 = 63 j 12 × 12 = 144 k 7 ÷ 1 = 7 l 7 × 6 = 42
4 a 6 × 4 = 24, 4 × 6 = 24, 24 ÷ 4 = 6, 24 ÷ 6 = 4 b 7 × 3 = 21, 3 × 7 = 21, 21 ÷ 3 = 7, 21 ÷ 7 = 3 c 7 × 8 = 56, 8 × 7 = 56, 56 ÷ 8 = 7, 56 ÷ 7 = 8
5 a 6 × 7 b 11 × 9 c 2 × 8 d 8 × 6 e 5 × 8 f 7 × 7 g 4 × 8 h 6 × 6 i 11 × 11
6 a 4 b 8 c 10 d 8 e 12 f 12 g 4 h 10 i 7
7 36 ÷ 3, 12 ÷ 1, 24 ÷ 2
8 a $6\overline{)42}$ b $9\overline{)27}$ c $9\overline{)81}$ d $2\overline{)127}$ e $4\overline{)464}$ f $3\overline{)721}$
9 a $\frac{25}{5}$ b $\frac{36}{4}$ c $\frac{24}{6}$ d $\frac{72}{8}$ e $\frac{9}{9}$ f $\frac{63}{7}$ g $\frac{40}{4}$ h $\frac{32}{8}$ i $\frac{10}{2}$
10 a 24 ÷ 3 b 42 ÷ 6 c 49 ÷ 7 d 121 ÷ 11 e 80 ÷ 8 f 90 ÷ 9
11 a 4 b 6 c 6 d 7 e 12 f 5 g 9 h 10
12 a 3 r 6 b 10 r 2 c 3 r 0 d 4 r 3 e 1 r 2 f 5 r 1 g 5 r 3 h 7 r 2 i 4 r 0 j 9 r 1
13 a 6 r 3 b 8 r 2 c 9 r 6 d 4 r 5
14 a 18 ÷ 4 = 4 r 2 b 134 ÷ 12 = 11 r 2 c 109 ÷ 10 = 10 r 9 d 79 ÷ 8 = 9 r 7 e 28 ÷ 6 = 4 r 4
15 a 31 b 12 c 24 d 21 e 43 f 12 g 13 h 17 i 15 j 29 k 16 l 14
16 a 17 b 20 c 14 d 18
17 a 5 r 2 b 15 r 4 c 9 r 3 d 9 r 5 e 14 r 0 f 8 r 1
18 a 421 b 213 c 122 d 111 e 234 f 332
19 a 321 r 1 b 116 r 6 c 119 r 3 d 181 r 1 e 343 r 1 f 129 r 4
20 a 127 r 2 b 118 r 1 c 106 r 1 d 127 r 4
21 a 4, 6, 8, 14, 22 b 6, 1, 7, 2, 11 c 1, 3, 6, 8, 5 d 1, 4, 8, 10, 9
22 a $155\frac{3}{4}$ b $106\frac{2}{5}$ c 126 d $156\frac{1}{6}$ e $95\frac{8}{9}$ f $193\frac{3}{4}$ g $83\frac{1}{5}$ h $109\frac{1}{3}$ i 347 j $121\frac{5}{7}$ k $64\frac{1}{4}$ l $48\frac{1}{2}$
23 a 197.5 b 156.8 c 168.75 d 281.5 e 224.75 f 124.33 g 52.25 h 64.8 i 196.2 j 158.5 k 151.33 l 60.5
24 a correct b correct c incorrect, 90 r 5 d incorrect, 110 r 4 e incorrect, 71 f correct g incorrect, 92 r 6 h correct

6 FRACTIONS

Parts of a Fraction

Page 194 – Your Turn

1 Adult to check
2 a $\frac{1}{3}$ b $\frac{11}{100}$ c $\frac{1}{2}$ d $\frac{1}{5}$ e $\frac{10}{100}$

Page 195 – Practice

1 a 2 b 3 c 4 d 7 e 1 f 1 g 2
2 a 3 b 100 c 8 d 10 e 8 f 5 g 8 h 4 i 10
3 a two-quarters b one-eighth c one-tenth d eighteen-hundredths e four-fifths f seven-eighths g two-thirds
4 a $\frac{3}{4}$ b $\frac{2}{5}$ c $\frac{10}{100}$ d $\frac{3}{5}$ e $\frac{1}{2}$
5 a three-quarters b two-fifths c ten-hundredths d three-fifths e one-half

Halves and Halves of Collections

Page 196 – Your Turn

1 Circle **c**
2 a Half of 8 = 4 b Half of 12 = 6 c Half of 20 = 10

Page 197 – Practice

1 Adult to check
2 Cross out **b**, **e** and **g**
3 a Half of 8 = 4 b Half of 18 = 9 c Half of 20 = 10
4 a 8, 16, 16, 8 b 7, 14, 14, 7 c 20, 40, 40, 20

Quarters and Eighths – Fractions and Collections

Page 198 – Your Turn

1 a b c

2 a 16, 2 b 24, 3

Page 199 – Practice

1 a b c
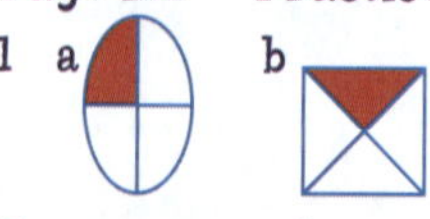

2 a b c
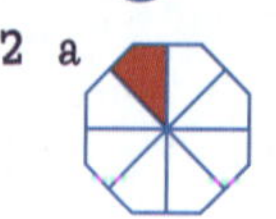
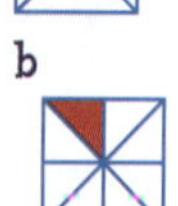

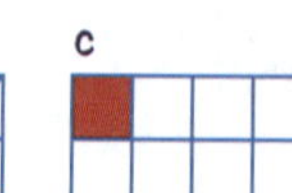

CATCH UP MATHS YEAR 5 BOOK A © PASCAL PRESS ISBN: 9781925726169

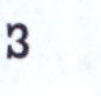

3

	Fraction name	Fraction	Picture
a	two-quarters	$\frac{2}{4}$	
b	seven-eighths	$\frac{7}{8}$	
c	three-quarters	$\frac{3}{4}$	
d	five-eighths	$\frac{5}{8}$	
e	two-eighths	$\frac{2}{8}$	

4 a 4 b 6 c 8 d 1 e 10 f 12 g 9

5 a 3 b 1 c 4 d 6 e 7 f 5 g 8

Thirds and Fifths – Fractions and Collections

Page 200 – Your Turn

1 a 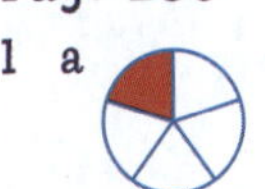 b  c 

2 a $\frac{1}{3}$ of 12 = 4

b $\frac{1}{3}$ of 15 = 5

Page 201 – Practice

1 a 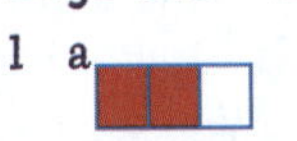 b 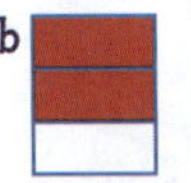 c 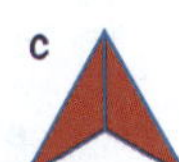

2 a 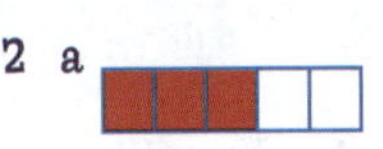 b 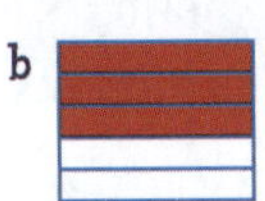 c 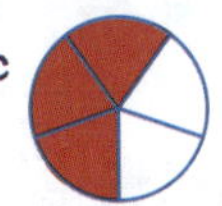

3

	Fraction name	Fraction	Picture
a	three-fifths	$\frac{3}{5}$	
b	one-third	$\frac{1}{3}$	
c	two-fifths	$\frac{2}{5}$	
d	three-thirds	$\frac{3}{3}$	
e	four-fifths	$\frac{4}{5}$	

4 a 6 b 11 c 12 d 9 e 1 f 2 g 4

5 a 5 b 1 c 8 d 12 e 10 f 2 g 7

Equivalent Fractions

Page 202 – Your Turn

a $\frac{4}{10}$ b $\frac{1}{2}$

Page 203 – Practice

1 a 2 b 6 c 1 d 1 e 3 f 1 g 5

2 a 6 b 15 c 1 d 2 e 2 f 4 g 1 h 20 i 24 j 25 k 16

Comparing Fractions

Page 204 – Your Turn

1 Adult to check

2 a 1 b 2 c 2

Page 205 – Practice

1 a $\frac{1}{4}$ (1), $\frac{2}{3}$ (4), $\frac{3}{5}$ (3), $\frac{5}{8}$ (5), $\frac{3}{4}$ (6), $\frac{1}{3}$ (2)

b $\frac{1}{3}$ (2), $\frac{1}{4}$ (1), $\frac{2}{3}$ (3), $\frac{3}{3}$ (6), $\frac{4}{5}$ (5), $\frac{6}{8}$ (4)

2 Adult to check

3 Adult to check

4 a > b = c > d < e = f < g = h < i > j < k <

Proper and Improper Fractions

Page 206 – Your Turn

Improper: $\frac{8}{2}$, $\frac{7}{3}$, $\frac{10}{4}$, $\frac{9}{6}$

Proper: $\frac{2}{8}$, $\frac{4}{9}$, $\frac{5}{10}$, $\frac{1}{4}$

Page 207 – Practice

1 a $\frac{1}{2}$ b $\frac{2}{3}$ c $\frac{2}{8}$ d $\frac{12}{24}$ e $\frac{53}{60}$

2 a $\frac{8}{5}$ b $\frac{3}{1}$ c $\frac{5}{3}$ d $\frac{81}{41}$ e $\frac{53}{24}$

3 Adult to check

4 Adult to check

Mixed Numbers

Page 208 – Your Turn

1 a 

b 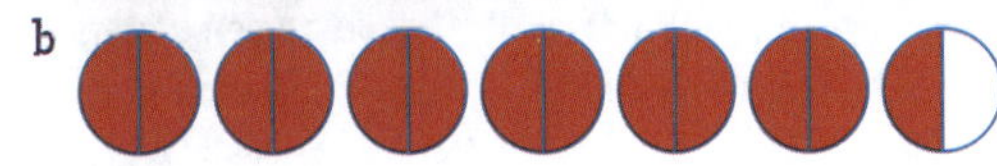

c 

Page 209 – Practice

1 a b 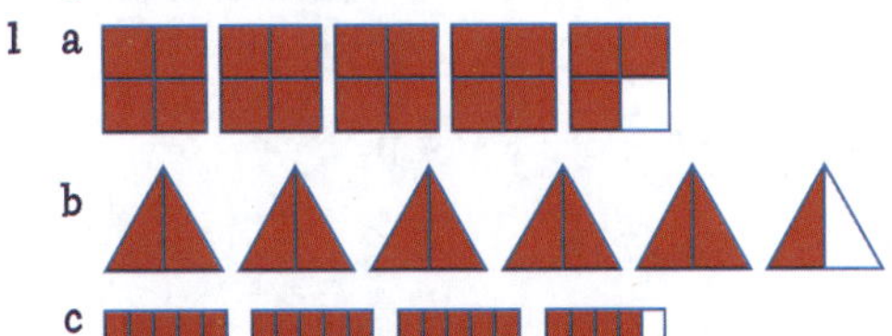

c 

2 a $4\frac{4}{5}$ b $5\frac{2}{8}$ c $1\frac{1}{3}$

Improper Fractions and Mixed Numbers

Page 210 – Your Turn

a 18, $4\frac{2}{4}$ b 13, $1\frac{5}{8}$ c 11, $5\frac{1}{2}$

6 FRACTIONS CONTINUED

Page 211 – Practice

1 a $\frac{7}{2}$ b $\frac{16}{6}$ c $\frac{11}{3}$ d $\frac{7}{4}$ e $\frac{17}{3}$ f $\frac{13}{2}$ g $\frac{9}{2}$ h $\frac{12}{5}$ i $\frac{33}{4}$ j $\frac{23}{8}$ k $\frac{14}{8}$

2 a $3\frac{1}{3}$ b $1\frac{2}{5}$ c $1\frac{2}{3}$ d $4\frac{1}{2}$ e $1\frac{2}{4}$ f $1\frac{2}{10}$ g $5\frac{1}{4}$ h $10\frac{2}{3}$ i $4\frac{2}{3}$ j $8\frac{1}{3}$ k $6\frac{1}{7}$

3 a $3\frac{4}{7}$ b $5\frac{5}{9}$ c $3\frac{5}{10}$ d $8\frac{3}{8}$ e $7\frac{3}{6}$

4 a $\frac{18}{5}$ b $\frac{39}{8}$ c $\frac{20}{3}$ d $\frac{12}{9}$ e $\frac{39}{7}$ f $\frac{36}{5}$ g $\frac{21}{8}$

Add and Subtract Fractions with the Same Denominator

Page 212 – Your Turn

a 3 b 4 c 4 d 7 e 4

Page 213 – Practice

1 a $\frac{2}{3}$ b $\frac{5}{8}$ c $\frac{5}{8}$ d $\frac{5}{10}$ e $\frac{2}{2}$ f $\frac{8}{10}$ g $\frac{4}{5}$ h $\frac{4}{6}$ i $\frac{9}{10}$ j $\frac{7}{8}$ k $\frac{8}{10}$

2 a $\frac{4}{10}$ b $\frac{3}{8}$ c $\frac{4}{8}$ d $\frac{5}{10}$ e $\frac{1}{3}$ f $\frac{1}{4}$ g $\frac{2}{8}$ h $\frac{2}{5}$ i $\frac{6}{10}$ j $\frac{4}{8}$ k $\frac{4}{12}$

3 a $\frac{7}{10}$ b $\frac{7}{8}$ c $\frac{2}{4}$ d $\frac{1}{10}$ e $\frac{3}{4}$

Fractions Review Page 214

1 Adult to check

2 a three-fifths b two-eighths c one-third d seven-tenths e three-quarters f one-half

3 Sample answers:

a b c d e

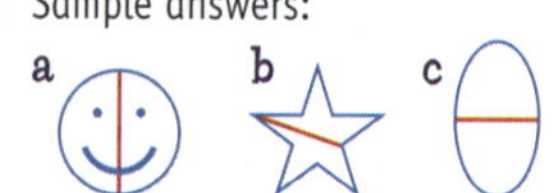

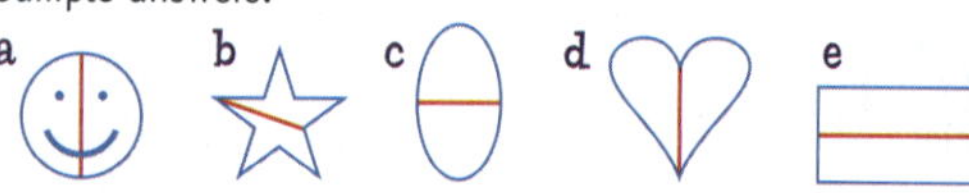

4 a Half of 12 = 6 b Half of 24 = 12 c Half of 30 = 15

5 a 8, 16, 16, 8 b 20, 40, 40, 20

6 a $\frac{3}{4}$ b $\frac{4}{5}$ c $\frac{5}{8}$ d $\frac{3}{6}$

7 a $\frac{5}{8}$ b $\frac{1}{5}$ c $\frac{5}{8}$ d $\frac{5}{10}$

8 a b c d

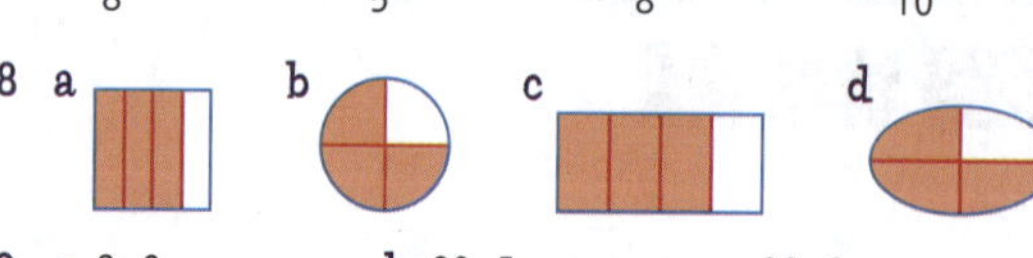

9 a 8, 2 b 20, 5 c 32, 8

10 a b c d

11 a 8, 1 b 24, 3 c 40, 5

12 a 6 b 9 c 1 d 4 e 11 f 3 g 7 h 10

13 a 2 b 4 c 7 d 8 e 9 f 12 g 10 h 6

14 Colour blue: a, c, d, f Colour red: b, e, g, h

15 a 3, 1 b 9, 3 c 27, 9

16 a 8 b 10 c 5

17 a 10, 2 b 20, 4 c 30, 6

18 a 3 b 7 c 8

19 a 6 b 5 c 4 d 2 e 1 f 6 g 1 h 1 i 4 j 1 k 8 l 2 m 8 n 4 o 5

20 a $\frac{3}{2}$ b $\frac{4}{2}$ c $\frac{4}{3}$ d $\frac{5}{1}$ e $\frac{3}{2}$ f $\frac{5}{4}$ g $\frac{45}{30}$ h $\frac{51}{30}$

21 a

b

c

d

22 a $2\frac{2}{3}$ b $3\frac{3}{4}$ c $2\frac{4}{8}$ d $3\frac{3}{5}$

23 a $\frac{10}{3}$ b $\frac{15}{6}$ c $\frac{13}{8}$ d $\frac{13}{3}$ d $\frac{21}{2}$ f $\frac{59}{8}$ g $\frac{18}{7}$ h $\frac{9}{4}$

24 a $2\frac{2}{4}$ b $2\frac{1}{7}$ c $4\frac{1}{5}$ d $3\frac{2}{10}$ e $7\frac{1}{9}$ f $10\frac{2}{3}$ g $9\frac{4}{5}$ h $8\frac{1}{2}$

7 DECIMALS

Writing Decimals

Page 218 – Your Turn

	Fraction	Decimal Fraction	Out of 100
a	$\frac{32}{100}$	0.32	32 out of 100
b	$\frac{57}{100}$	0.57	57 out of 100
c	$\frac{93}{100}$	0.93	93 out of 100
c	$\frac{22}{100}$	0.22	22 out of 100
h	$\frac{138}{100}$	1.38	138 out of 100
j	$\frac{365}{100}$	3.65	365 out of 100

Page 219 – Practice

1 a 124.35 b 20.34 c 82.10 d 64.39 e 913.26 f 1.34 g 52.45 h 142.53 i 106.50

CATCH UP MATHS YEAR 5 BOOK A © PASCAL PRESS ISBN: 9781925726169

2

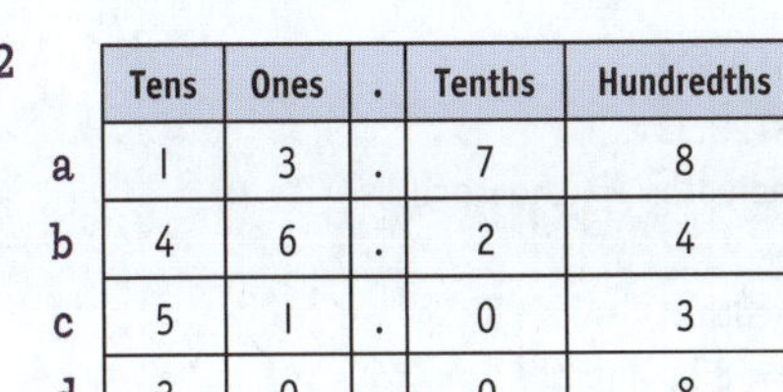

	Tens	Ones	.	Tenths	Hundredths
a	1	3	.	7	8
b	4	6	.	2	4
c	5	1	.	0	3
d	3	0	.	0	8

3 a 8 ones + 2 tenths + 3 hundredths
b 4 ones + 5 hundredths
c 7 tenths + 6 hundredths

4 a 1.27, 2.17, 2.71, 7.12, 7.21
b 0.30, 0.33, 3.03, 3.30, 3.31
c 1.8, 1.85, 5.18, 8.15, 8.51
d 3.67, 3.76, 6.37, 6.73, 7.63

Relating Tenths to Hundredths

Page 220 – Your Turn

1 a 9, 0.90 b 64, 0.64 c 33, 0.33

Page 221 – Practice

1 a 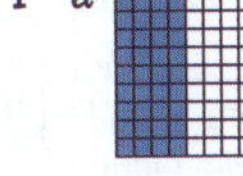 b 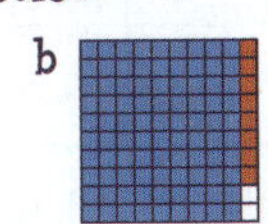 c 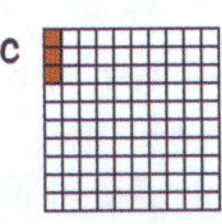

2 a 0.13 b 0.80 c 0.07

3

	Words	Picture	Fraction	Decimal
a	sixty hundredths		$\frac{60}{100}$	0.60
b	forty-two hundredths		$\frac{42}{100}$	0.42
c	ninety-three hundredths		$\frac{93}{100}$	0.93
d	one hundredth		$\frac{1}{100}$	0.01
e	eighteen hundredths		$\frac{18}{100}$	0.18

Decimal Fractions and Fractions in Words

Page 222 – Your Turn

	Fraction	Fraction in Words	Decimal Fraction	Decimal in words
a	$\frac{84}{100}$	eighty-four hundredths	0.84	zero point eight four
b	$\frac{36}{100}$	thirty-six hundredths	0.36	zero point three six
c	$\frac{139}{100}$	one hundred and thirty-nine hundredths	1.39	one point three nine

Page 223 – Practice

1 a two point nine four
b zero point eight two
c eight point nine three
d one point three seven
e two point four zero
f ten point seven three
g twenty point zero two
h zero point one six
i twenty-nine point zero six
j forty-three point four four

2 a $\frac{6}{100}$ b $\frac{23}{100}$ c $\frac{42}{100}$ d $\frac{89}{100}$ e $\frac{102}{100}$ f $\frac{712}{100}$ g $\frac{963}{100}$ h $\frac{890}{100}$ i $\frac{701}{100}$ j $\frac{547}{100}$ k $\frac{800}{100}$ l $\frac{320}{100}$

Thousandths

Page 224 – Your Turn

1 a 7, 3, 8, 2
b 5, 4, 1, 0
c 9, 7, 3, 4,
d 8, 5, 3, 2
e 6, 5, 9, 3
f 7, 1, 1, 3

Page 225 – Practice

1 a 6 b 2 c 4
2 a 9 b 3 c 7
3 a 8.691 b 4.385 c 6.702
4 a 1.672 b 5.381 c 9.035 d 8.354 e 7.438 f 1.085 g 1.006 h 6.952 i 2.010 j 2.340 k 5.000 l 6.100 m 7.467

Place Value

Page 226 – Your Turn

1 a 62.493 b 25.620 c 7.358
2 a 1.246 b 29.368 c 84.721
3 a 32.739 b 1.435 c 0.583
4 a 73.282 b 1.122 c 40.475
5 a 10.650 b 74.009 c 21.436

Page 227 – Practice

1 a thousandths
b tens
c hundredths
d tenths
e tenths
f thousandths
g hundredths
h tenths
i tens
j hundredths
k ten thousandths

2 a 21.350 b 71.4583 c 0.359 d 54.9252 e 99.9999 f 64.381 g 82.563 h 0.5034 i 17.32 j 53.1 k 80.0019

3

	Decimal	Tens	Ones	.	Tenths	Hundredths	Thousandths	Ten Thousandths
a	43.246	4	3	.	2	4	6	
b	74.9783	7	4	.	9	7	8	3
c	87.903	8	7	.	9	0	3	
d	95.0902	9	5	.	0	9	0	2
e	16.10	1	6	.	1	0		
f	24.649	2	4	.	6	4	9	
g	30.2587	3	0	.	2	5	8	7
h	61.082	6	1	.	0	8	2	
i	53.942	5	3	.	9	4	2	
j	60.4365	6	0	.	4	3	6	5

Partitioning Decimals

Page 228 – Your Turn

a 2, 7
b 3, 2, 7
c 43, 4, 3
d 103, 4, 1, 0, 3

7 DECIMALS CONTINUED

Page 229 – Practice

1

	Decimal	Mixed Number	Wholes	Tenths	Hundredths	Thousandths
a	4.175	$4\frac{175}{1000}$	4	$\frac{1}{10}$	$\frac{7}{100}$	$\frac{5}{1000}$
b	6.157	$6\frac{157}{1000}$	6	$\frac{1}{10}$	$\frac{5}{100}$	$\frac{7}{1000}$
c	1.493	$1\frac{493}{1000}$	1	$\frac{4}{10}$	$\frac{9}{100}$	$\frac{3}{1000}$
d	8.459	$8\frac{459}{1000}$	8	$\frac{4}{10}$	$\frac{5}{100}$	$\frac{9}{1000}$
e	3.237	$3\frac{237}{1000}$	3	$\frac{2}{10}$	$\frac{3}{100}$	$\frac{7}{1000}$
f	7.368	$7\frac{368}{1000}$	7	$\frac{3}{10}$	$\frac{6}{100}$	$\frac{8}{1000}$
g	9.945	$9\frac{945}{1000}$	9	$\frac{9}{10}$	$\frac{4}{100}$	$\frac{5}{1000}$

2

	Decimal	Mixed Number	Wholes	Thousandths
a	5.324	$5\frac{324}{1000}$	5	$\frac{324}{1000}$
b	6.173	$6\frac{173}{1000}$	6	$\frac{173}{1000}$
c	4.159	$4\frac{159}{1000}$	4	$\frac{159}{1000}$
d	3.438	$3\frac{438}{1000}$	3	$\frac{438}{1000}$
e	6.805	$6\frac{805}{1000}$	6	$\frac{805}{1000}$
f	7.590	$7\frac{590}{1000}$	7	$\frac{590}{1000}$
g	9.720	$9\frac{720}{1000}$	9	$\frac{720}{1000}$

3 a 90 + 386 thousandths = $90 + \frac{386}{1000}$
b 38 + 4 tenths + 9 hundredths + 1 thousandths $= 38 + \frac{4}{10} + \frac{9}{100} + \frac{1}{1000}$
c 46 + 93 hundredths + 7 thousandths = $46 + \frac{93}{100} + \frac{7}{1000}$
d 354 + 25 thousandths = $354 + \frac{25}{1000}$
e 460 + 3 tenths + 2 hundredths + 1 thousandth $= 460 + \frac{3}{10} + \frac{2}{100} + \frac{1}{1000}$
f 46 + 93 hundredths + 7 thousandths = $46 + \frac{93}{100} + \frac{7}{1000}$
g 90 + 386 thousandths = $90 + \frac{386}{1000}$

Decimals Review Page 231

1 a 24.20 c 0.30 e 70.59
b 63.58 d 58.37

2

	Tens	Ones	.	Tenths	Hundredths	Thousandths
a	3	5	.	6	5	1
b	0	8	.	5	6	2
c	6	9	.	8	3	9
d	7	6	.	5	8	6
e	3	9	.	1	5	6
f	1	0	.	4	9	7

3 a 4.32, 3.42, 3.24, 2.43, 2.34
b 7.317, 4.371, 3.174, 1.734, 1.473
c 8.965, 8.695, 6.895, 5.968, 5.698
d 9.34, 9.304, 9.034, 4.093, 3.904
e 9.529, 9.501, 5.243, 3.781, 3.331

4 a 8 ones + 2 tenths + 3 hundredths + 4 thousandths
b 5 ones + 4 tenths
c 2 tens + 4 ones + 7 hundredths
d 5 tens + 1 one + 7 tenths + 8 hundredths
e 4 ones + 7 tenths + 3 hundredths + 8 thousandths
f 2 tens + 4 ones + 8 tenths + 2 hundredths + 5 thousandths

5 a 0.80 b 0.73 c 0.04 d 0.13 e 2.38 f 1.34

6 a c e
b d f

7

	Fraction	Fraction in words	Decimal fraction	Decimal in words
a	$\frac{29}{100}$	twenty-nine hundredths	0.29	zero point two nine
b	$\frac{38}{100}$	thirty-eight hundredths	0.38	zero point three eight
c	$\frac{197}{100}$	one hundred and ninety-seven hundredths	1.97	one point nine seven
d	$\frac{254}{100}$	two hundred and fifty-four hundredths	2.54	two point five four
e	$\frac{6348}{1000}$	six thousand, three hundred and forty-eight thousandths	6.348	six point three four eight
f	$\frac{241}{1000}$	two hundred and forty-one thousandths	0.241	zero point two four one
g	$\frac{6846}{1000}$	six thousand, eight hundred and forty-six thousandths	6.846	six point eight four six
h	$\frac{483}{100}$	four hundred and eighty-three hundredths	4.83	four point eight three

8 a six point five nine three
b zero point seven four two
c three point five six one
d zero point eight zero three
e two point four one

9 a 5 b 9 c 3 d 0 e 6
10 a 4 b 7 c 6 d 3 e 8
11 a 4 b 2 c 9 d 8
12 a 3.768 b 69.178 c 52.834 d 80.927 e 64.085
13 a 5.935 c 8.423 e 4.731 g 3.7 i 4.907
b 6.710 d 2.872 f 6.481 h 5.49 j 8.034
14 a 0.2 c 1.47 e 7.33 g 8.43
b 3.75 d 0.9 f 0.6 h 1.25

15

	Decimal	Tens	Ones	.	Tenths	Hundredths	Thousandths
a	15.723	1	5	.	7	2	3
b	74.978	7	4	.	9	7	8
c	73.098	7	3	.	0	9	8
d	3.432	0	3	.	4	3	2
e	5.24	0	5	.	2	4	0
f	6.13	0	6	.	1	3	0
g	8.354	0	8	.	3	5	4
h	91.407	9	1	.	4	0	7
i	70.320	7	0	.	3	2	0
j	90.004	9	0	.	0	0	4

CATCH UP MATHS YEAR 5 BOOK A © PASCAL PRESS ISBN: 9781925726169

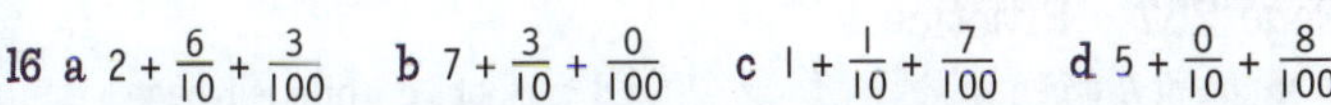

16 a $2 + \frac{6}{10} + \frac{3}{100}$ b $7 + \frac{3}{10} + \frac{0}{100}$ c $1 + \frac{1}{10} + \frac{7}{100}$ d $5 + \frac{0}{10} + \frac{8}{100}$

17 a $6 + \frac{28}{100}$ b $8 + \frac{48}{100}$ c $4 + \frac{3}{100}$ d $3 + \frac{50}{100}$

18 a $2 + \frac{6}{10} + \frac{2}{100} + \frac{4}{1000}$ d $4 + \frac{9}{10} + \frac{2}{100} + \frac{3}{1000}$

b $8 + \frac{1}{10} + \frac{8}{100} + \frac{3}{1000}$ e $15 + \frac{4}{10} + \frac{0}{100} + \frac{9}{1000}$

c $9 + \frac{7}{10} + \frac{5}{100} + \frac{9}{1000}$ f $10 + \frac{7}{10} + \frac{9}{100} + \frac{0}{1000}$

19 a $3 + \frac{729}{1000}$ c $5 + \frac{745}{1000}$ e $20 + \frac{130}{1000}$

b $8 + \frac{430}{1000}$ d $14 + \frac{657}{1000}$ f $39 + \frac{429}{1000}$

20 a 32 + 849 thousandths = $32 + \frac{849}{1000}$

b 54 + 98 hundredths + 1 thousandth = $54 + \frac{98}{100} + \frac{1}{1000}$

c 84 + 7 tenths + 8 hundredths + 1 thousandth = $84 + \frac{7}{10} + \frac{8}{100} + \frac{1}{1000}$

d 43 + 5 thousandths = $43 + \frac{5}{1000}$

21

	Decimal	Mixed Number	Wholes	Tenths	Hundredths	Thousandths
a	5.632	$5\frac{632}{1000}$	5	$\frac{6}{10}$	$\frac{3}{100}$	$\frac{2}{1000}$
b	7.365	$7\frac{365}{1000}$	7	$\frac{3}{10}$	$\frac{6}{100}$	$\frac{5}{1000}$
c	18.792	$18\frac{792}{1000}$	18	$\frac{7}{10}$	$\frac{9}{100}$	$\frac{2}{1000}$
d	24.839	$24\frac{839}{1000}$	24	$\frac{8}{10}$	$\frac{3}{100}$	$\frac{9}{1000}$
e	16.516	$16\frac{516}{1000}$	16	$\frac{5}{10}$	$\frac{1}{100}$	$\frac{6}{1000}$
f	31.126	$31\frac{126}{1000}$	31	$\frac{1}{10}$	$\frac{2}{100}$	$\frac{6}{1000}$
g	14.203	$14\frac{203}{1000}$	14	$\frac{2}{10}$	$\frac{0}{100}$	$\frac{3}{1000}$
h	17.041	$17\frac{41}{1000}$	17	$\frac{0}{10}$	$\frac{4}{100}$	$\frac{1}{1000}$
i	5.278	$5\frac{278}{1000}$	5	$\frac{2}{10}$	$\frac{7}{100}$	$\frac{8}{1000}$
j	1.435	$1\frac{435}{1000}$	1	$\frac{4}{10}$	$\frac{3}{100}$	$\frac{5}{1000}$
k	12.349	$12\frac{349}{1000}$	12	$\frac{3}{10}$	$\frac{4}{100}$	$\frac{9}{1000}$
l	49.636	$49\frac{636}{1000}$	49	$\frac{6}{10}$	$\frac{3}{100}$	$\frac{6}{1000}$

8 PATTERNS AND ALGEBRA

Number Patterns

Page 238 – Your Turn

a 16, 32, 64, 128 c 48, 24, 12, 6 e 18, 54, 162, 486
b 132, 120, 108, 96 d 73, 79, 85, 91

Page 239 – Practice

1 a 2000, 1000, 500, 250, 125 c 188, 191, 194, 197, 200
b 2002, 2004, 2006, 2008, 2010 d 1017, 1025, 1033, 1041, 1049

2 a − 3 b × 2 c × 5 d + 11 e × 6 f ÷ 2

3 a 14, 26, 50, 98, 194 c 68, 36, 20, 12, 8
b 13, 25, 49, 97, 193 d 38, 62, 110, 206, 398

4 a 54, 45, 36. Rule: − 9 c 61, 57, 53. Rule: − 4
b 500, 2500, 12 500. Rule: × 5

5 a ● − ■ = ▲ d G ÷ H = I
b D × E = F e X + A = ⬢
c ▲ ÷ ■ = ★

Pattern Grids

Page 241 – Your Turn

a 8, 16, 24, 32, 40 b 0, 5, 10, 15, 20 c 3, 6, 9, 12, 15

Page 242 – Practice

1 a 12, 14, 16, 18, 20, 24, 26

2 a D = B ÷ 4 b K = C × 11 c ▲ = ■ ÷ 3

3 a 46, 106, 226, 466 c 68, 260, 1028, 4100
b 20, 28, 44, 76

4 a

÷ 3	9	6	1	3	4	21
+ 7	34	25	10	16	19	70

b

÷ 5	5	1	13	2	12	4
+ 8	33	13	73	18	68	28

c

+ 12	19	20	22	23	25	21
− 7	0	1	3	4	6	2

Equivalent Number Sentences

Page 243 – Your Turn

a 20 c 20 e 100 g 50 i 36
b 3 d 4 f 25 h 8

Page 244 – Practice

1 a 4 d 4 g 9 j 20
b 16 e 34 h 10 k 5
c 3 f 7 i 4

2 Adult to check

3 a 3 c 45 e 2 g 26
b 22 d 3 f 5

Number Sentences and Patterns with Fractions and Decimals

Page 245 – Your Turn

a 1.1 b 12 c 16 d 3.6

Page 246 – Practice

1 a 3.2 d 1.1 g 2 j 4.8 m 2
b 1.2 e 4 h 2.2 k 8 n 4.1
c 2.4 f 1.5 i 3 l 0.9

2 a 4 d 6 g 4 j 8
b 16 e 12 h 1 k 9
c 6 f 12 i 18

3 a 3 b 12 c 20 d 6 e 10

4 a 1.8, 1.5, 1.2 c 1.0, 1.2, 1.4 e 4.9, 4.5, 4.1 g 8.0, 7.7, 7.4
b 3.5, 3.7, 3.9 d 2.5, 2.7, 2.9 f 9.4, 9.0, 8.6

Terms in Sequences

Page 247 – Your Turn

1 a 9 b 10 c 18 d 80 e 3.2

2 a 8 b 10 c 18 d 14 e 56

Page 248 – Practice

1 a 4th b 2nd c 5th d 3rd e 4th

2 a 24, 28 b 63, 70 c 93, 85 d 93, 101 e 35, 42

8 PATTERNS AND ALGEBRA CONTINUED

3 a Increase, 4   b Increase, 7   c Decrease, 8   d Increase, 4   e Increase, 7
4 a 34, 32, 30   b 8, 10, 12   c 32, 40, 48
5 a 16, 25, 36   b 100   c 144

Patterns and Algebra Review Page 249

1 a 18, 54, 162, 486   b 104, 106, 108, 110   c 250, 50, 10, 2   d 85, 79, 73, 67
2 a + 7   b − 3   c × 2   d − 4
3 a 10, 18, 34, 66, 130   b 45, 18, 9, 6, 5   c 16, 36, 76, 156, 316
4 a A × B = C   b ▲ − ● = ■   c D × F = E
5 a 45, 35, 25, 15, 5, 0
b 12, 32, 44, 60, 80, 88
c

÷ 5	12	10	5	2	3
+ 9	69	59	34	19	24

d

× 7	35	49	63	77	42
− 4	1	3	5	7	2

6 a 3   b 13   c 12   d 7   e 40   f 8
7 Adult to check
8 a 12   b 16   c 20   d 48   e 96   f 3
9 a 6   b 1.2   c 9   d 8   e 2   f 4.8
10 a 2.8, 3.1, 3.4   b 9.4, 9.0, 8.6   c 10.5, 13, 15.5   d 4.2, 5.2, 6.2   e 8.5, 10, 11.5   f 5.4, 4.3, 3.2
11 a 4th   b 2nd   c 5th   d 1st   e 2nd   f 4th
12 a increase by 2   b increase by 9   c increase by 2   d decrease by 0.3   e decrease by 6   f increase by 3
13 a 12, 15   b 24   c 30   d 36

9 CHANCE

Likely and Unlikely Events

Page 252 – Your Turn

a likely   b unlikely   c likely   d unlikely

Page 253 – Practice

1 a likely   b likely   c likely   d unlikely   e unlikely
2 Adult to check

Certain and Uncertain Events

Page 254 – Your Turn

a certain   b uncertain   c certain   d uncertain

Page 255 – Practice

1 a uncertain   b uncertain   c certain   d certain
2 a uncertain   b uncertain   c uncertain   d uncertain   e certain   f certain   g uncertain   h uncertain   i uncertain

Probability

Page 256 – Your Turn

a Bernard arrives at work at 6:00 am.

Page 257 – Practice

1 a paint a green car   b pick out a black ball   c land on the number 30   d pick out a white jellybean   e Tiana's birthday tomorrow

Related Events

Page 258 – Your Turn

a Rosie eats healthy food

Page 259 – Practice

1 a False   b True   c False
2 a Jason's friend loves skiing.
b Adrianna's mum has a red bike.
c Roisin will buy a new dress.
d Jonah enjoys watching movies.
e Paul just learned a new skateboard trick.

Outcomes

Page 260 – Your Turn

a Blue, Blue, Blue; Blue, Blue, Orange; Blue, Orange, Orange; Orange, Orange, Orange
b impossible

Page 261 – Practice

1 1. pink shirt, pink shorts
2. pink shirt, orange shorts
3. pink shirt, blue shorts
4. green shirt, pink shorts
5. green shirt, orange shorts
6. green shirt, blue shorts
7. red shirt, pink shorts
8. red shirt, orange shorts
9. red shirt, blue shorts

2 1. Yellow hat, yellow shirt, orange shorts
2. Yellow hat, red shirt, orange shorts
3. Yellow hat, yellow shirt, white shorts
4. Yellow hat, red shirt, white shorts
5. Red hat, yellow shirt, orange shorts
6. Red hat, yellow shirt, white shorts
7. Red hat, red shirt, orange shorts
8. Red hat, red shirt, white shorts
9. Orange hat, yellow shirt, orange shorts
10. Orange hat, yellow shirt, white shorts
11. Orange hat, red shirt, orange shorts
12. Orange hat, red shirt, white shorts

Even Chance Events

Page 262 – Your Turn

a no   b no   c yes

Page 263 – Practice

1 Sample answers:

a b 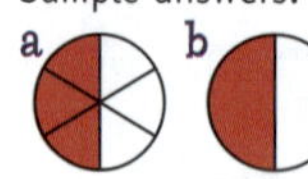 c  d 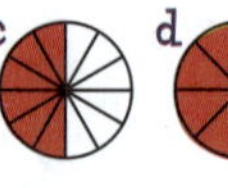 e  f g 

2 a Draw 2 more aqua marbles.
b Draw 4 more red marbles.
c Draw 4 more blue marbles.
d Draw 4 more blue marbles.
e Draw 4 more purple marbles.
f Already an equal chance.
g Draw 4 more green marbles.

3 a ✗   b ✓   c ✓   d ✗

Unequal Chances

Page 264 – Your Turn

a Unequal   b Equal

Page 265 – Practice

1 a 7   b 5   c 20   d 20   e pink   f yellow

CATCH UP MATHS YEAR 5 BOOK A © PASCAL PRESS ISBN: 9781925726169

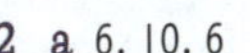

2 a 6, 10, 6 b 4, 10, 4 c unequal

3 Adult to check

Probability From 0 To 1

Page 266 – Your Turn

[From left to right] impossible, unlikely, even chance, likely, certain

Page 267 – Practice

1 a 0.5 b 1 c 0.8 d 0.2 e 0 f 0.5

2 Adult to check

Chance Review Page 268

1 a unlikely b likely c unlikely d unlikely e likely

2 a certain b uncertain c uncertain d certain

3 Adult to check

4 a Max cannot serve a customer at 6:00 am.
b Marissa cannot draw an orange boat.
c Matthew cannot land on number 20.

5 a False b False c True d True

6 a I love eating pasta for dinner. b It is Tuesday tomorrow.

7 orange, orange, orange; orange, orange, blue; orange, blue, blue; blue, blue, blue.

8 1. red shirt, green shorts
2. red shirt, red shorts
3. red shirt, pink shorts
4. orange shirt, green shorts
5. orange shirt, red shorts
6. orange shirt, pink shorts
7. blue shirt, green shorts
8. blue shirt, red shorts
9. blue shirt, pink shorts

9 Sample answers:

a 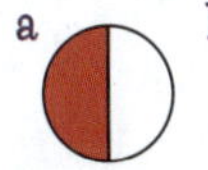 b 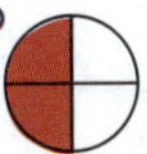 c 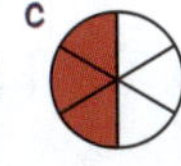 d 

10 a Draw 4 the same colour but not green.
b Draw 3 the same colour but not purple.
c Draw 5 the same colour but not red.
d Draw 6 the same colour but not orange.

11 a 8, 4, 4, 1, 2, $\frac{1}{2}$, 1, 2, $\frac{1}{2}$, equal
b 9, 6, 3, 6, 9, $\frac{2}{3}$, 3, 9, $\frac{1}{3}$, unequal
c 13, 6, 7, 7, 13, $\frac{7}{13}$, 6, 13, $\frac{6}{13}$, unequal

12 a 0.1, 0.3, 0.4, 0.5, 0.6, 0.8, 0.9
b impossible, unlikely, even chance, likely, certain

13 a Answers will vary.
b 0
c 1
d Answers will vary.
e 0.5

10 DATA

Data Tables

Page 273 – Your Turn

a

Items Sold at Port Hacksville High	
Item	**Total**
Nachos	15
Hot chips	21
Hamburgers	28
Hot dogs	32
Spring rolls	21
	117

b hot dogs
c spring rolls, hot chips
d 13
e 32

Page 274 – Practice

1 a

5T's Favourite Food	
Food	**Total**
Pasta	8
Pizza	9
Sushi	7
Tacos	3
Hot dogs	2
	29

b

5M's Favourite Colour		
Colour	**Tally**	**Total**
Red	||||	4
Blue	~~||||~~ ||||	9
Green	~~||||~~	5
Yellow	~~||||~~ |||	8
		26

2 a Ice hockey
b Football
c 9
d 8
e 4

3

Colours of Tennis Balls		
Colour	**Tally**	**Total**
Green	||||	4
Pink	~~||||~~	5
Yellow	~~||||~~ ||	7
Orange	~~||||~~ |||	8
Blue	|||	3
		27

Picture Graphs

Page 276 – Your Turn

Number of Ice Creams Sold in One Week	
Day	**Total**
Sunday	24
Monday	18
Tuesday	16
Wednesday	10
Thursday	20
Friday	28
Saturday	34
	150

Page 277 – Practice

1 a 110 b 2021 c 2017 d 40 e 1070

2 a eggs
b toast
c 5
d 2
e 4
f 2

Column Graphs

Page 279 – Your Turn

a 11 b 4 c zebras d 1 e 8

Page 280 – Practice

1 a March b 80 c 55 d 50

2 a 2
b 5, 1
c 13
d 9
e 0
f 1

Bar Graphs

Page 281 – Your Turn

a cherry b more c more d more

Page 282 – Practice

1 a Sample answer: Animals Rehomed in November Last Year
b cats
c less
d 40
e 4
f less

10 DATA CONTINUED

2 Money Spent on Petrol

January	February	March	April	May

Dot Plots

Page 284 – Your Turn

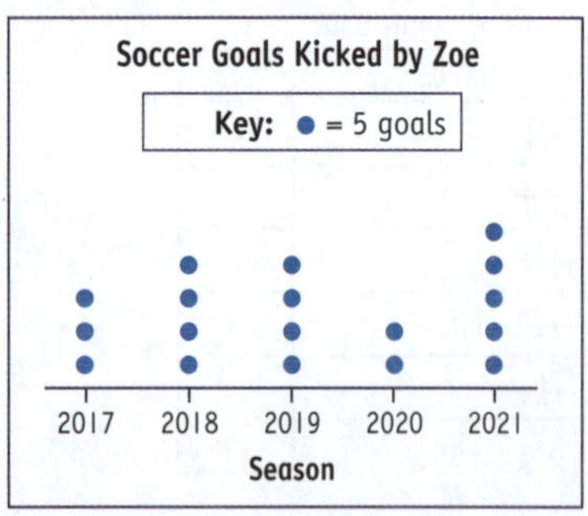

Page 285 – Practice

1 a 25 b 20 c Enda High d 15

2 a Chris b Dani c Mai d Zoe e 5 f Lu g 29

3 a 10 b 2 c 59 d 65 e 6

4 a 6 b 9 c 10 d 31 e 3

5

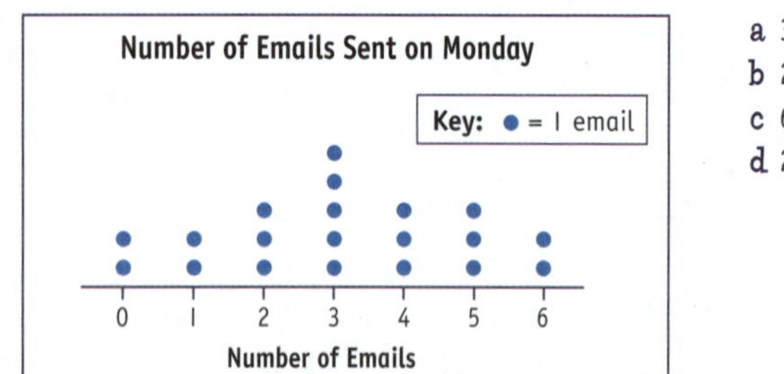

a 3
b 2
c 62
d 2

6

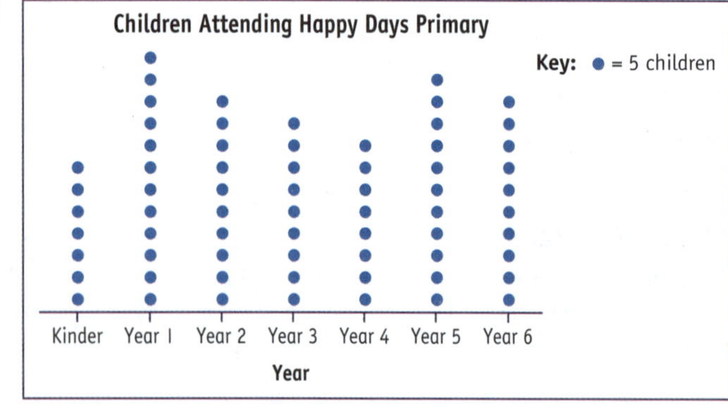

Line Graphs

Page 289 – Your Turn

a

Day of the Week	Mon	Tue	Wed	Thu	Fri	Sat	Sun
Number of Visitors	2	4	3	3	6	7	6

b Thursday c 6 d 1 e 13 f 18

Page 290 – Practice

1 a 12 pm b 5 c 7 d 27

2 hours, minutes, 1, 60

3 a 120 minutes b 300 minutes c 240 minutes

4 a 4.5 hours b 5 hours c 2.5 hours

5 a 4 minutes b 50 litres c 20 litres

6 a

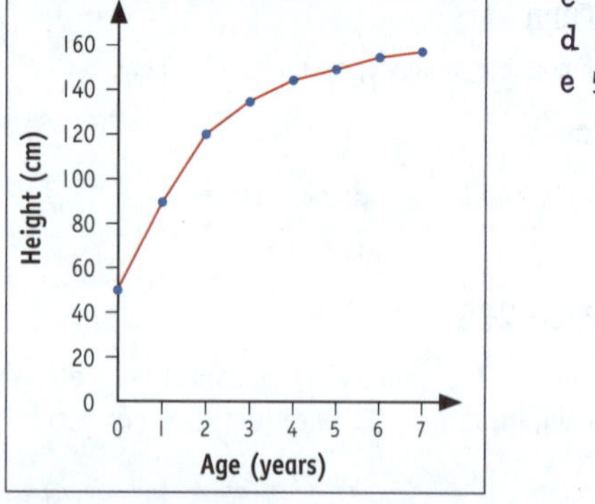

b 0 and 1
c 108 cm
d 10 cm
e 50 cm

7 a

Population (thousands)	2	2.5	4	4.5	5	6.5	8
Year	2015	2016	2017	2018	2019	2020	2021

b

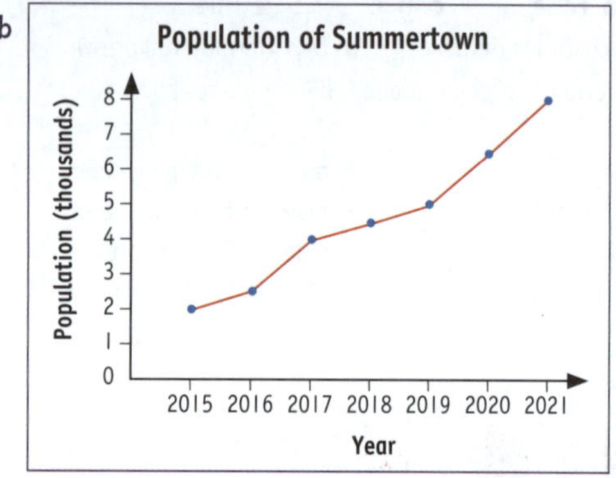

Spreadsheets

Page 294 – Your Turn

a $9175.00
b $1550.00
c 30 June
d $1850.00
e 6
f $4800.00
g $14 625.00

Page 295 – Practice

1 a D3 = $3835.50, D4 = $5258.50, D5 = $6578.50, D6 = $7178.50, D7 = $8433.50, D8 = $9753.50

b $2435.50 c 24 September d $1255.00 e $1400.00

2

	A	B	C	D
6	17 MAR	Bought Balls	$150.00	$7100.00
7	21 MAR	Bought Trophies	$295.00	$6805.00
8	28 MAR	Bought Awards	$560.00	$6245.00
9	31 MAR	Bought Shirts	$1526.00	$4719.00

Data Review Page 296

1

Jellybean Colour

Colour	Tally	Total
Red	|||	3
Blue	||||	4
White	~~||||~~ |||	8
Green	~~||||~~ |	6
Black	~~||||~~ ||||	9
		30

a 30
b 8
c 9
d 5
e red

CATCH UP MATHS YEAR 5 BOOK A © PASCAL PRESS ISBN: 9781925726169

2 a **Books Read in Book Week**

Class	Total
5A	12
5T	10
5P	9
5S	11
5D	8
	50

b

3 a **Number of Planes Taking Off**

Day	Total
Monday	4
Tuesday	5
Wednesday	10
Thursday	4
Friday	6
Saturday	12
Sunday	13
	54

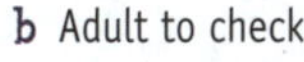
b Adult to check

c
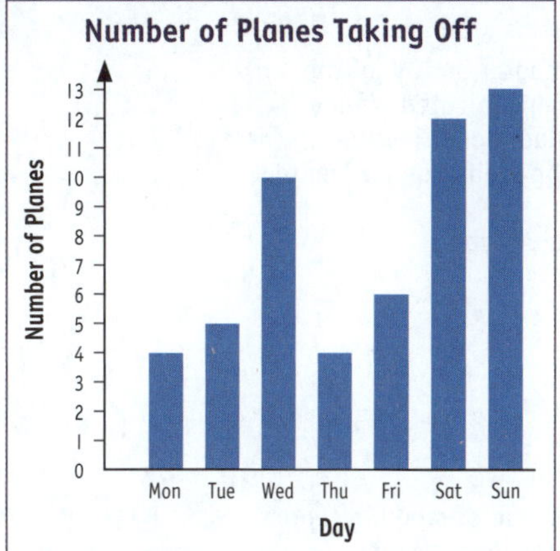

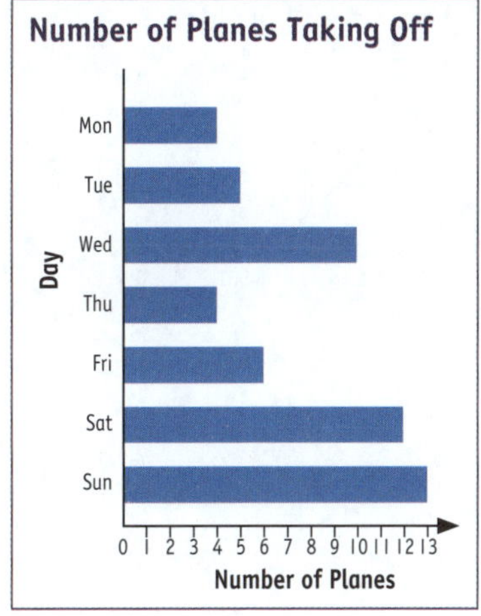

4 a 7
b True
c Alyssa, 5
d 24
e 51
f 4
g 5
h 2
i Alyssa
j Tanya

5 a 4
b 30
c goats
d chickens
e 15

6 **Favourite Snack Food**

Chocolate	Chips	Lollies	Pretzels	Cake

a 25
b 15
c most
d 5
e potato chips

7 a 24
b 3
c 4
d 5
e 18
f 21
g 100
h 30

8
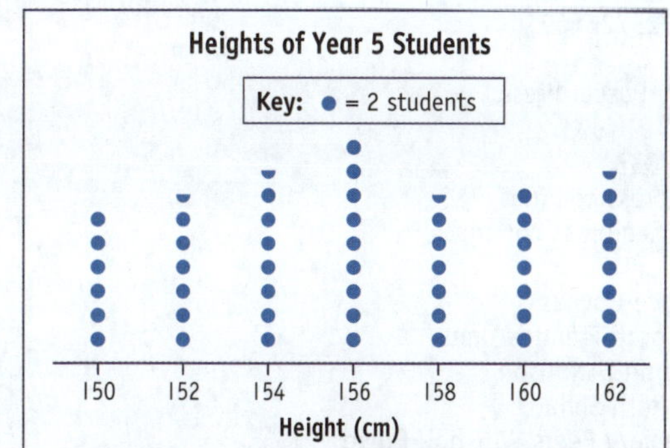

9 a
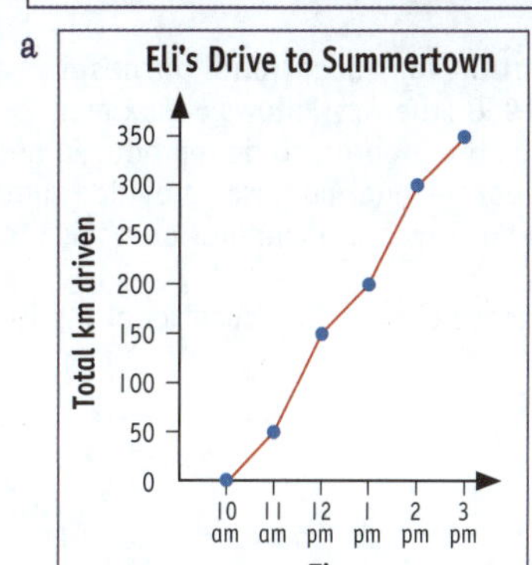

b 50, 100, 50, 350

10 a D3 $670.00, D4 $570.00, D5 $370.00, D6 $940.00, D7 $1115.00, D8 $615.00
b $1195
c $800
d $570
e $200
f 5 Feb

  ISBN: 9781925726169

Catch-Up Maths Book 5A

ISBN: 9781925726169

Published by Pascal Press
PO Box 250
Glebe NSW 2037
www.pascalpress.com.au
contact@pascalpress.com.au

Design: Janice Bowles
Author: Deborah Frendo-Toman
Publisher: Lynn Dickinson
Typesetter: Ruth Schultz
Editor: Rosemary Peers, Vaishali Batra

Printed by Wai Man Book Binding (China) Ltd.